Principles of Soldering and Brazing

Giles Humpston
David M. Jacobson

Acquisitions Editor - Mary Thomas Haddad
Production Project Manager - Suzanne E. Hampson
Production/Design - Randall Boring

**Copyright 1993
by
ASM International®**
All rights reserved

No part of this book may be reproduced, stored in a retrieval system, or transmitted, in any form or by any means, electronic, mechanical, photocopying, recording, or otherwise, without the written permission of the copyright owner.

First printing, March 1993
Second printing, August 1994

This book is a collective effort involving hundreds of technical specialists. It brings together a wealth of information from worldwide sources to help scientists, engineers, and technicians solve current and long-range problems.

Great care is taken in the compilation and production of this Volume, but it should be made clear that NO WARRANTIES, EXPRESS OR IMPLIED, INCLUDING, WITHOUT LIMITATION, WARRANTIES OF MERCHANTABILITY OR FITNESS FOR A PARTICULAR PURPOSE, ARE GIVEN IN CONNECTION WITH THIS PUBLICATION. Although this information is believed to be accurate by ASM, ASM cannot guarantee that favorable results will be obtained from the use of this publication alone. This publication is intended for use by persons having technical skill, at their sole discretion and risk. Since the conditions of product or material use are outside of ASM's control, ASM assumes no liability or obligation in connection with any use of this information. No claim of any kind, whether as to products or information in this publication, and whether or not based on negligence, shall be greater in amount than the purchase price of this product or publication in respect of which damages are claimed. THE REMEDY HEREBY PROVIDED SHALL BE THE EXCLUSIVE AND SOLE REMEDY OF BUYER, AND IN NO EVENT SHALL EITHER PARTY BE LIABLE FOR SPECIAL, INDIRECT OR CONSEQUENTIAL DAMAGES WHETHER OR NOT CAUSED BY OR RESULTING FROM THE NEGLIGENCE OF SUCH PARTY. As with any material, evaluation of the material under end-use conditions prior to specification is essential. Therefore, specific testing under actual conditions is recommended.

Nothing contained in this book shall be construed as a grant of any right of manufacture, sale, use, or reproduction, in connection with any method, process, apparatus, product, composition, or system, whether or not covered by letters patent, copyright, or trademark, and nothing contained in this book shall be construed as a defense against any alleged infringement of letters patent, copyright, or trademark, or as a defense against liability for such infringement.

Comments, criticisms, and suggestions are invited, and should be forwarded to ASM International.

Library of Congress Cataloging Card Number: 93-70224
ISBN: 0-87170-462-5

ASM International®
Materials Park, OH 44073-0002

Preface

Assemblies of all shapes and sizes frequently contain soldered or brazed joints, and the functional operation and integrity of these assemblies usually rely on the soundness and stability of the joints. Therefore, most engineers and technicians, whether in the chemical, mechanical, electrical, or electronic fields, are necessarily exposed to soldering and brazing technologies and are required to have an understanding if not a mastery of the subject.

This book is intended to address this need as far as possible, bearing in mind that practicing engineers usually want the necessary information in a form that is at once readily accessible and easy to assimilate. Therefore, priority is given to the fundamental principles that underlie this field of technology. The largely artificial distinctions between soldering and brazing are removed, and aspects of their commonality are highlighted. Much of the detail to be found in other technical publications on the subject is deliberately omitted, because this often crowds out or obscures the key features of a particular joining process. A large proportion of the literature on soldering and brazing may be charged with being heavy on description and light on critical analysis. We have endeavored to redress the balance, while striving to avoid being unduly simplistic in our approach, although admittedly we may not always have succeeded in this aim.

In order to keep the focus on the fundamental aspects of soldering and brazing, we have deliberately avoided entering into specific joining technologies in detail. Thus, little coverage is given to surface mount technology. This burgeoning subject is a field of technology in its own right and is addressed specifically in the literature: a representative bibliography to 1988 is given in Klein Wassink [1989], Minges [1989] and Tummala and Rymaszewski [1989], supplemented by Manko [1991]. However, from the perspective of our study, the metallurgy of microscopic soldered joints to surface-mounted components is essentially the same as that of larger joints; the process parameters, such as cooling and heating rates and the phases that form, may represent some of the more extreme conditions encountered, but the basic principles are the same. If there is an evident bias toward electronic applications, this reflects the professional orientation of the authors. Some topics are inevitably not accorded due consideration, although it is hoped that sufficient references are provided to enable the reader to pursue these further.

One of the features of soldering and brazing that makes this an exciting subject is that it cuts right across many branches of science and engineering. These include chemical thermodynamics; mechanics; physical, inorganic, and organic chemistry; electronics; thermal management; as well as many of the subject areas of metallurgy. Soldering and brazing also embrace virtually the entire periodic table of elements, aside from the radioactive series.

Some final words are needed regarding bibliography. No attempt has been made to gather a comprehensive list of references. Those that are included have been selected because they are useful basic texts, cover important subject matter, or relate to exemplary pieces of work, whether in respect to methodology, technique, or other innovative features. It was felt that if the value of the book depended on its bibliography, it would become dated more rapidly. The advent of computer search facilities and databases should enable the reader who wishes to locate references on a specific topic to obtain further information without too much difficulty.

The reader should be aware that all compositions given in this book are in weight percentage in accordance with standard industrial practice (except where otherwise stated), and these have, for the most part, been rounded to the nearest integer. Certain conventions have been

been followed when citing eutectic compositions of solders; for example, eutectic tin-lead solder is, for historical reasons, referred to as Pb-60Sn, whereas in fact it is closer to Pb-62Sn. In practical terms, this is a very minor discrepancy, because such small differences in the proportion of the major constituents of solders and brazes will have only a marginal effect on the joining process and the characteristics of the resulting joints.

Specific references are given with each chapter. For those wishing to read more generally on particular topics, the authors recommend the texts listed below. Because they are frequently cited throughout the text, these sources are not listed in the bibliographies appended to the individual chapters.

Soldering and Brazing

Aluminum Association, 1991. *Welding Aluminum: Theory and Practice*

Aluminum Association, 1991. *Soldering and Brazing of Aluminum*

Cieslak et al., M.J., 1991. *The Metal Science of Joining*, TMS

International Organization for Standardization (ISO), 1990. *Welding, Brazing and Soldering Processes—Vocabulary*

Johnson, C.C. and Kevra, J., 1990. *Solder Paste Technology*, TAB Books

Klein Wassink, R.J., 1989. *Soldering in Electronics*, 2nd ed., Electrochemical Publications

Kumar, P. and Greenhut, V.A. 1990. *Metal-Ceramic Joining*, TMS

Lea, C. 1988. *A Scientific Guide to Surface Mount Technology,* Electrochemical Publications Ltd.

Lieberman, E., 1988. *Modern Soldering and Brazing Techniques*, Business News

Manko, H.H., 1986. *Soldering Handbook for Printed Circuits and Surface Mounting*, Van Nostrand Reinhold

Manko, H.H., 1992. *Solders and Soldering*, 3rd ed., McGraw-Hill

Milner, D.R. and Apps, R.L., 1968. *Introduction to Welding and Brazing*, Pergamon Press

Minges, M.L., 1989. *Electronic Materials Handbook.* Vol 1: Packaging, ASM International

Schwartz, M.M., 1987. *Brazing*, ASM International

Schwartz, M.M., 1990. *Ceramic Joining*, ASM International

Schwartz, M.M., 1969. *Modern Metal Joining Manual*, John Wiley & Sons

Thwaites, C.J., 1983. *Capillary Joining: Brazing and Soft-Soldering*, Books Demand UMI

Tummala, R.R. and Rymaszewski, E.J., 1989. *Microelectronics Packaging Handbook,* Van Nostrand Reinhold

Alloy Constitution

John, V.B., 1974. *Understanding Phase Diagrams*, Macmillan

Prince, A., 1966. *Alloy Phase Equilibria*, Elsevier

West, D.R.F., 1982. *Ternary Equilibrium Diagrams*, Chapman & Hall

Thermodynamics, Kinetics, and Physical Metallurgy

Haasen, P., 1986. *Physical Metallurgy*, 2nd ed., Cambridge University Press

Machlin, E.S., 1991. *An Introduction to Aspects of Thermodynamics and Kinetics Relevant to Materials Science*, Giro Press

Rollason, E.C., 1973. *Metallurgy for Engineers*, Edward Arnold

Swalin, R.A., 1972. *Thermodynamics of Solids*, 2nd ed., John Wiley & Sons

Van Vlack, L.H., 1970. *Materials Science for Engineers*, Addison-Wesley

Warn, J.R.W., 1979. *Concise Chemical Thermodynamics*, Van Nostrand Reinhold

Testing and Assessment of Joints

Anderson, R.C., 1988. *Inspection of Metals*, Vol 2, *Destructive Testing*, ASM International

Collins, J.A., 1981. *Failure of Materials in Mechanical Design: Analysis, Prediction, Prevention*, John Wiley & Sons

Dieter, G.E., 1986. *Mechanical Metallurgy*, 3rd ed., McGraw-Hill

Ensminger, D. 1988. *Ultrasonics (Fundamentals, Technology, Applications)*, 2nd ed. Mercel Dekker

Frear, D.R., Jones, W.B., and Kinsman, K.R., 1990. *Solder Mechanics—A State of the Art Assessment*, TMS

Halmshaw, R., 1988. *Non-destructive Testing*, Edward Arnold

Hull, B. and John, V., 1989. *Non-destructive Testing*, Macmillan Education

Lau, J.H., 1991. *Solder Joint Reliability*, Van Nostrand Reinhold

Lau, J.H., 1991. *Solder Joint Reliability: Theory and Applications,* Van Nostrand Reinhold

Minges, M.L. 1989. *Electronic Materials Handbook.* Vol. 1: Packaging, ASM International

Scully, J.C., 1990. *The Fundamentals of Corrosion*, Pergamon Press

Shreir, L.L., 1963. *Corrosion*, Vol 1 and 2, George Newnes Ltd.

Bibliographic Sources

Information on new developments in soldering and brazing is scattered throughout a wide range of periodicals, as reflected in the sources cited in the references appended to the individual chapters. However, the authors have found especially useful the following publications:

Welding Journal (particularly its Research Supplement); *Soldering and Surface Mount Technology*; *IEEE Transactions on Components, Hybrids and Manufacturing Technology*

Abstract publications of value include:

Metals Abstracts (refer to appropriate classification headings); *Weldasearch Abstracts*, produced by The Welding Institute on behalf of the British Association for Brazing and Soldering (issued quarterly)

Acknowledgments

We wish to thank our many colleagues at the Hirst Research Centre, GEC-Marconi Ltd., for their helpful advice and encouragement, and particularly Brian J. Isherwood, Norman J. Iungius, and Satti P.S. Sangha. External readers to whom we are also indebted for their constructive comments and suggestions are A. Arun Junai, Pr. Chidambaram, J.A. DeVore, G.R. Edwards, M. Fletcher, S. Liu, D. Olson, and, not least, Alan Prince, who has championed our undertaking from its inception.

About The Authors

Giles Humpston took a first degree in Metallurgy from Brunel University in 1982, followed by a Ph.D. on the constitution of solder alloys in 1985. He has since been employed at the Hirst Research Centre of GEC-Marconi Ltd., where he has been involved with determining alloy phase diagrams and developing procedures for producing high integrity soldered and brazed joints for a wide variety of metallic and nonmetallic materials. He now leads the Metals Technology Group, with a staff of six scientists, which is responsible for projects involving soldering and brazing, magnetic materials, metallurgical failure analysis, environmental testing of components, and the production and sale of testing equipment. He is cited inventor on 9 patents involving joining processes and the author of 18 papers, several of which are related to the constitution of alloy systems of relevance to soldering and brazing.

Dr. Humpston is a licensed amateur radio enthusiast and has published several articles and reviews on electronics, radios, and computing. His other interests include exploring alternative energy sources, gardening, and wine making. He lives with his wife Jacqueline and their two children in a small village in Buckinghamshire, England.

David M. Jacobson graduated in Physics from the University of Sussex in 1967 and obtained his doctorate in Materials Science there in 1972. Between 1972 and 1975 he lectured in Materials Engineering at the Ben Gurion University, Beer-Sheva, Israel, returning as Visiting Senior Lecturer in 1979-1980. Having gained experience in brazing development with Johnson Matthey Ltd. (U.K.), he extended his range of expertise to soldering at the Hirst Research Centre, GEC-Marconi Ltd., which he joined in 1980. He currently manages the Materials Fabrication Division, which encompasses research groups specializing in the fields of electrochemistry, metal joining, magnetics, polymer technology, and the testing of materials, components, and electronic systems. He is the author of more than 50 scientific and technical publications covering several subjects, including ancient metallurgy, and is a co-inventor with Giles Humpston of several joining materials and processes.

Dr. Jacobson's principal outside interests are archaeology and architectural history, focusing on the Near East in the Graeco-Roman period. He has published extensively in these fields, which extend to the numismatics and early metallurgy of that region. He is completing a Ph.D. thesis on Herodian Architecture at King's College, London. Dr. Jacobson is married, with two children, and lives in Wembley, England, fairly close to the internationally famous football stadium. His leisure pursuits include classical music and art appreciation.

Contents

Chapter 1: Introduction
1.1 Joining Methods 1
    1.1.1 Mechanical Fastening 1
    1.1.2 Adhesive Bonding 3
    1.1.3 Soldering and Brazing 3
    1.1.4 Welding 4
    1.1.5 Diffusion Bonding 4
1.2 Comparison of Solders and Brazes 5
1.3 Key Parameters of Soldering and Brazing 8
    1.3.1 Surface Energy and Surface Tension 8
    1.3.2 Wetting and Contact Angle 9
    1.3.3 Fluid Flow 12
    1.3.4 Filler Spreading Characteristics 13
    1.3.5 Surface Roughness of Components 14
    1.3.6 Dissolution of Parent Materials by Molten Fillers 16
    1.3.7 Significance of the Joint Gap 18
1.4 The Design and Application of Soldering and Brazing Processes 19
    1.4.1 Functional Requirements and Design Criteria 19
        1.4.1.1 Metallurgical Stability 19
        1.4.1.2 Mechanical Integrity 20
        1.4.1.3 Environmental Durability 20
        1.4.1.4 Electrical and Thermal Conductivity 20
    1.4.2 Processing Aspects 20
        1.4.2.1 Jigging of the Components 21
        1.4.2.2 Form of the Filler Metal 22
        1.4.2.3 Heating Method 23
        1.4.2.4 Temperature Measurement 23
        1.4.2.5 Joining Atmosphere 24
        1.4.2.6 Coatings Applied to Surfaces of Components 26
        1.4.2.7 Cleaning Treatments 26
        1.4.2.8 Heat Treatments Prior to Joining 26
        1.4.2.9 Heating Cycle of the Joining Operation 27
        1.4.2.10 Postjoining Treatments 28
References 28

Chapter 2: Filler Alloys and Their Metallurgy
2.1 Introduction 31
2.2 Survey of Filler Alloy Systems 32

2.2.1 Brazes 33
    2.2.1.1 Pure Silver 33
    2.2.1.2 Pure Copper 33
    2.2.1.3 Silver-Copper Eutectic 34
    2.2.1.4 Copper-Zinc and Silver-Zinc Alloys 36
    2.2.1.5 Silver-Copper-Zinc 36
    2.2.1.6 Silver-Copper-Zinc-Cadmium 39
    2.2.1.7 Silver-Copper-Zinc-Tin 42
    2.2.1.8 Gold-Bearing Filler Metals 43
    2.2.1.9 Palladium-Bearing Filler Metals 46
    2.2.1.10 Nickel-Bearing Filler Metals 47
    2.2.1.11 Other Brazing Alloy Families 50
  2.2.2 Aluminum- and Zinc-Bearing Filler Metals 50
    2.2.2.1 Aluminum-Bearing Brazes 51
    2.2.2.2 Zinc-Bearing Solders 52
  2.2.3 Solders 54
    2.2.3.1 Gold-Bearing Filler Metals 54
    2.2.3.2 Other Solder Alloy Families 58
    2.2.3.3 Lead-Tin Solders 66
References 69

Chapter 3: Application of Phase Diagrams to Soldering and Brazing
3.1 Introduction..71
3.2 Application of Phase Diagrams to Soldering and Brazing 72
  3.2.1 Examples Drawn From Binary Alloy Systems.............................73
  3.2.2 Examples Drawn From Ternary Alloy Systems 81
  3.2.3 Complexities Presented by Higher-Order and Nonmetallic Systems..........................91
3.3 A Methodology for Determining Phase Diagrams 91
  3.3.1 Literature Search 92
  3.3.2 Metallographic Examination 93
  3.3.3 Subdivision of High-Order Systems 93
  3.3.4 Predicting the Composition of Eutectic Points in High-Order Systems......................95
  3.3.5 Thermal Analysis 96
    3.3.5.1 Preparation of Alloys for Thermal Analysis 96
    3.3.5.2 Selection of Alloys for Thermal Analysis 97
  3.3.6 Quantitative Metallography 100
Appendix 3.1: Conversion Between Weight and Atomic Fraction of Constituents
  of Alloys...103
Appendix 3.2: Methods of Thermal Analysis 104
  Differential Thermal Analysis 104
  Differential Scanning Calorimetry 105
  Calorimetric Thermal Analysis 106
  Thermogravimetric Analysis 106
  Thermomechanical Analysis 107
References 108

Chapter 4: The Role of Materials in Defining Process Constraints
4.1 Introduction..111

4.2 Metallurgical Constraints and Solutions 113
    4.2.1 Wetting of Metals 114
    4.2.2 Wetting of Nonmetals 115
    4.2.3 Erosion 117
    4.2.4 Phase Formation 118
4.3 Mechanical Constraints and Solutions 118
    4.3.1 Interlayers 120
    4.3.2 Compliant Structures 121
4.4 Constraints Imposed by the Components and Solutions 123
    4.4.1 Strength as a Function of Joint Area 123
        4.4.1.1 Trapped Gas 124
        4.4.1.2 Solidification Shrinkage 127
    4.4.2 Diffusion Soldering and Brazing 128
    4.4.3 Joints to Strong Materials 132
        4.4.3.1 Joint Design To Minimize Concentration of Stresses 133
        4.4.3.2 Reinforced Filler Alloys To Enhance Joint Strength 138
Appendix 4.1: A Brief Survey of the Main Metallization Techniques 140
    Physical Vapor Deposition 140
    Chemical Vapor Deposition 140
    Wet Plating 140
    Thick Film 140
References 143

Chapter 5: The Joining Environment

5.1 Introduction 145
5.2 Joining Atmospheres 145
    5.2.1 Atmospheres and Reduction of Oxide Films 147
    5.2.2 Thermodynamic Aspects of Oxide Reduction 147
    5.2.3 Practical Application of the Ellingham Diagram 150
        5.2.3.1 Joining in Inert Atmospheres and Vacuum 150
        5.2.3.2 Joining in Reducing Atmospheres 151
        5.2.3.3 "Fluxless" Brazing of Aluminum 155
5.3 Chemical Fluxes 156
    5.3.1 Soldering Fluxes 157
    5.3.2 Brazing Fluxes 159
    5.3.3 Fluxes for Aluminum and Its Alloys 160
        5.3.3.1 Aluminum Soldering Fluxes 161
        5.3.3.2 Aluminum Brazing Fluxes 161
        5.3.3.3 Aluminum Gaseous Fluxes 162
    5.3.4 Self-Fluxing Filler Alloys 162
    5.3.5 Ultrasonic Fluxing 163
5.4 Reactive Filler Alloys 164
    5.4.1 Wetting of Ceramics 164
    5.4.2 Influence of Concentration of the Reactive Constituent 166
    5.4.3 Formation and Nature of the Reaction Products 168
    5.4.4 Mechanical Properties of Joints 169
5.5 Health, Safety, and Environmental Aspects of Soldering and Brazing 171

Appendix 5.1: Thermodynamic Equilibrium and the Boundary Conditions
   for Spontaneous Chemical Reaction 175
     The First Law of Thermodynamics and Internal Energy 175
     Entropy and the Second Law of Thermodynamics...175
     Dependence of Gibbs Free Energy on Pressure..176
  References...178

Chapter 6: Assessment of Joint Quality

6.1 Introduction..181
6.2 Evaluation of Metallization Quality...183
    6.2.1 Tests for Adhesion 184
    6.2.2 Tests for Surface Cleanliness and Wettability 184
    6.2.3 Tests for Thickness and Thickness Uniformity 184
6.3 Assessment of Wetting 186
    6.3.1 Spreading Tests..188
    6.3.2 Wetting Balance Testing 191
    6.3.3 Accelerated Aging of Components..193
6.4 Microscopic Examination of Joints...194
    6.4.1 Metallographic Examination 195
    6.4.2 Scanning Acoustic Microscopy 196
    6.4.3 Scanning Electron Microscopy 198
6.5 Mechanical Testing of Joints 201
    6.5.1 Tensile Strength...206
    6.5.2 Shear Strength 208
    6.5.3 Impact Resistance, Bending, and Peeling Strength 209
    6.5.4 Creep Rupture Strength 211
    6.5.5 Fatigue Strength 212
    6.5.6 Other Mechanical Tests 215
6.6 Nondestructive Evaluation of Joints and Subassemblies 216
    6.6.1 Visual Inspection and Metrology 216
    6.6.2 Hardness Measurements 217
    6.6.3 Liquid Penetrant Inspection 218
    6.6.4 X-Radiography..219
    6.6.5 Ultrasonic Inspection 222
    6.6.6 Technique Selection 225
6.7 Evaluation of Fabricated Products 225
Appendix 6.1: Relationships Among Spread Ratio, Spread Factor, and Contact
   Angle of Droplets..226
     Spread Ratio and Contact Angle 226
     Spread Factor and Contact Angle 226
     Contact Angle and Dimensions of the Solidified Pool of Filler....................................226
References...227

Chapter 7: Characterization and Process Development in Soldering and Brazing: Selected Case Studies

7.1 Introduction..229
7.2 Illustration of Factors That Can Influence Joint Integrity 229
    7.2.1 Effect of Impurity Elements on the Properties of Filler Metals and Joints.................229

7.2.1.1 Formation of Embrittling Interfacial Phases in Brazed Joints 229
7.2.1.2 Generation of Voids That Compromise the Strength of Soldered Joints 230
7.2.1.3 Adverse Effect on Wetting and Spreading Characteristics 230
7.2.1.4 Beneficial Effect in Preventing "Tin Pest" in Solders 231
7.2.2 Corrosion of Soldered and Brazed Joints in Tap Water 232
7.2.2.1 Water Installations of Copper 232
7.2.2.2 Water Installations of Stainless Steel 233
7.2.3 Effect of Filler/Component Reactions on Properties and Performance of Joints 234
7.2.3.1 Changes Introduced to the Melting Point of Soft Solders 234
7.2.3.2 Metallurgical Reactions That Vary With Solder Composition 235
7.2.3.3 Influence of Solder Composition on the Mechanical Properties of Soldered Joints 236
7.3 Practical Examples of Process Design 236
7.3.1 Mitigating Thermal Expansion Mismatch in Bonded Assemblies 237
7.3.1.1 Background 237
7.3.1.2 Formulating the Problem 238
7.3.1.3 Solution 238
7.3.2 Designing Soldered Joints for Hermetic Seals 241
7.3.2.1 Specifying the Functional Requirements 241
7.3.2.2 Design Guidelines 242
7.3.2.3 Solution 242
7.3.2.4 Postscript: Sputtered Chromium Metallizations 244
7.3.3 Development of an Improved Solder for Die Attachment in Electronics Fabrication 244
7.3.3.1 Background 244
7.3.3.2 Formulating the Problem 245
7.3.3.3 Improving the Spreading Characteristics of the Au-2Si Solder 246
7.3.3.4 Improving the Wetting of Silicon by Gold Solders 247
7.3.4 Problem Solving: Improving Reliability by Modifying Joint Geometry 247
7.3.4.1 Introduction 247
7.3.4.2 Defining the Problem 247
7.3.4.3 Identification of a Viable Solution 248
7.3.5 Product Development: Improving the Fabrication of Silicon Power Device Assemblies 250
7.3.5.1 Introduction 250
7.3.5.2 Defining the Problem 250
7.3.5.3 Solutions by Incremental Changes to the Fabrication Process 252
7.3.5.4 Solutions by Radical Changes to the Fabrication Process 252
7.3.5.5 Outcome 254
References 254

Index 257

Chapter 1

Introduction

1.1 *Joining Methods*

Soldering and brazing represent one of several types of methods for joining solid materials. The five methods primarily used are:

- Mechanical fastening
- Adhesive bonding
- Soldering and brazing
- Welding
- Diffusion bonding

Other methods, such as glass/metal sealing, electrostatic welding, and so forth, are dealt with elsewhere [Bever 1986].

Schematic illustrations of these joining methods are shown in Fig. 1.1. These methods have a number of common features but also certain significant differences. For example, soldering and brazing are the only joining methods that can produce smooth and rounded fillets at the periphery of the joints. The joining methods have been listed above in the order in which they lead to fusion of the joint surfaces and tend towards a "seamless" joint.

Because soldering and brazing lie in the middle of this sequence, they share several of the features of the other methods. For example, soldered and brazed joints can be endowed with the advantageous mechanical properties of welded and diffusion-bonded joints; at the same time, they can be readily disassembled, usually without detriment to the components, like mechanically fastened joints. These features make soldering and brazing highly versatile.

The principal characteristics of the various joining methods are summarized below.

1.1.1 Mechanical Fastening

Mechanical fastening involves the clamping together of components without fusion of the joint surfaces. This method often, but not always, relies on the use of clamping members, such as screws and rivets. In crimping, the components are keyed together by mechanical deformation.

Characteristic features of mechanical fastening include:

- A heating cycle is generally not applied to the components being joined. A notable exception is riveting, where the rivets used for clamping are heated immediately prior to the fastening operation. On subsequent cooling the rivets shrink, causing the components to be clamped tightly together.
- The reliance on local stressing to effect joining requires thickening or some other means of reinforcement of the components in the joint region. This places a severe restriction on the joint geometries that can be used and imposes a weight penalty on the assembly. Another constraint on permissible joint configu-

Principles of Soldering and Brazing

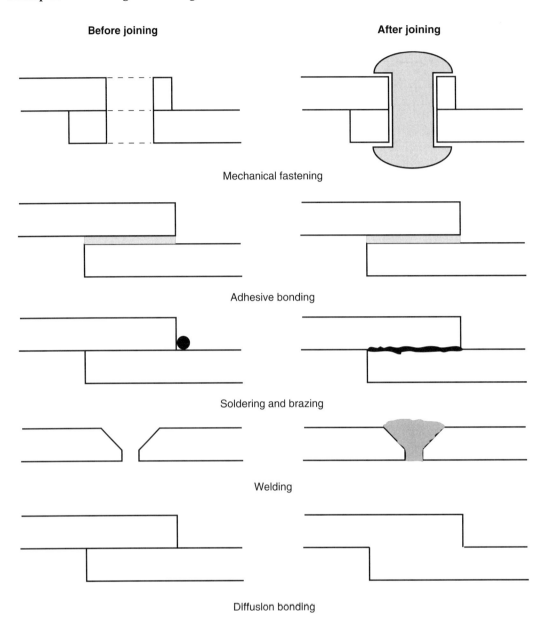

Fig. 1.1 Principal methods for joining engineering materials, represented schematically

rations is the need for access to insert the clamping member.
- The method usually requires special mechanical preparation, such as drilling holes, machining screw threads, or perhaps chamfering of abutting surfaces in the case of components to be crimped.
- The choice of suitable joint configurations is highly dependent on service conditions—for example, whether or not leak tightness is re-

quired. Joints may be designed to accommodate thermal expansion mismatch between the components in the assembly. In the extreme case, joints can be made to permit complete freedom of movement in the plane perpendicular to the clamping member, as applied to the fishplates used to couple train rails.
- The electrical and thermal conductance across the joint is a function of the effective area that is in contact. This depends on many other pa-

rameters, such as the clamping force and the materials used, and in service is unlikely to be constant.

1.1.2 Adhesive Bonding

Adhesive bonding involves the use of a polymeric material, often containing various additives, to "stick" components together. The process involves a chemical reaction, which may simply comprise exposure of the adhesive to air, leading to the formation of a hydrogen-type bond between the cured adhesive and the respective components. The original interfaces of the joint are often preserved in this type of bonding process.

Characteristic features of adhesive bonding include:

- It is inherently a low-stress joining method, because it is carried out at relatively low temperatures and most adhesives have high compliance.
- A diverse range of methods is available for curing adhesives.
- The geometry of the components tends not to be critical.
- Constraints apply to the geometry of the actual joint; in particular, large areas and very narrow gaps are necessary to ensure mechanical integrity.
- Joints tend to be weak when subjected to forces that cause peeling. For this reason, adhesive joints are frequently used in combination with mechanical fastening—for example, in airframe assembly.
- Joint integrity tends to be sensitive to the atmosphere of the service environment and to the state of cleanliness of the mating surfaces.
- The service temperature range of adhesively bonded joints is usually limited.
- Joints usually possess poor electrical and thermal conductivity, although loading the organic adhesive with metal particles allows moderate conductance to be achieved.

1.1.3 Soldering and Brazing

Soldering and brazing involve the use of a molten filler metal to wet the mating surfaces of a joint, with or without the aid of a fluxing agent, leading to the formation of metallurgical bonds between the filler and the respective components. In these processes, the original surfaces of the components are "eroded" by virtue of the reaction occurring between the molten filler metal and the solid components, but the extent of this "erosion" is usually at the microscopic level (<100 µm, or 4000 µin.). Joining processes of this type, by convention, are defined as soldering if the filler melts below 450 °C (840 °F) and as brazing if it melts above this temperature.

Characteristic features of soldering and brazing include:

- All brazing operations and most soldering operations involve heating the filler and joint surfaces above ambient temperature.
- In most cases, the service temperature of the assembly must be lower than the melting temperature of the filler metal.
- It is not always necessary to clean the surfaces of components prior to the joining operation, because fluxes are available that are capable of removing most oxides and organic films. However, there are penalties associated with the use of fluxes—for example, the residues that they leave behind, which are often corrosive and can be difficult to remove.
- The appropriate joint and component geometries are governed by the filler/component material combination and by service requirements (need for hermeticity, stress loading, positional tolerances, etc.). Complex geometries and combinations of thick and thin sections can usually be soldered and brazed together.
- Intricate assemblies can be produced with low distortion, high fatigue resistance, and good resistance to thermal shock.
- Joints tend to be strong if well filled, unless embrittling phases are produced by reaction between the filler metal and the components.
- Soldered and brazed joints can be endowed with physical and chemical properties that approximately match and, in some cases, even exceed those of the components.
- Fillets are formed under favorable conditions. These can act as stress reducers at the edges of joints, improving the overall mechanical properties of the joined assembly.
- Soldering and brazing can be applied to a wide variety of materials, including metals, ceramics, plastics, and composite materials. For many materials, and plastics in particular, it is

necessary to apply a surface metallization prior to joining.

1.1.4 Welding

Welding involves the fusion of the joint surfaces by controlled melting through heat being specifically directed toward the joint. Commonly used heating sources are plasma arcs, electron beams, lasers, and electrical current through the components and across the joints (electrical resistance). Filler metals may be used to supplement the fusion process for components of similar composition, as for example when the joint gap is wide and of variable width. In that situation, the filler is often chosen to have a marginally lower melting point than the components in order to ensure that it completely melts.

Characteristic features of welding include:

- Welding invariably involves a heating cycle, which tends to be rapid.
- Welding cannot be used to join metals to nonmetals or materials of greatly differing melting points. There are exceptions to this, but these are generally limited to precise combinations of materials and highly specific welding methods.
- Joint geometries are limited by the requirement that all joint surfaces be accessible to the concentrated heat source.
- Welded joints may approach the physical integrity of the components, but are often inferior in their mechanical properties, particularly fatigue resistance. This is due to stress concentrations produced by the high thermal gradients developed during joining and the relatively rough surface texture of welds.
- The heating cycle usually affects the microstructure and hence the properties of the components over a macroscopic region around the joint, called the heat-affected zone (HAZ). The HAZ is often influential in determining the properties of welded joints.
- Welding tends to distort the components in the region of the HAZ. This is associated with the thermal gradients developed through the use of a concentrated heat source to fuse the joint surfaces.

1.1.5 Diffusion Bonding

Diffusion bonding involves the removal of joint interfaces through the migration of atoms toward and across the joint interfaces, induced by thermal activation. Because no filler materials are used, it is necessary for the interfaces to be placed in intimate contact with one another. This may be achieved by compressively loading to the assembly either before or during the bonding operation. Alternatively, joint formation may be promoted by the generation of a liquid phase through reaction of the migrating species with constituents of the components, in which case the pressure requirement may be relaxed and the other features of the process also tend to those of soldering and brazing. For very small components, usually in the microelectronics industry, the static load may be replaced by frictional agitation, which simultaneously raises the temperature of the joint, enabling ultrasonic vibration to be used as the energy source to produce diffusion-bonded joints.

Characteristic features of solid-state diffusion bonding include:

- This method generally involves heating of the joint to a temperature below the melting point of the components.
- The bonding process is relatively slow, because it relies on solid-state diffusion. Ultrasonic agitation applied to small-area joints (<1 mm^2, or 0.0016 in.^{2}) can reduce the bonding time significantly (<0.1 s per joint).
- The joints have no fillets, although the edges are often rounded due to local creep of the components in response to the temperature/stress conditions of the bonding cycle.
- The service temperature of joined assemblies can be higher than the joining temperature and tend toward the melting point of the components.
- Diffusion bonding is limited in application to specific combinations of materials that provide adequate diffusion without a consequential formation of voids (Kirkendall porosity) or embrittling phases in the joint.
- Of all the joining methods, it is the least tolerant to poor mating of the joint surfaces.
- Joint surfaces must be scrupulously clean, because diffusion bonding is a fluxless process.
- The properties of diffusion-bonded joints can approach those of the parent materials.

1.2 Comparison of Solders and Brazes

In many respects it is fruitful to consider solders together with brazes. This integrated treatment can be justified on metallurgical grounds. These two classes of filler cannot be demarcated by a temperature boundary as is generally done: conventionally, solders are defined as filler metals with melting points below 450 °C (840 °F) and brazes as having melting points above this temperature. This distinction has a historical origin. The earliest solders were based on alloys of tin, while brazes were based on copper-zinc alloys. Indeed, the word "braze" is a derivative of the Old English *braes*, meaning to cover with brass. On the other hand, the term "solder" is an adaptation of the Old French *soudure*, meaning to make solid.

The type of metallurgical reaction between a filler and parent metal is sometimes used to differentiate soldering from brazing. Solders usually react to form intermetallic phases, that is, compounds of the constituent elements that have different atomic arrangements than the elements in solid form. By contrast, most brazes form solid solutions, which are mixtures of the constituents on an atomic scale. However, this distinction does not have universal validity. For example, silver-copper-phosphorus brazes react with steels to form the interfacial phase of Fe_3P in a similar manner to the reaction of tin-base solders with iron or steels to form $FeSn_2$. On the other hand, solid solutions form between silver-lead solders and copper just as they do between the common silver-base brazes and copper.

Soldering and brazing involve the same bonding mechanism: that is, reaction with the parent material, usually alloying, so as to form metallic bonds at the interface. In both situations, good wetting promotes the formation of fillets that serve to enhance the strength of the joints. Similar processing conditions are required, and the physical properties are comparable provided that the same homologous temperature (the temperature at which the properties are measured as a fraction of the melting temperature expressed in degrees Kelvin) is used for the comparison.

The perpetuation of the distinction of solders from brazes on the basis of the 450 °C (840 °F) boundary has arisen from the significant gap that exists between the melting points of available solder alloys, the highest being that of Au-3Si, which melts at 363 °C (685 °F), and the lowest temperature standard braze, the Al-4Cu-10Si alloy, which melts at 524 °C (975 °F) but which, being a noneutectic alloy, is fully liquid only above 585 °C (1085 °F). Eutectic alloys are defined in Chapter 3; for the present, it shall suffice to state that eutectic alloys are akin to pure metals in that they melt and freeze at the same temperature. The temperature ranges of the principal solder and braze alloy families are shown in Fig. 1.2 and 1.3.

For most purposes, the temperature gap between solders and brazes is substantially wider than 160 °C (290 °F). This is because the gold-base solders are expensive and are largely limited to use in the high-added-value manufacturing of the electronics industry. Removing the high-gold-content alloys from consideration, the highest-melting-point solders are the lead-rich alloys, which melt at about 300 °C (570 °F). The lowest-melting-point brazes used commercially in significant quantities are the reasonably ductile alloys based on aluminum-silicon, which melt at 577 °C (1070 °F). A selection of eutectic alloys with melting points in the temperature range of 300 to 550 °C (570 to 1020 °F) that have been promoted as solders and brazes are listed in Table 1.1. They are, without exception, brittle and often contain one or more volatile constituents, notably magnesium, cadmium, or zinc.

The dearth of filler metals with melting points in the range of 300 to 550 °C (570 to 1020 °F) is not necessarily a handicap; techniques are available for making joints using molten filler metal with effective melting points in this temperature interval. Transient-liquid phase diffusion bonding is one such example and is discussed in Chapter 4. Moreover, multicomponent alloys are under development that melt in this temperature interval. Some of these are discussed in Chapter 2.

From the "maps" of solders and brazes presented in Fig. 1.2 and 1.3, it might appear that there are many more solders than brazes. In fact, the contrary is true. The alloys that are specifically indicated in these figures are mostly eutectic compositions or those characterized by narrow melting ranges. Most commercially used solders are included, because these are almost all of eutectic composition. However, entire families of brazes have been omitted based on the fact that there is no eutectic in the alloy system and that they instead exhibit complete intersolubility. Examples are the copper-nickel, silver-gold, silver-palladium, and silver-gold-palladium alloys. Alloys in such systems melt over a temperature range that varies with composition.

Principles of Soldering and Brazing

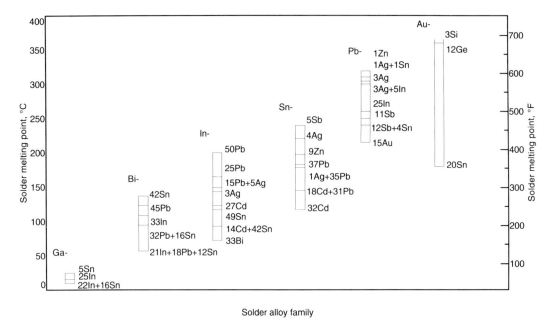

Fig. 1.2 Principal solder alloy families and their melting ranges

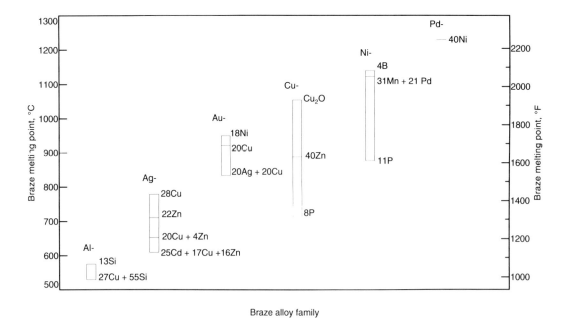

Fig. 1.3 Principal braze alloy families and their melting ranges

Table 1.1 Selected eutectic alloys promoted as high-melting-point solders and low-melting-point brazes

Solder composition, wt%	Melting point °C	°F	Problems
5Ag-95Cd	340	644	Toxic fumes, volatile
75Au-25Sb	356	680	High cost, brittle
88Au-12Ge	361	682	High cost, brittle
97Au-3Si	363	685	High cost, brittle
6Al-94Zn	381	718	Volatile, brittle
48Al-52Ge	424	795	High dross, brittle
36Al-37Mg	450	842	Volatile, brittle
75Pb-25Pd	454	849	Poor fatigue resistance, brittle
56Ag-44Sb	485	905	Volatile, brittle
58Au-42In	495	923	High cost, brittle
68Al-27Cu-5Si	524	975	Difficult to clean, brittle
23Ag-53Cd-24Cu	525	977	Toxic fumes, volatile, brittle
24Cu-76Sb	526	979	Volatile, brittle
62Cd-38Cu	549	1020	Toxic fumes, volatile, brittle

The higher process temperatures needed to make brazed joints have important consequences, because more thermal activation energy is present. These are:

- More extensive metallurgical reaction between the filler metal and the substrate. Solders typically do not dissolve more than a few microns of the component surfaces, whereas brazes often dissolve tens of microns. Larger changes in the composition of the filler metal therefore occur during brazing, which in turn significantly affect the fluidity of and wetting by the molten filler as well as the properties of the joint.
- Greater reactivity with the atmosphere surrounding the workpiece. All other factors being equal, brazes are less tolerant to oxidizing atmospheres than solders but, for the same reasons, are also better suited to cleaning by reducing atmospheres. When joints are made in air with the aid of a flux, the greater reactivity of brazes means that a higher proportion of flux to filler metal is generally required. Consequently, flux-cored solders are adequate for soldering in air, while brazing rods intended for use in ambient atmosphere must be provided with a thick external coating of flux. Fluxes are discussed in Chapter 5.

Most, but not all, soldering and brazing processes are performed at small excess temperatures above the melting point of the filler metal (commonly referred to as the superheat). Much higher process temperatures are occasionally used when it is desirable to exploit thermal activation. For example, tin-containing solders can wet and join nonmetallized ceramics provided that the solder incorporates an active ingredient, such as titanium, and the alloy is heated above about 900 °C (1650 °F) [Kapoor and Eagar 1989]. Although the freezing point of the solder is unchanged at about 250 °C (480 °F), such "activated" solders have several of the characteristics of brazes at the process temperature.

Several general features distinguish the majority of solders from brazes, namely:

- Most commercial solders are of eutectic composition, because there is usually a need to minimize the processing temperature while maintaining reasonable fluidity of the molten filler. Also, these alloys, which are intrinsically soft and therefore weak, must be conferred with optimum mechanical properties; these generally are achieved by a fine-grained microstructure, which is a characteristic feature of a true eutectic alloy.
- Most brazes possess mutual solid solubility between their constituents and are therefore offered with a wide range of compositions and melting ranges. The low degree of intersolubility and the propensity to form intermetallic compounds possessed by solder alloys are related to the noncubic crystal symmetry of their constituent elements.
- Solders find application at temperatures that are between 50 and 90% of their melting point in degrees Kelvin, under strain levels that

often exceed 10%. At these relatively high temperatures, the alloys are not metallurgically stable and the joint microstructure tends to change with time. Brazes tend to be used at temperatures that are relatively much lower, usually below half their melting point in degrees Kelvin.

These points are discussed in further detail in Chapters 2 and 6. Notwithstanding the differences, solders and brazes operate on similar principles; hence, the frequent use of the collective term "filler" throughout this book has some justification.

1.3 Key Parameters of Soldering and Brazing

The quality of soldered and brazed joints depends strongly on the combination of filler and component materials, including surface coatings that may be applied to the components, and also on the processing conditions that are used. It is precisely for this reason that a sound understanding of the metallurgical changes accompanying the sequence of events that occurs in making soldered and brazed joints is so vital for developing reliable joining processes.

Soldering and brazing technology has generally evolved in an empirical manner, largely by trial and error. Theoretical principles have helped to furnish insights, guidelines, and qualitative explanations for this technology, but have rarely provided reliable data for use in the design of joining processes. The basic difficulty is that the real situation is highly complex, as it brings into play a large number of variables, some of which may not be easy to recognize. Among the relevant factors are the condition of the solid surfaces (i.e., the nature of any oxides or other coatings, surface roughness, etc.), the temperature gradients that develop during the joining operation, the metallurgical reactions involving the filler and parent materials, and the chemical reactions with fluxes when these are used.

Another key aspect of joining with fillers is the manner and extent of flow of the molten filler into the joint. This is influenced by the following features:

- Dimensions of the joint
- Spread characteristics of the filler metal
- Surface condition of the components

The limitation of theory in accounting for observed behavior is well illustrated by the classical model of wetting and spreading. This model nevertheless does provide useful concepts and insights. It is given a detailed treatment by Harkins [1952] and will not be repeated here. For the purposes of the present discussion, it will suffice to outline the main features of this model.

In the classical model of wetting, the surface of the solid is taken to be invariant as a liquid droplet spreads over it. That is to say, the reaction between the liquid and the solid components across their common interface is considered to be negligible. It is also assumed that the composition and other characteristics of the solid and liquid components, likewise, do not change with time. This assumption is not generally valid, as shall be shown.

1.3.1 Surface Energy and Surface Tension

At this juncture, the concepts of surface energy and surface tension will be briefly reviewed. Figure 1.4 provides a simplified representation of the atomic structure of a solid close to one of its free surfaces. The atom at position A in the bulk of the solid has a balanced array of neighboring atoms, whereas atom B at the surface of the solid is lacking in neighbors above it, apart from the occa-

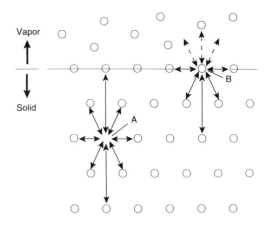

Fig. 1.4 Simplified diagram of surface energies. Atom B, at the surface, has unsaturated bonds and thus a higher energy than atom A. This difference in energy is the origin of surface energy γ_{SV}.

sional vapor molecule, and thus, has some unsaturated bonds.

The potential energy of atoms at the free surface, such as B, is higher than the energy of atoms within the bulk of the solid, such as A, by the energy of the unsaturated bonds. The aggregate of this excess energy that is possessed by atoms in the vicinity of the free surface constitutes the surface energy of the solid. In a similar manner, a liquid also possesses a surface energy, which is directly manifested in the tendency to draw up into drops. If small, the droplets are perfect spheres. Because a sphere has the smallest surface-to-volume ratio of any shape, it is clear that the surface energy of a liquid is greater than its volume energy. In the classical model, when a liquid spreads over a surface, the volume remains constant, because evaporation and reaction with the substrate are excluded. Therefore, only surface energy changes must be considered.

A surface of a liquid acts like an elastic skin covering the liquid; in other words, the surface is in a state of tension. The tensile force (F), known as surface tension (γ), is defined as the force acting at right angles to a line of unit length (L) drawn in the surface. The relationship between surface tension and surface energy under specific conditions can be seen as follows.

Consider a liquid film of length L and width W. Apply a force F at a barrier AB, as shown in Fig. 1.5, parallel to one surface of the film, so as to extend the liquid film a distance x.

The increase in area of the film is xL.

The work done in obtaining this increase is the mathematical product of the force applied times the distance moved, or Fx.

Work done by the liquid film in opposing the increase in area, under isothermal conditions (i.e., constant temperature), is $2\gamma xL$, where γ is the surface tension force acting on each surface at the prescribed temperature.

At a fixed temperature (under isothermal conditions),

$$Fx = 2\gamma xL$$

Rearranging, $F/L = 2\gamma$, or $F/L = \gamma$ for each surface.

Thus, surface energy is equivalent to surface tension under isothermal conditions. The MKS (meter-kilogram-second) unit of surface energy is $J \cdot m^{-2}$ and that of surface tension is $N \cdot m^{-1}$. Because these parameters are properties of an interface (e.g., between liquid and air), surface energy and tension must be defined with reference to the appropriate pair of materials that meet at the interface, and the test conditions, such as temperature and atmosphere, also must be specified.

1.3.2 Wetting and Contact Angle

According to the classical model of wetting, the liquid will spread over a solid surface until the three surface tensions—between the liquid droplet and the solid substrate, the liquid droplet and the atmosphere, and the substrate and the atmosphere—are in balance as shown in Fig. 1.6.

According to the balance of forces,

$$\gamma_{SL} = \gamma_{SV} - \gamma_{LV} \cos \theta \qquad \text{(Eq 1.1)}$$

where γ_{SL} is the surface tension between the solid and liquid, γ_{SV} is the surface tension between the solid and the vapor, γ_{LV} is the surface tension between the liquid and the vapor, and θ is the contact angle of the liquid droplet on the solid surface. Equation 1.1, known as the wetting equation, shows that $\theta < 90°$ corresponds to the condition $\gamma_{SV} > \gamma_{SL}$. This imbalance in surface tension (i.e., surface energy) provides the driving force for the

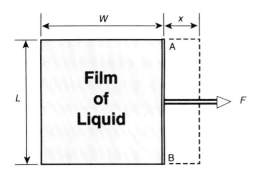

Fig. 1.5 Schematic diagram used to explain the relationship between surface energy and surface tension

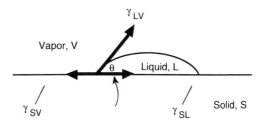

Fig. 1.6 Surface tension forces acting when a liquid droplet wets a solid surface, according to the classical model

spreading of liquid over the solid surface and diminution of the unwetted surface area.

The contact angle θ provides a measure of the quality of wetting. Thus, if $90° < \theta < 180°$, some wetting is said to occur, but a liquid droplet will not spread on the surface with which it is in contact. On the other hand, if $\theta < 90°$, a liquid droplet will wet the substrate and also spread over an area defined by the contact angle θ. Clearly, the area of spreading will increase with decreasing contact angle. For further details of the interrelationship between these two parameters, refer to the appendix to Chapter 6.

Rewriting Eq 1.1 in terms of $\cos \theta$,

$$\cos \theta = \frac{\gamma_{SV} - \gamma_{SL}}{\gamma_{LV}}$$

Thus, wetting is improved by decreasing θ. This can be achieved by:

- Increasing γ_{SV}
- Decreasing γ_{SL}
- Decreasing γ_{LV}

The term γ_{SV} can be maximized for a given solid by cleaning the surfaces. The presence of adsorbed material, such as water vapor, dust, or other nonmetallic surface films on a metal surface, markedly reduces γ_{SV} and correspondingly raises the contact angle θ. Therefore, it is important in soldering and brazing that joint surfaces be clean and metallic—hence the need for fluxes or protective atmospheres to achieve and then maintain this condition.

The term γ_{SL} is a constant at a fixed temperature for a particular solid-liquid combination, according to the classical model of wetting. This parameter can be reduced by changing the materials system. However, this is not usually easy to achieve in practice, because component materials are specified to fulfill certain functional requirements. Fortunately, γ_{SL} is highly temperature dependent and usually decreases rapidly with increasing temperature [Schwartz 1987, Table 1.4], thereby providing a ready means of controlling spreading.

The term γ_{LV} is a constant at a fixed temperature and pressure for a particular liquid-vapor combination, but can be varied by altering the composition and pressure of the atmosphere. Although the composition of the atmosphere used for the joining operation is known to affect the contact angle, in practice it is often easier to promote spreading by reducing the pressure of the atmosphere. This is one of the reasons for the popularity of vacuum-based joining processes.

In general, the relative magnitudes of the surface energies are $\gamma_{SV} > \gamma_{SL} > \gamma_{LV}$. For water wetting on mica, subjected to an atmosphere of water vapor, the following values have been measured [Tabor 1969, p 215-216]:

$\gamma_{SV} = 0.183 \text{ N} \cdot \text{m}^{-1}$
$\gamma_{SL} = 0.107 \text{ N} \cdot \text{m}^{-1}$
$\gamma_{LV} = 0.073 \text{ N} \cdot \text{m}^{-1}$

Thus, $\cos \theta = (0.183 - 0.107)/0.073 = 1$ (within the limits of experimental error), and the contact angle, $\theta = 0°$.

The wetting equation (Eq 1.1) applies when the liquid is practically insoluble in the solid over which it spreads (i.e., the solubility is less than 0.1%). For binary metal systems where this condition is satisfied (e.g., silver-iron, copper-iron, tin-aluminum), it has been shown that the wetting equation can be reduced to:

$$\cos \theta = 1 + k \left[\frac{T_m^s}{T_m^l} - 1 \right]$$

where k is a constant equal to approximately 0.3, T_m^s is the melting point of the solid metal, and T_m^l is the melting point of the liquid metal. This expression has been verified experimentally [Eustathopoulos and Coudurier 1979]. Higher-order metal systems (ternary, quaternary, etc.) are considerably more complex, and the wetting equation cannot be formulated in such a simple form. A more sophisticated analysis of wetting that takes into account the influence of certain microscopic features, including the influence of local defects and van der Waal forces, is provided by de Gennes [1985]. However this is still a continuum analysis and does not consider the local atomic environment.

So far, we have considered idealized filler spread over a single surface. In a joint there are always two facing surfaces. If the contact angle θ is less than $90°$, the surface energies will give rise to a positive capillary force that will act to fill the joint. For a pair of vertical parallel plates, D mm apart and partly immersed in a liquid, the capillary force per length (in millimeters) of joint is equal to $2\gamma_{LV} \cos \theta$. Under this force, the liquid will rise to an equilibrium height h at which the capillary

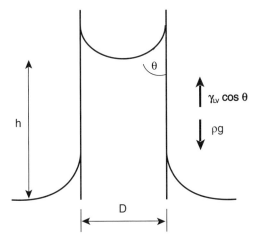

Fig. 1.7 Rise of a liquid between two parallel plates by capillary force

force balances the hydrostatic force (as shown schematically in Fig. 1.7), such that:

$$h = \frac{2\gamma_{LV} \cos \theta}{\rho g D} \quad \text{(Eq 1.2)}$$

where ρ is the density of the liquid and g is the acceleration due to gravity.

The actual situation in soldering and brazing is much more complex than that represented by Eq 1.2 and the classical wetting model. The irreversible nature of spreading and the time dependence of contact angle that is commonly observed are at variance with this simple model. These and other departures from the classical model occur because the joining process almost invariably involves a degree of chemical reaction between the filler metal and the solid surface, which is neglected in the conventional model. This is clearly demonstrated in a study [Schwartz 1987, p 10-11] that showed that the contact angle for various liquid metals on freshly cleaned beryllium generally decreased with time, over a period of several minutes, at a fixed temperature. This is consistent with the known reaction of beryllium with constituents of most brazing alloys. More predictably, perhaps, was the finding that the contact angle decreased with increasing temperature and that the atmosphere in which the test was conducted also made a difference.

Reactions between a filler metal and the substrates often result in dissolution of the surface of the substrate; this process usually leads to the growth of new phases. Frequently these phases are intermetallic compounds that appear either distributed throughout the joint or form as layers adjacent to the surface of the solid substrate.

The energy of formation of an intermetallic layer by reaction between a molten solder and a solid substrate has been calculated by Yost and Romig [1988]. The energy of formation considered is the thermodynamic function known as the Gibbs free energy. This function and its properties are briefly explained in the appendix to Chapter 5. In order to simplify the analysis, Yost and Romig limited their consideration to the clean surfaces of pure metals, wetted by liquids of elemental metals, in the absence of fluxes, to form binary interfacial phases. It was demonstrated that the free energy of formation of intermetallic phases by reaction of liquid antimony, cadmium, and tin with solid copper was approximately two orders of magnitude larger than the energy release created by the surface energy imbalance during the advance of a spreading solder droplet, as exclusively considered in the classical model.

Therefore, in these cases, and probably more generally in soldering and brazing processes, the Gibbs free energy change that occurs on reaction by a filler with the substrate is demonstrably the dominant driving force for wetting. Empirical evidence for this is provided, for example, by the fact that the measured contact angle of molten germanium on silicon carbide at 1430 °C (2600 °F) is approximately 120°, whereas that of molten silicon on this ceramic at the same temperature is 38° [Li and Hausner 1991]. The substantial difference in the two contact angles cannot be accounted for by the difference in γ_{LV} in the wetting equation (Eq 1.1), because the values of γ_{SL} are likely to be similar. It can only be due to the greater intersolubility of silicon with silicon carbide. This example clearly demonstrates that the classical wetting equation cannot be relied on for a quantitative description of wetting, contact angle, or spreading.

The effect of metallurgical interaction between filler and the component material in promoting wetting is exploited in active filler metals: the addition of a small fraction of a reactive metal, such as titanium, hafnium, or zirconium, to fillers enables them to wet and spread over ceramic materials. In this instance, wetting of and reaction with the ceramic are inextricably linked. Activated filler alloys are discussed further in Chapter 5.

Although a low contact angle is used as an index for judging the quality of wetting, there are situations where higher contact angles are preferred. This is illustrated by Fig. 1.8, which shows two joints, one between two component surfaces

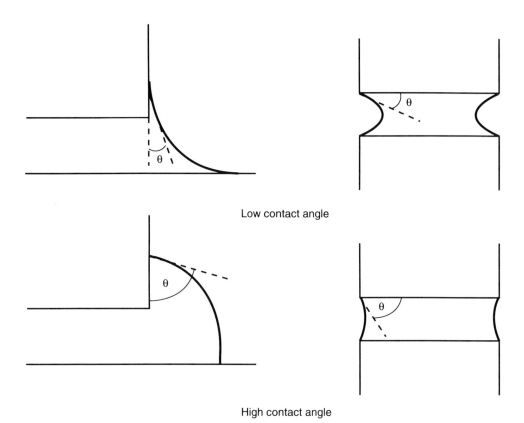

Fig. 1.8 Effect of contact angle on fillet formation and joint filling. Low contact angles tend to be preferred when external fillets can form. In other joint geometries, high contact angles result in lower stress concentrations.

of unequal area and the other between component surfaces that entirely correspond. In the first case, a low contact angle serves to form a gentle concave fillet, which enhances the mechanical properties of the joint. In the other configuration, a low contact angle encourages the formation of a neck in the joint, which can be a source of weakness. A contact angle close to 90° will eliminate this problem.

A further point in connection with wetting is that a situation can arise where the molten filler is physically prevented from achieving its equilibrium contact angle—as, for example, when a solder droplet is confined to a small metallized area. This is commonly encountered on electronic circuit boards, where solder droplets are constrained to individual metal pads. In this situation, the pad is often too narrow to accommodate the spherical metal cap that would form if this restriction did not apply. The enforced wetting angle imposes a pressure on the droplet that is often adequate to cause the solder to flow along the length of the conductor that leads away from the pad and, more seriously, to lift and flow under the solder resist that surrounds the pad. An analysis of the pressure arising from a nonequilibrium contact angle, using the classical wetting model, is given in Klein Wassink [1989, p 56].

1.3.3 Fluid Flow

According to the classical model, the force that acts to fill the joint gap with liquid is given by Eq 1.2. The liquid will flow into the joint under this force at a rate that is governed by its viscosity. Simple fluid flow theory assumes that:

- There is no interaction between the liquid and the solid surfaces with which it is in contact.
- All surfaces are smooth and perfectly clean.
- Flow is laminar, not turbulent.

For a detailed treatment of this subject, the reader is referred to a paper by Milner [1958]. Here, we shall merely quote the expression (given as Eq 8 in Milner's paper) for the volume rate of liquid flow, dV/dt, between a pair of horizontal parallel plates, length l, separated a distance D, under a pressure P per unit area transverse to the plates. The viscosity of the liquid is η.

$$\frac{dV}{dt} = \frac{PD^3}{12\eta l} \quad \text{(Eq 1.3)}$$

It is assumed that the liquid front will advance at a rate (dl/dt) equal to the mean velocity of flow, that is:

$$\frac{dl}{dt} = \left(\frac{1}{D}\right)\left(\frac{dV}{dt}\right) = \frac{PD^2}{12\eta l}$$

From the wetting equation (Eq 1.1), under isothermal conditions the change in surface energy as a unit area of a surface becomes wetted by the liquid is:

$$\gamma_{SL} - \gamma_{SV} = -\gamma_{LV} \cos \theta$$

Therefore, the change in surface energy when the pair of parallel plates becomes wetted is:

$$2l(\gamma_{SV} - \gamma_{SL}) = 2l\gamma_{LV} \cos \theta$$

It follows that the force acting on the liquid to cause it to wet the plates is:

$$F = \frac{2l\gamma_{LV} \cos \theta}{l}$$

so that the pressure is:

$$P = \frac{2\gamma_{LV} \cos \theta}{D}$$

and the velocity of flow of the liquid into the space between two parallel surfaces, of separation D, according to this simple model is given by:

$$\frac{dl}{dt} = \frac{\gamma_{LV} D \cos \theta}{6\eta l} \quad \text{(Eq 1.4)}$$

Equation 1.4 shows that the rate of liquid flow increases when:

- The liquid-vapor surface tension, γ_{LV}, increases
- The joint gap, D, increases
- The contact angle θ decreases

Rates of flow calculated from this expression for molten solders and brazes in joints 50 μm (2000 μin.) wide are typically 0.3 to 0.7 m/s (1 to 23 ft/s). In other words, a joint 5 mm (0.2 in.) long will be filled in a time on the order of 0.01 s. This implies that joint filling by the molten solder or braze occurs virtually instantaneously and that transient effects associated with fluid flow can generally be neglected in joining processes. De Gennes [1985] offers a more developed model of the dynamics of liquid spreading, in which the surface energy driving force is opposed by viscous drag and surface irregularities. Joint filling times of the order of 0.1 s are routinely measured on a meniscograph, an industry standard instrument used for determining solderability. This instrument and its capabilities are described in Chapter 6.

It should be noted that, although the rate of filling is proportional to the joint gap D, the driving force for filling, according to the classical model, is inversely proportional to D, that is, these two aspects of filling act in opposition.

1.3.4 Filler Spreading Characteristics

Molten filler metals do not all have the same spreading characteristics, although, with few exceptions, the degree of spread over an "ideal" substrate increases as the temperature is raised, the atmosphere is lowered and is made more reducing. In this context, an "ideal" substrate, suitable for reference purposes, needs to be defined. This is understood to possess a perfectly clean metal surface that is highly wettable by the filler metal under consideration, but with which it does not significantly alloy. Any alloying reactions will be highly specific to the combination of materials in question, so that the substrate will lose its ideal characteristics.

An example of a substrate that approximates the ideal, and which has been used by the authors in comparative soldering assessments, comprises a flat glass plate sputter-coated with 0.1 μm (4 μin.) of chromium and overlaid with 0.1 μm (4 μin.) of gold. The chromium represents a metal that is essentially insoluble in most solders, and the gold layer provides this reactive metal with protection against oxidation. The gold layer is sufficiently thin to not significantly alter the composition of a solder pellet as it spreads over the substrate [Humpston and Jacobson 1990].

Eutectic composition alloys are often regarded as having the best spreading characteristics, and this is frequently one of the reasons cited for their selection in preference to hypo- and hypereutectic compositions. The superior spreading of alloys of eutectic composition in comparison with off-eutectic alloys of the same system can be explained by the different melting characteristics in the two cases. An alloy of eutectic composition melts instantly. Spreading of the molten alloy is then driven by interaction with the substrate [Ambrose, Nicholas, and Stoneham 1992]. In the case of a noneutectic filler, melting, wetting, and spreading commence before the alloy is entirely molten, when it tends to be somewhat viscous. Under these conditions, movement of the filler will be relatively sluggish. By the time the alloy is completely molten, the filler will have partly alloyed with the substrate, and the driving force for spreading will have been diminished. Eutectic composition alloys also have lower viscosity than adjacent compositions when completely molten; further details are given in Chapter 3.

Whether or not the filler alloy is of eutectic composition is of much less importance to the phenomenon of spreading than the composition *per se*. The spreading of a filler metal depends greatly on the elemental constituents present and their relative proportions. The authors have compared the spreading characteristics, as a function of excess temperature above the melting point ("superheat"), of all combinations of the elements bismuth, indium, lead, silver, and tin when used as eutectic solders on "ideal" substrates [Humpston and Jacobson 1990]. The results presented in Fig. 1.9 show that the area of spreading increases at an accelerating rate as a function of the excess temperature above the melting point of the solder. Furthermore, this study has demonstrated that there is a consistent ranking order for these elements in their ability to promote spreading—namely, tin > lead > silver > indium > bismuth. This ranking order is maintained even for ternary and quaternary solders and when these are applied to a range of substrates, in air, using mild fluxes.

The spreading characteristics of brazes tend to be less sensitive to composition, because the constituent elements are usually extensively soluble in both the solid and liquid states. Although some comparative data on the influence of the composition of brazes on spreading characteristics are available [e.g., McDonald *et al.* 1989], the detailed picture remains to be established because comprehensive data for brazes on "ideal" substrates are unavailable at present.

Although high fluidity of a filler metal is a desirable property when it is required to flow into the joint gap of a heated assembly by capillary action, it is not quite so important when the preferred method of applying the filler is to sandwich a thin foil preform or layer of paste between the components, which are then joined together in an appropriate heating cycle. For this type of configuration, a high degree of spreading is detrimental to joint filling, as the filler tends to flow out of the joint. Placement of the filler metal and its influence on joint filling are discussed in Chapter 4.

In a vacuum or neutral protective atmosphere, the spreading of a filler will tend to be inferior to that obtained in air in the presence of a chemical flux. This is to be expected in view of the limited effectiveness of these environments: neither a vacuum nor a protective atmosphere is usually capable of removing oxides that form on the surface of components or the filler while exposed to air before the joining operation. In both cases, the spreading is inferior to that achieved in the presence of an active flux that can remove the surface oxide [Humpston and Jacobson 1991].

1.3.5 Surface Roughness of Components

The roughness of joint surfaces can have a significant effect on the wetting and spreading behavior of a filler. It is well known that for each parent material there is an optimum surface roughness for maximizing the spreading of a filler. For example, when brazing aluminum alloys with the Al-12Si filler alloy in high vacuum and in the absence of fluxes, the best results in terms of spreading of the molten filler metal and of fillet formation have been obtained when the surface of the components were prepared by dry grinding with SiC papers of grit size between 400 and 600 [Okamoto, Takemoto, and Den 1976].

Surface roughness reduces the effective contact angle θ^*, where θ^* is related to θ, the contact angle for a perfectly flat surface, through the relation:

$$\cos \theta^* = r \cos \theta$$

where

$$r = \frac{\text{actual area of rough surface}}{\text{plan area}}$$

At the same time, by producing a network of fine channels, the texturing may increase the capillary force acting between the filler and the component surfaces. Both phenomena will tend to aid spreading. A directionally orientated surface texture promotes preferential flow parallel to the channeling [Nicholas and Crispin 1986].

Introduction

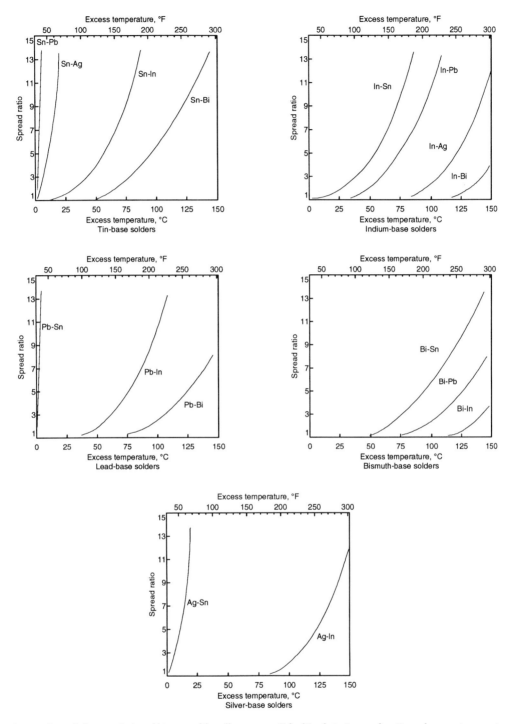

Fig. 1.9 Spread characteristics of binary solder alloys on an "ideal" substrate as a function of excess temperature above the melting point. The substrate is a flat, glass microscope slide, sputter metallized with 0.1 μm (4 μin.) of chromium overlaid with 0.1 μm (4 μin.) of gold. Spread ratio is defined in Appendix 6.1.

For maximum effect, the surface texture should be as jagged as possible. A surface prepared by grit blasting or abrading with SiC-impregnated paper is therefore preferable to a shot-

Principles of Soldering and Brazing

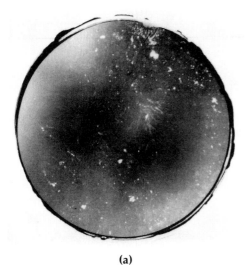
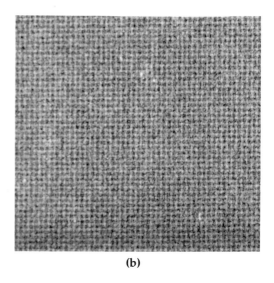

(a)   (b)

Fig. 1.10 (a) Radiograph of a 50 mm (2 in.) diameter component soldered using two 50 μm (2000 μin.) thick foils of Ag-96Sn solder, in high vacuum, and incorporating a # 400 gauze. (b) High-resolution radiograph that reveals the true nature of the joint filling in (a), with a void present at the center of each aperture in the mesh. 640×.

peened surface. The reason for this is as follows. At sharp reentrant angles that exist on jagged surfaces, there are sudden changes in the crystallographic orientation of the exposed parent metal. The adhesion of native oxides at these discontinuities will tend to be relatively weak and provide sites at which the oxide layer can be more readily undermined or penetrated. This is likely to have been one of the contributory factors in the example of aluminum brazing cited above.

There is a limit to the roughness of surface that can be used to promote spreading by a molten filler. If the texturing is too deep, then capillary dams can be formed and these will impede the spreading of the filler metal [Funk and Udin 1952]. Another factor that should be considered in connection with texturing is the extent of alloying between the filler and the parent material. For example, when lead-tin eutectic solders are used to join gold-coated components, there is a limit of about 5% to the concentration of gold that can be accommodated before the solder is embrittled (see Chapter 3). If the solder spreads over a gold-coated surface the critical thickness of the gold coating will be reduced by a factor related to the surface roughness, r, and which must be considered when calculating the total volume of gold corresponding to a given plan area of spread.

Attempts have been made to improve joint filling by introducing capillarity enhancers to the joint gap. Such enhancers include finely divided powders and fine meshes that are wetted by the filler but that are effectively inert. This type of approach has been explored by the authors and has not been found to radically improve joint filling. A volume fraction of powder that was calculated to significantly increase capillary forces had an adverse effect on the fluidity of the filler metal. On the other hand, meshes provided stable traps for air and evolved gases in the joint. This led to the formation of an array of voids corresponding to each aperture in the mesh, as revealed by a high-resolution radiograph of a joint with a #400 gauze that was soldered using a foil of Ag-96Sn alloy (Fig. 1.10).

1.3.6 Dissolution of Parent Materials by Molten Fillers

It is frequently observed that a filler metal will continue to spread beyond an initially wetted surface area over an extended period of time (>10 s), which would not be expected from classical fluid flow theory. Clearly, classical expressions for fluid flow, exemplified by Eq 1.4, do not strictly apply in such cases. Indeed, this type of flow can usually be associated with solid-liquid interfacial reactions, which are neglected in the model described in Milner's paper [1958]. Where joint filling is sluggish because of reactions occurring between the filler and the solid surface, increasing the temperature to reduce the viscosity of the molten filler is unlikely to enhance filling, because the reactions that are occurring transverse to the flow directions will accelerate [Tunca, Delamore, and Smith 1990].

Dissolution of the substrate and resulting growth of intermetallic compounds both follow Arrhenius-type rate relationships, represented by the expression:

$$\text{rate} = \exp\left[\frac{-Q}{kT}\right]$$

where Q is an activation energy that characterizes the reaction taking place at temperature T (in degrees Kelvin) and k is the Boltzmann constant. The alternative of increasing the joint gap is not usually an option, because this is likely to lead to a reduction in joint filling and/or joint strength, as discussed in Chapter 6. The solution then is to change the materials system; several means by which this can be achieved without changing the parent materials are described in Chapter 4.

Interfacial reactions are important, not only in determining the flow characteristics of the filler and its wetting behavior, but also the properties of the resulting joints. When a molten filler wets the parent materials, there is normally some intersolubility between them. It is usually manifested as dissolution of the surfaces of the parent materials in the joint region and the formation of new phases at either interface between the parent materials and the molten filler or within the filler itself when it solidifies. The effects of dissolution of the parent materials and compound formation on joints are discussed in detail in Chapter 3.

The following expression describes the rate of dissolution of a solid metal in a molten metal [Weeks and Gurinsky 1958, p 106-161; Tunca, Delamore, and Smith 1990]:

$$\frac{dC}{dt} = \frac{KA(C_S - C)}{V} \quad \text{(Eq 1.5)}$$

where C is the instantaneous concentration of the dissolved metal in the melt, C_S is the concentration limit of the dissolved metal in the melt at the temperature of interest, t is the time, K is the dissolution rate constant, A is the wetted surface area, and V is the volume of the melt. This equation is known as both the Nernst-Shchukarev and the Berthoud equation. In the integral form, Eq 1.5 becomes:

$$C = C_s\left[1 - \exp\left(\frac{-KAt}{V}\right)\right] \quad \text{(Eq 1.6)}$$

assuming initial conditions of $C = 0$, $t = 0$.

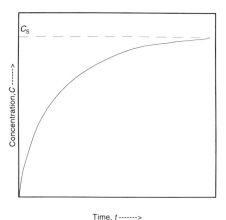

Fig. 1.11 The concentration of a solid metal in a liquid metal wetted by it changes in an inverse exponential manner with respect to time and is limited by the saturation concentration of the solid constituent in the liquid at that temperature.

Equation 1.6 reflects the fact that, in general, the concentration of dissolved metal in the molten filler increases in an inverse exponential manner with respect to time. That is, the dissolution rate is initially very fast, but then slows as the concentration of the dissolved parent material tends toward its saturation limit (i.e., equilibrium), as shown in Fig. 1.11. Substituting measured values into Eq 1.6 shows that, for a soldered or brazed joint of typical geometry, the equilibrium condition is reached within seconds at the process temperature. Thus, it is possible to use an equilibrium phase diagram to predict the change in the composition of the filler metal that will occur in typical joining operations and the associated depth of erosion of the joint surfaces. Equilibrium phase diagrams and their use in soldering and brazing are considered more fully in Chapter 3.

In some materials systems, the product of reaction between a molten filler and the parent materials is a continuous layer of an intermetallic compound over the joint interface. Once formed, the rate of erosion greatly decreases, because it is then governed by the rate at which atoms of the parent material can diffuse through the solid intermetallic compound. As a rough guide, solid-state diffusion processes are two orders of magnitude slower than solid-liquid reactions, and thus continued dissolution of the parent materials effectively ceases, within the timescales of typical joining processes. The practical implications of this behavior are also discussed in Chapter 3.

1.3.7 Significance of the Joint Gap

The joint gap at the process temperature influences both the joint filling and the mechanical properties of the resulting joint. The relationship between joint dimensions and mechanical properties is discussed in Chapter 6. In summary, the thinner a joint, the greater its load-bearing capability, until a limiting condition is reached.

Contact angle, surface tension, and viscosity all reduce with increasing temperature, making good joint filling in narrow joints more readily achievable as the joining temperature is raised above the melting point of the filler metal. A lower practical limit to the joint gap is imposed by the three factors discussed below:

i. **The need to provide a path for vapors to escape.** Flux vapors evolved within the joint and pockets of air must be allowed to escape if the formation of voids through gas entrapment is to be prevented (see Chapters 4 and 5). At the same time, any reducing agent needs to gain access to all joint surfaces and be present in sufficient concentration to work effectively.
ii. **Reaction with the components.** The metallurgical reaction that occurs between a molten filler metal and the surfaces of the components can take one of two forms.

    The surface region of the workpiece has limited solubility in the molten filler. This is the preferred situation. The dissolution of metal from the surface of the components can result in either compound formation at the interface, which may prevent further dissolution, or alloying with the filler, which will change its composition and hence its melting point.

    On the whole, solders tend to form interfacial compounds with parent materials, while brazes usually exhibit more extensive alloying between the materials. This can be partly explained by the fact that most solder alloys are based on elements with crystal structures that differ from those of most common parent metals. Consequently, intermetallic compounds tend to form in preference to solid solutions.

    A reaction that depresses the melting point of the filler metal is desirable for narrow joints, because its fluidity will be enhanced by such a reaction at a constant temperature. A reaction that raises the melting point of the filler metal will tend to increase its viscosity and holds out the danger that the filler will solidify at the process temperature before it has filled the entire joint. Wider joints mitigate this effect because the alloying will tend to be diluted.

    Dissolution of the filler in the parent metal. In this situation, the volume of filler will shrink as the reaction progresses; therefore, a larger volume of filler metal accommodated in a wider joint gap is again preferred and for similar reasons. However, absorption of the filler is generally undesirable, because its constituents will tend to penetrate into the parent materials preferentially along grain boundaries, generally to the detriment of the mechanical properties of the assembly and sometimes resulting in embrittlement and/or hot shortness.
iii. **Control of the joint gap.** The width of the joint gap must be predictable and stable during the bonding cycle. The size of the gap will be influenced by the coefficients of thermal expansion of the respective components, and allowances must be made for different expansivities of the mating components. The expansivities of a representative range of engineering materials at room temperature (25 °C, or 77 °F) are listed in Table 1.2. Temperature gradients along the joint must be considered from the same viewpoint. Variations of joint gap should be avoided wherever possible, as this can have a serious effect in impeding flow of the filler by capillary action. An upper practical limit to the joint gap is determined by the following two factors:
    i. **Mechanical properties of the joint.** As the gap is increased, the mechanical properties of the joint decline progressively to those of the bulk filler metal, which in the case of solders are particularly poor in relation to most structural materials. This aspect is discussed further in Chapter 6.
    ii. **Joint filling requirements.** As noted in section 1.3.2, the capillary force decreases as the joint gap increases, and this will place a practical upper limit on the joint gap. At the same time, a sufficient quantity of filler must be supplied to the joint to entirely fill it. Hydrostatic forces will promote the flow of low-viscosity filler metals out of wide gap joints.

The optimum balance of these factors is achieved when the joint gap is about 10 to 100 μm (400 to 4000 μin.), depending on the type of reaction that occurs between the filler and the component. Generally, when components rest freely on

Table 1.2 Typical thermal expansivity of common engineering materials at room temperature(a)

Material	Linear expansivity, 10^{-6} K^{-1}
Polymers	
Rubbers	150-300
Semicrystalline	100-200
Amorphous	50-100
Metals	
Zinc alloys	25-30
Aluminum alloys	20-23
Copper alloys	16-19
Stainless steels	15-17
Iron alloys	13-15
Nickel alloys	12-15
Cast irons	10-13
Titanium alloys	8-10
Tungsten/molybdenum alloys	5-7
Low-expansion alloys (Fe-Ni-base)	1-5
Graphite	7-9
Ceramics	
Glass	6-10
Oxide	4-8
Porcelain/clay	3-7
Nitride/carbide	2-6
Diamond/silica/carbon fiber	–1 to 1

(a) The values given are representative of the most widely used materials, rather than providing absolute limits for the different classes listed. The thermal expansivity will depend not only on elemental composition but also on microstructure and temper. Composite materials can have expansivities that effectively range between those of the constituents and depend on the relative proportions of the matrix and reinforcement phases.

one another and the assembly is heated until the filler is molten, the joint will tend to self-regulate to widths between these values. Indeed, it has been demonstrated that for a fixed combination of filler metal, component materials and process conditions the joint gap will tend to a fixed value specific to the combination. This value must be determined by experiment. If there is insufficient filler metal to fill this gap then the joint will contain voids of if too much filler is applied the excess will spill out [Bakulin, Shorshorov and Shapiro 1992]. Where thinner or wider joints are required, it is necessary to insert spacers (such as wires) of the desired width between the components and, for thin joints, to apply pressure during the bonding cycle to overcome the hydrostatic forces that will act to levitate the upper component.

1.4 *The Design and Application of Soldering and Brazing Processes*

A soldered or brazed joint usually must satisfy a specific set of requirements. Most frequently, it must achieve a certain mechanical strength, which it must retain to the highest service temperature in the intended application. The joint must also endure a particular service environment, which may be corrosive, and it may have to provide good electrical and thermal conductance. In addition, the joint must be capable of being formed in a cost-effective manner.

The principal aspects to be addressed can be divided as follows:

- The functional requirements of the application and the means of satisfying these through appropriate structural design.
- The achievement of the specified assembly through successful processing

1.4.1 Functional Requirements and Design Criteria

All soldered and brazed joints used in manufactured products must remain solid in service and retain their associated components in fixed positions when subjected to stress. These requirements are usually satisfied by suitable design of the geometry and the metallurgy of the joint, but there are also other aspects to consider. Several factors that affect the functional integrity of soldered and brazed joints are discussed below.

1.4.1.1 *Metallurgical Stability*

For a joint to remain solid, the melting point (solidus temperature) of the filler metal must exceed the peak temperature that the component is likely to experience. There are exceptions to this rule, which will be discussed in Chapter 4. Because the strength of all metals decreases rapidly as the melting point is approached, the peak operating temperature should not exceed about 70% of the melting point of the filler, in degrees Kelvin, if the joint is required to sustain a load.

The phases that form on solidification in soldered joints are frequently unstable at elevated service temperatures. Instability of phases present in the joint at the service temperature may be undesirable, because it can affect integrity. The effects of continued reactions between the filler and the components also must be considered, as explained in Chapters 2 and 3.

Because most filler metals are softer than the materials that are commonly joined, the mechanical properties of the joint are generally limited by those of the filler. Brazes tend to be harder than solders, a feature that is related to their higher melting point. For this reason, brazed joints are usually stronger at room temperature.

1.4.1.2 Mechanical Integrity

The durability of engineering and consumer products often depends on joints maintaining their mechanical integrity for the duration of their expected service life. The mechanical integrity of a soldered or brazed joint depends on a number of factors, including:

- The mechanical properties of the bulk filler metal (Chapter 2)
- The joint geometry—namely, area, width, and shape (Chapters 4 and 7)
- The mechanical properties of any new phases formed in the joint by reaction between the filler and the components, either during the joining operation or subsequently in service (i.e., there is an interdependence with the microstructure) (Chapter 3)
- The number, size, shape, and distribution of voids within the joint (Chapter 4)
- The quality of fillets formed between the filler and the surface of the components at the edge of the joint (i.e., their radius of curvature and extent of continuity) (Chapter 4)

The mechanical properties of joints, taking into account the influence of joint geometry, are extensively reviewed elsewhere. The reader is particularly referred to Frear [1990], Manko [1992], and Schwartz [1987].

1.4.1.3 Environmental Durability

Joints are normally expected to be robust in relation to the service environment. This most commonly involves exposure to corrosive gases, including sulfur dioxide and other constituents of a polluted atmosphere; to moisture, perhaps laden with salt; and to variable temperature. The corrosion and stress-corrosion characteristics of the joint are then of relevance. Corrosion mechanisms are generally very complex and specific to a given combination of materials, chemical environment, and joint geometry. Therefore, each situation should be determined empirically. A typical example is provided by the case study of corrosion in brazed stainless steel pipes conveying drinking water, an environment in which extensive corrosion would not normally be expected [Jarman, Linekar, and Booker 1975].

The temperature of a joint can be shifted well beyond a normal ambient range, especially in aerospace applications and in situations where heat is generated within the assembly itself. Then, thermal fatigue and other changes to the metallurgical condition of the joint, such as the growth of phases, can occur, and these invariably affect the properties of the joint. In other words, there is an interdependence between environmental stability and microstructural stability. An appropriate choice of the materials combination used should enable these changes to be constrained within predictable and acceptable limits. In this regard, solders tend to be used at service temperatures that are proportionately much closer to their melting points than are brazes. Hence, they are metallurgically less stable, and microstructural changes take place more readily.

1.4.1.4 Electrical and Thermal Conductivity

In certain applications, the soldered or brazed joint is required to perform the function of providing electrical and/or thermal contact between components. Generally, thin, well-filled soldered and brazed joints amply satisfy this requirement. Only in a few extreme situations are the thermal and electrical properties of such joints close to the allowed limits. A case in point is silicon high-power device assemblies, where the joint between the silicon device and the metal backing plate is required to conduct away up to 1 W/mm^2 of thermal power. Here it is crucial to ensure that the joints are kept thin (<30 μm, or 1200 $\mu in.$) and essentially void-free (<5% by volume). Further details of such components and the joining methods used in their fabrication are discussed in Chapter 7.

1.4.2 Processing Aspects

An important aspect that must be considered when designing a joint is the practicality of the process involved. Among the relevant issues are:

- Jigging of the components
- Form of the filler metal
- Heating method
- Temperature measurement

Introduction

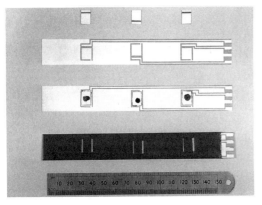 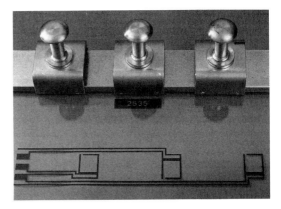

Fig. 1.12 (a) Sensor comprising three piezoelectric ceramic elements. These are metallized and soldered onto a metallized substrate. The component is then encapsulated in a polymer to provide protection against the environment. (b) To ensure the designed degree of acoustic coupling between the piezoelectric elements and the substrate, the soldered joints must be of a specified thickness. This was achieved using tungsten spacer wires in each joint and a spring-loaded jig to apply a compressive stress to each element during the process cycle. Also shown is the mask used to apply the metallization pattern to the substrate.

- Joining atmosphere
- Coatings applied to surfaces of components (as necessary)
- Cleaning treatments
- Heat treatments prior to joining
- Heating cycle of the joining operation
- Postjoining treatments

1.4.2.1 *Jigging of the Components*

The components being joined normally must be held in the required configuration until the filler metal has solidified. Even if the components can be preplaced without a fixture, the use of some form of jig is still frequently beneficial to ensure that the components are not disturbed by capillary forces originating from the molten filler metal. On the other hand, it is also possible to exploit the capillary forces to make the assembly self-align during the joining process; this is widely practiced in the fabrication of electronic circuits using lightweight surface-mounted components.

The jig can be used to fulfill more than a holding function. For example, it can serve as a heat sink or as a heat source. A jig should be constructed from a nonporous material to prevent contamination of the atmosphere surrounding the workpiece. Moreover, the jig material should not be wettable by the molten filler metal, in case they come into contact through accidental spillage. Materials such as brass should be avoided as zinc readily volatilizes. Care should be taken in designing jigs so as not to stress the constrained components through thermal expansion mismatch.

A note of caution needs to be sounded when joining to stressed components, as this can lead to brittle failure of the components through a mechanism known as liquid metal embrittlement (LME), which can be understood as follows. When a solid is wetted by a liquid, the interfacial energy between the atmosphere and the liquid, γ_{LV}, must be less than that between the solid and the atmosphere, γ_{SV} (see section 1.3.2), so that the threshold stress for incipient cracks in the stressed component to grow is correspondingly reduced. The degradation in the mechanical properties of an engineering steel, when stressed in tension and wetted by Pb-4Sn solder, has been studied by Watkins, Johnson, and Breyer [1975].

Graphite is often favored as a jig material. It is inexpensive, easy to machine, a good thermal conductor and absorber of radiant heat, and is not wetted by the majority of molten filler metals. Graphite also has the merit of "mopping-up" oxygen in an oxidizing atmosphere to form CO and CO_2. However, if the oxygen partial pressure is already low, then its role as a reducing agent is negligible. Care should be taken to ensure that the graphite used for jigs is of a dense grade so that it will be mechanically robust and have a low porosity to minimize outgassing. The desorption of water vapor, in particular, frequently determines the quality of a gas atmosphere in the vicinity of the workpiece.

Jigs are sometimes used to apply a controlled pressure to a joint in an assembly. One component can then be deliberately and elastically distorted to bring it into close and uniform contact with its mating part. This is an advantage when very nar-

Fig. 1.13 Production of foil directly from a molten charge by strip casting. Courtesy of Vacuumschmelze GmbH, Germany

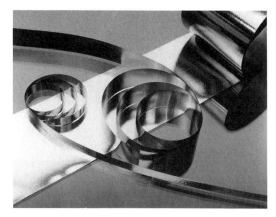

Fig. 1.14 Examples of foil strip produced by rapid solidification casting technology. Fleetwood et al. [1988]

row joints are required and when solid-state diffusion constitutes an important part of the joining process. Another application is outlined in Fig. 1.12. Jigs that exert a compressive stress are also used for vacuum brazing with aluminum-base filler metal preforms. Here the applied load serves a dual function. First, it helps puncture the oxide films on the surfaces of the filler. Second, when the filler metal melts, the applied force acts against any dewetting capillary force of the liquid to ensure flow. The combination of these two factors leads to improved joint filling.

1.4.2.2 Form of the Filler Metal

Filler metals are available in many different forms. These include configurations that normally can be produced from an ingot by mechanical working—for example, wire, rings, and foil. Such geometries are not restricted to ductile alloys. The development of rapid solidification processes has led to the availability of foils and wire of joining alloys that are inherently brittle. These foils are produced directly from the melt: the process involves forcing molten metal through a hole or slot onto a rapidly spinning, water-cooled metal wheel. Figure 1.13 shows such a strip casting process in operation; Fig. 1.14, some typical foils produced by this route. The high rate of heat extraction that occurs causes the molten metal to solidify almost immediately on striking the wheel, resulting in the formation of a strip of the alloy with a fine crystalline, or occasionally amorphous, microstructure. The dimensions of the cast material can be controlled by varying the nozzle dimensions, the ejection pressure, the speed of rotation of the wheel, and other parameters of the casting process [Jones 1982]. The refined microstructure of the rapidly solidified alloys generally gives them greater ductility than the same alloys produced by conventional casting and mechanical working. A good example is the Bi-43Sn eutectic solder, which is normally brittle. However, when prepared as a foil by rapid solidification, the ductility of this alloy is comparable to that of other solders.

Solders and brazes are also available as finely divided powders that can be mixed with a binder to form a paste capable of being screen printed onto a substrate or applied to the workpiece, via a dispenser, in an automated production line. However, powders, and pastes containing powders, have an extremely high ratio of surface area to volume of filler, which generally produces high oxide fractions and therefore tends to be detrimental to the quality of the resulting joints.

Some of the more common filler metals are available as wire or rod that incorporates a flux. Examples are flux-cored solder wire and flux-coated brazing rods. In this form, the filler metals can be readily used in air without any additional precautions.

In more specialized joining processes, the solder or braze can be deposited as a coating on the components by electroplating and by vapor deposition techniques such as evaporation. Where it is not possible to deposit the actual alloy, sequential layers of the constituent elements can be applied. The former method is generally preferred because the melting point of an alloy is well defined, whereas there is no guarantee that melting will take place at the desired temperature in the case of the composite layers, unless significant solid-state diffusion has occurred first to form the appropriate low-melting-point phases.

The use of some form of preplaced filler metal has a number of advantages. Most particularly, because the thickness and area of filler metal are predetermined, the volume of molten filler may be carefully controlled. This can help contribute to a well-defined and reproducible process that gives consistent results.

When the filler is applied as a coating, further benefits are gained. The number of free surfaces is reduced from four (corresponding to a foil preform sandwiched in the joint) to just two, thereby considerably reducing the proportion of oxides and other impurities deriving from exposed surfaces. Another advantage is that the thickness of the filler can be precisely tailored to produce well-filled joints without undue edge spillage.

1.4.2.3 *Heating Method*

Heat must be supplied to the joint to raise the temperature of the filler metal and joint surfaces above the melting point of the filler. The joint surfaces need to be heated; otherwise the filler metal will be incapable of wetting them and therefore will "ball up." To prevent this situation, it is good practice always to heat the filler metal via the components to be joined and never vice versa. The available methods of heating are: (1) local heating, in which only those parts of the components in the immediate vicinity of the joint are heated to the desired temperature, and (2) diffuse heating, where the temperature of the entire assembly is raised.

Common local heat sources include soldering irons, gas torches, and resistance heating using the assembly as the resistive element. More sophisticated heating techniques, such as induction heating and laser heating, also fall within this category. Although some methods of local heating are applicable to joining in a controlled atmosphere, this is not usually the case with a soldering iron or torch, and a flux must then be used.

In local heating, the rate of heat energy input must be high to swamp the heat conducted away by the components and jigging. A high rate of heat input can ensure fast heating and cooling of the joint. Fast heating coupled with short heating cycles minimizes erosion of substrate surfaces and therefore restricts the formation of undesirable phases, while rapid cooling ensures a fine grain size to the solidified filler and thereby superior mechanical properties. However, these potential benefits can be offset by the thermal distortion that might be produced in the components being joined. Local heating can be used to create specified temperature gradients that will restrict the flow of the molten filler metal to the immediate vicinity of the joint.

Diffuse heating sources include furnaces (both resistance and optical), hot plates, and induction coils. The features of diffuse heating methods are the opposite of those of the local heating methods. For example, the total energy requirement is higher, as the temperature of the entire assembly has to be raised, which also significantly increases the process cycle time. On the other hand, there is less risk of thermal distortion, and accurate control of temperature is easier to achieve. Diffuse heating methods tend to impose fewer constraints on the atmosphere surrounding the workpiece, because the source of heat is relatively remote from the components. For example, a torch is not generally compatible with a special atmosphere.

If diffuse heating is to be used in the fabrication of complex assemblies, the designer must ensure that all of the component parts are able to withstand the peak process temperature. With local heating, heat sinks can be used to protect sensitive areas from excessive thermal excursions. A related consideration when using diffuse heating in a situation where several joints must be made is that the melting point of the filler metal used for the preceding joining operations must be higher than the peak process temperature used in the succeeding cycle. Several different filler metals will therefore be required to fabricate a multijointed component in a step joining process.

1.4.2.4 *Temperature Measurement*

The liquid-solid metallurgical reactions that occur during soldering and brazing operations are highly temperature dependent. Therefore, reliable measurement of temperature is essential. Thermocouples and pyroelectric elements are the most common types of temperature sensors.

A number of precautions should be taken when employing thermocouples. Regular *in situ* calibration checks should be made to determine whether the thermoelectric characteristics of the thermocouple materials have been altered and to test for electrical interference affecting the display system. Correct temperature measurement requires good thermal contact between the thermocouple and the object being monitored. This tends to present a problem in vacuum joining processes, where thermal contact by mechanical means—namely, resting the thermocouple against a surface—tends to be inadequate. The thermal mass of the thermocouple and its protective sheath further impedes the thermocouple junction from sensing the true temperature of the component surface.

These effects can be minimized by embedding the thermocouple within the workpiece to improve thermal contact.

Even when thermocouples are used for temperature measurement in gas atmospheres, where the thermal coupling is better than it is in a vacuum, a change in the measured temperature will lag behind that actually occurring. This delay, which is not usually less than several seconds, is difficult to measure accurately, but it must be taken into account if a thermocouple is being used to monitor the temperature of assemblies exposed to high heating and cooling rates.

Pyrometers have one important advantage over thermocouples: they are noncontacting sensors of temperature. Measurements may be made remotely from the workpiece, and the response time of the instrument can be accurately determined. Traditional pyrometers are primarily designed for operation above about 750 °C (1380 °F). However, recent advances in pyroelectric technology have led to the commercial development of thermal imaging bolometers that are capable of measuring radiated energy down to and even below ambient temperature with a fast response time. Bolometers are now worth considering not only for high-temperature brazing but also for general application to soldering and brazing processes. Apart from their flexibility, bolometers are capable of rapidly measuring temperature over a large surface area. By contrast, a thermocouple provides only a highly localized measurement of temperature at its tip; this may not be representative of the entire joint region. Nonetheless, thermocouples have a greater sensitivity than bolometers, being capable of resolving smaller temperature differences.

1.4.2.5 *Joining Atmosphere*

For a molten filler metal to wet and bond to a metal surface, the latter must be free from nonmetallic surface films. Although it is possible to ensure that this condition is met at the beginning of the heating cycle, by prescribed cleaning treatments, significant oxidation will generally occur if the components are heated in air. Steps must therefore be taken to either prevent oxidation or remove the oxide film as fast as it forms.

The approach adopted will depend largely on the atmosphere surrounding the workpiece. Brazing and soldering processes are conducted in one of three types of atmospheres, defined according to the reaction that occurs between the atmosphere and the constituent materials, as follows:

- Oxidizing (e.g., air)
- Essentially inert (e.g., nitrogen, vacuum)
- Reducing (e.g., carbon monoxide, halogen containing)

The implications associated with the use of these atmospheres are considered below.

Oxidizing Atmospheres. Air is the most common oxidizing atmosphere. The principal advantages of joining in air are that no special gas-handling measures are required and that there are no difficulties associated with access to the workpiece during the joining process. However, because most component surfaces and those of the filler metal are likely to form oxide scale when heated in air, fluxes normally must be applied to the joint region.

An active flux is capable of chemically and/or physically removing an oxide film. The flux may be applied either as a separate agent or may be an integral constituent of the joining alloy. Fluxes are discussed in detail in Chapter 5.

Gold and some of the platinum group metals do not oxidize when heated in air. These metals are therefore sometimes applied as surface metallizations to the surfaces of the components being joined in fluxless processes. The use of wettable metallizations is discussed in Chapter 4. Solders and brazes that contain significant proportions of the precious metals are generally less susceptible to oxidation, enabling mild fluxes to be used.

An oxidizing atmosphere is occasionally desirable during soldering and brazing. Not only do some fluxes require the presence of oxygen in order to work, but it is a prerequisite for successful joining that oxygen be present in some instances. An example is provided by the copper-copper oxide eutectic brazing process in which copper is brazed to ceramic materials, such as alumina, by a eutectic that is formed *in situ* between copper and Cu_2O just below the melting point of copper [Schwartz 1990, p 84].

Inert Atmospheres. From a practical point of view, an atmosphere is either oxidizing or reducing. This is because it is not possible to remove and then totally exclude oxygen from the workpiece, except perhaps under rigorous laboratory conditions. Thus, when defining an atmosphere as inert, it must be taken as meaning that the residual level of oxygen present is not sufficient to adversely affect the joining process under consideration. An atmosphere that might be suitable for soldering to copper is likely to be inadequate for brazing aluminum.

Because the "inertness" of an atmosphere is judged relative to the specific application, it is necessary to define a quantitative measure of the oxygen present. This parameter is the oxygen partial pressure. Partial pressure provides a measure of the concentration of one gas in an atmosphere containing several gases. The partial pressure of a gas in a mixture of gases is defined as the pressure it would exert if it alone occupied the available volume. Thus, dry air at atmospheric pressure (0.1 MPa, or 14.5 psi) contains approximately 20% O_2 by volume, so that the oxygen partial pressure in air is 0.02 MPa (2.9 psi).

Typical inert atmospheres among the common gases include nitrogen, argon, and hydrogen. Hydrogen is included here because it is not capable of reducing the oxides present on the majority of metals at normal soldering and brazing temperatures. The oxygen partial pressure in standard commercial-grade bottled gases is on the order of 1 mPa (1.5×10^{-7} psi). Higher-quality grades are available, but their cost is usually too prohibitive to permit their use in most industrial applications.

Vacuum is frequently used as a protective environment for filler metal joining processes. Vacuum offers several advantages compared with a gas atmosphere, particularly the ability to measure and control the oxygen partial pressure more readily. In a substantially leak-free system, the oxygen partial pressure is a fifth of the vacuum pressure, which is relatively easy to determine, as compared with direct measurement of oxygen partial pressure. Although a roughing vacuum of 10 mPa (1.5×10^{-6} psi) will provide an atmosphere with the same oxygen partial pressure as a standard inert gas, it is possible to improve on this value, by several orders of magnitude, by using a high-vacuum pumping system. Alternatively, a low oxygen partial pressure may be achieved by obtaining a roughing vacuum, backfilling with an inert gas, and then roughing-out again. The effect of the second pumping cycle will be to reduce the oxygen partial pressure to less than typically one-thousandth of that in the inert gas, that is, approximately 10 μPa (1.5×10^{-9} psi). This estimate assumes that the furnace chamber is completely leaktight and does not outgas from interior surfaces, nor does any oxygen or water vapor backstream through the pump.

The disadvantages of using a vacuum system for carrying out a joining process are, principally, restricted access to the workpiece and the inadvisability of using either fluxes or filler metals with volatile constituents, such as cadmium, as the vapors can corrode the vacuum chamber, degrade its seals, and contaminate the pumping oils. This problem is not limited to the well-known volatile elements. Many metals that have a negligible vapor pressure at normal ambient temperatures will volatilize during high-temperature brazing processes (>1000 °C, or 1830 °F), particularly when these entail using reduced pressure atmospheres. Manganese-containing brazes and base materials fall into this category as the vapor pressure of this element is 1 Pa (1.5×10^{-4} psi) at 1000 °C (1830 °F).

A frequently overlooked consideration in reduced-pressure atmospheres is adsorbed water that naturally exists on surfaces exposed to ambient atmospheres. The continuous streaming of water vapor that desorbs from surfaces and flows past the workpiece as the pressure in a vacuum chamber is reduced is a source of oxidation; this is discussed in quantitative detail in Chapter 5. In a vacuum system operating at 10 mPa (1.5×10^{-6} psi), the desorbing water vapor constitutes the major proportion of the residual atmosphere. An adsorbed monolayer of water vapor of just 100 mm^2 (0.16 $in.^2$) in area desorbs to a gas pressure of 4 mPa (6×10^{-7} psi) per liter of chamber volume. The surfaces of the chamber should therefore be smooth to minimize the surface area, and also dry. In order to reduce this problem further, the walls of the vacuum chamber should be heated and the system should be vented to a dry atmosphere. To effectively desorb water vapor, the bakeout temperature should be at least 250 °C (480 °F), which may be difficult to achieve in practice owing to design constraints and the employment of rubber and other organic seals.

Another source of oxidizing contamination in a vacuum system is oil vapor mixed with air and water vapor backstreaming from a rotary pump. This can occur whenever the pressure inside the vacuum chamber drops below 1 Pa (1.5×10^{-4} psi), but can be largely eliminated by employing a foreline trap or by isolating the pump from the chamber once the required pressure reduction has been obtained.

The widespread practice of relying on an open gas shroud to provide an inert atmosphere is often unsatisfactory, because it is extremely difficult to control such an atmosphere reliably. For example, turbulence in the inert gas shroud can result in a supply of air actually being directed at the workpiece.

A reducing atmosphere is one that is capable of chemically removing surface contamination from metals. Gases that provide reducing conditions are generally proprietary mixtures that liberate halogen radicals. Specific gas-handling systems are usually needed for these in order to satisfy health and safety legislation.

For a few metals, hydrogen is satisfactory as a reducing atmosphere. No less important for meeting its functional requirement than the oxygen partial pressure of the gas is its water content. Hydrogen is a relatively difficult gas to dry, and the water vapor present can present a serious problem. A frost point of -70 °C (-95 °F) is equivalent to a water content of 0.0002% by volume—that is, an oxygen partial pressure of about 10 mPa (1.5×10^{-6} psi). There is also the risk of explosion when dealing with hydrogen at high temperatures, and hydrogen can embrittle some materials. A more detailed treatment of reducing atmospheres and their use is presented in Chapter 5, which deals with fluxing agents.

1.4.2.6 *Coatings Applied to Surfaces of Components*

Only occasionally is the desired joining alloy (chosen on the basis of melting temperature and physical properties) metallurgically compatible with the substrate, in the sense that the filler will wet the substrate uniformly, without the consequential formation of embrittling phases by reaction. A solution is to apply a surface coating that reacts in a benign manner with the filler. Coatings can be applied by a variety of techniques and to thicknesses that suit the particular application.

On metals, it is usual to apply coatings by wet plating methods, which are quick, economical, and flexible with regard to the coating thickness. Electroplating cannot often be used directly for most nonmetals, and it is usual instead to rely on vapor-deposited coatings. If the substrate is refractory in character, adhesion of metal coatings tends to be poor unless the metal is itself sufficiently refractory that it will form a strong reactive bond to the substrate. Widely used metallizations are chromium or titanium as the reactive layer, overlaid by gold or a platinum group metal to provide protection from the atmosphere. These and other metallizations and the principles on which they are designed are described in Chapter 4.

1.4.2.7 *Cleaning Treatments*

The surfaces of the components to be joined and the filler metal preforms must be free from any nonmetallic films, such as organic residues and metal oxides, to enable the molten filler metal to wet and alloy with the underlying metal. Fluxes are often capable of removing surface oxides, provided that they are reasonably thin.

Organic films can be removed with solvents, which obviously should not react with the underlying materials. Thick oxides and other nonmetallic surface layers can be removed chemically. However, mechanical cleaning is generally preferable because chemicals tend to leave residues, which then also have to be removed.

Dry mechanical abrasion exposes a fresh metal surface. The roughness of the abraded surface can be readily controlled, and this can be used to advantage in promoting the spreading of the molten filler metal. The rougher the surface, the better the wetting and spreading of the molten filler tends to be, for the reasons given in section 1.3.5. However, rough surfaces create problems when it is required to cover them with thin metal coatings. For example, thickness uniformity of thin metallic films is difficult if not impossible to achieve on a rough-textured substrate.

Soft components, such as solder foils, can be difficult to mechanically clean using abrasive particles because these tend to get embedded. Scraping with a sharp blade is usually a satisfactory alternative solution. Thin vapor-deposited and electroplated metallizations should be protected against atmospheric corrosion by the application of a noble metal overcoat. If correctly stored and handled, such components will not require cleaning prior to bonding and, for obvious reasons, must never be abraded.

1.4.2.8 *Heat Treatments Prior to Joining*

Prejoining heat treatments are occasionally useful in providing stress relief and thereby preventing unpredictable distortion during heating of the components to the bonding temperature. Other situations where prejoining heat treatments can be beneficial include those involving components with nonmetallic surface films that are thermally unstable. In the case of silver, for example, the oxide will readily dissociate when heated above 190 °C (375 °F) in an ambient atmosphere. Likewise, silver sulfide dissociates on heating above 842 °C (1548 °F).

Heating cycles may be used to produce solder alloys *in situ* from layers of the constituent elements applied to surfaces by screen printing, electroplating, or vapor deposition. By heating the substrate above the melting point of the constituent with the lowest melting point, alloying will oc-

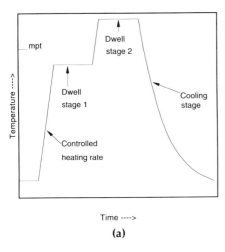 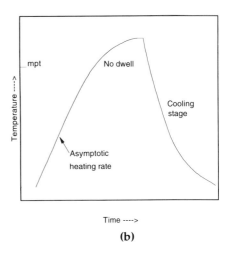

Fig. 1.15 Profiles of typical temperature cycles. (a) Heating cycle with a controlled profile incorporating dwell stages to reduce thermal gradients. (b) Heating cycle defined solely by attainment of a peak temperature

cur by solid-liquid interaction. The joint can then be formed in a subsequent heating cycle that is usually referred to as the reflow stage.

1.4.2.9 Heating Cycle of the Joining Operation

The prepared components, possibly mounted in jigs, are joined by applying heat. The heating cycle involves four important processing parameters: the heating rate, the peak bonding temperature, the holding time above the melting point of the filler, and the cooling rate.

In general, it is desirable to use a fast heating rate to limit reactions that can occur below the prescribed bonding temperature. However, the maximum heating rate is normally constrained by adverse temperature gradients developing in the assembly. These can produce distortions in the components and give rise to nonuniform reactions between the filler and the two joint surfaces. Also, temperatures are difficult to measure reliably during fast heating schedules. A better practice, when joining in a vacuum or special atmosphere furnace, is to heat the assembly rapidly to a preset temperature that is just below the melting point of the filler metal and then hold at this temperature for sufficient time (which can range from a few seconds to more than one hour, depending on the size of the assembly and the heating method) to allow the assembly to thermally equilibrate and for water vapor to flush out of the joint. Following this dwell, the assembly may then be rapidly heated to the bonding temperature. Profiles of typical temperature cycles are shown schematically in Fig. 1.15.

The bonding temperature should be such that the filler is guaranteed to melt, but at the same time should not be so high that the filler degrades through the loss of constituents or by reaction with the furnace atmosphere. The optimum temperature is normally determined by metallurgical criteria, most importantly the nature and extent of the filler-substrate interaction. The peak process temperature is frequently set at between 50 and 100 °C (90 and 180 °F) above the melting point, because accurate temperature measurement and control are not always readily achievable, especially in reduced-pressure atmospheres where conduction between the heat source and the workpiece is poor. Moreover, the reported melting temperatures of some fillers are not based on accurate measurements, and it is prudent to make some allowance for this uncertainty.

The minimum time that the assembly is held above the melting point must be sufficient to ensure that the filler has melted over the entire area of the joint. The maximum time is usually a compromise based on practical and metallurgical considerations. Extended dwell times tend to result in excessive spreading by the molten filler, possible oxidation gradually taking place, and deterioration of the properties of the parent materials.

The cooling stage of the cycle is seldom controlled by the operator, but tends to be governed by the thermal mass of the assembly and jig. Forced cooling can lead to problems, such as exacerbating mismatch stresses. Occasionally one or more dwell stages are required, either to provide stress relief to the bonded assembly or to induce some requisite microstructural change. An example of

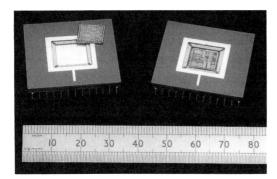

Fig. 1.16 Large-area silicon chip soldered into a metallized ceramic package using an alloy based on the Au-2Si composition (melting point of 355 °C, or 671 °F). The expansion mismatch between silicon and the ceramic makes it necessary to cool the assembly at a controlled rate in order to prevent fracture of the components.

the latter would be the aging treatment of a precipitation-hardening alloy. In this case, the solution treatment and quenching stages are carried out in tandem with the actual reflow operation.

For many steel components, controlled cooling is vital in order to achieve the desired microstructure and hence the correct temper. Stainless steels are susceptible to surface cracking if a martensitic transition is allowed to occur, owing to the associated volumetric strain. This phase change can be prevented by carrying out an isothermal anneal above the transition temperature; the component is thereafter cooled sufficiently slowly.

Stress relief of bonded assemblies tends to be difficult to achieve in practice. This is because the higher the temperature at which the heat treatment is performed, the greater the ability of the filler metal and the components to relieve the stress by creep; however, the residual stress in the assembly is correspondingly reduced. A heat treatment temperature of about 75% of the freezing point of the filler metal (in degrees Kelvin) usually provides the optimal stress relief. An example of an assembly that required a stress-relief treatment in order to avoid catastrophic cracking is illustrated in Fig. 1.16.

There are situations where a particular bonding cycle is designed to dovetail with other thermal processes that are required in the fabrication of the assembly—for example, spheroidization or paint curing. In such instances, the economics of production of the complete assembly often need to be taken into account when specifying the parameters of the process cycle.

1.4.2.10 *Postjoining Treatments*

Various types of postjoining treatments can be applied. A cleaning schedule is generally used to remove the flux residues and tarnishing that are produced when joining in air. Flux residues must be removed as they are usually corrosive, especially when moist, and can affect the long-term reliability of the component in service. Both chemical and mechanical means of flux removal are employed. Tarnishing tends to be removed chemically, often with acids, followed by thorough rinsing. If a heat treatment is not integrated into the cooling stage of the bonding cycle, a separate heat treatment may be carried out subsequently.

References

Ambrose, J.C., Nicholas, M.G., and Stoneham, A.M., 1992. Kinetics of Braze Spreading, *Proc. Conf. British Association for Brazing and Soldering*, 1992 Autumn Conference, Coventry, U.K.

Bakulin, S.S., Shorshorov, M.Kh., and Shapiro, A.E., 1992. A Thermodynamic Approach to Optimising the Width of the Brazing Gap and the Amount of Brazing Alloy, *Welding International,* 6(6), 473-475

Bever, M.B., Ed., 1986. *Encyclopedia of Materials Science and Engineering*, Pergamon Press, p 2458-2463

Eustathopoulos, N. and Coudurier, L., 1979. Wettability and Thermodynamic Properties of Interfaces in Metallic Systems, Paper 5, *Proc. Conf. British Association for Brazing and Soldering*, 3rd Int. Conf., London

Fleetwood, M.J., *et al.*, 1988. Control of Thin Strip Casting, *Proc. 2nd Int. Conf. Rapidly Solidified Materials*, ASM International, San Diego, 7-9 March

Funk, E.R. and Udin, H., 1952. Brazing Hydromechanics, *Weld. Res. Suppl.*, Vol 6, p 310s-316s

de Gennes, P.G., 1985. Wetting: stastics and dynamics, *Reviews of Modern Physics,* 57(3), 827-863

Harkins, W.D, 1952. *Physical Chemistry of Surface Films*, Van Nostrand Reinhold

Humpston, G. and Jacobson, D.M., 1990. Solder Spread: A Criterion for Evaluation of Soldering, *Gold Bull.*, Vol 23 (No. 3), p 83-95

Humpston, G. and Jacobson, D.M., 1991. The Relationship Between Solder Spread and Joint Filling, *GEC J. Res.*, Vol 8 (No. 3), p 145-150

Jarman, R.A., Linekar, G.A.B., and Booker, C.J.L., 1975. Interfacial Corrosion of Brazed Stainless Steel Joints in Domestic Tap Water, *Br. Corros. J.*, Vol 8, p 33-37; Vol 10, p 150-154

Jones, H., 1982. *Rapid Solidification of Metals and Alloys*, Monograph No. 8, The Institution of Metallurgists, London

Kapoor, R.R. and Eagar, T.W., 1989. Tin-Based Reactive Solders for Ceramic/Metal Joints, *Metall. Trans. B*, Vol 20B (No. 6), p 919-924

Li, J.G. and Hausner, H., 1991. Wettability of Silicon Carbide by Gold, Germanium and Silicon, *J. Mater. Sci. Lett.*, Vol 10, p 1275-1276

McDonald, M.M., *et al.* 1989. Wettability of Brazing Filler Metals on Molybdenum and TZM, *Weld. J.*, Vol 10, p 389s-395s

Milner, R.D., 1958. A Survey of the Scientific Principles Related to Wetting and Spreading, *Br. Weld. J.*, Vol 5, p 90-105

Nicholas, M.G. and Crispin, R.M., 1986. Some Effects of Anisotropic Roughening on the Wetting of Metal Surfaces, *J. Mater. Sci.*, Vol 21, p 522-528

Okamoto, I., Takemoto, T., and Den, K., 1976. Vacuum Brazing of Aluminum Using Al-12%Si System Filler Alloy, *Trans. Jpn. Weld. Res. Inst.*, Vol 5 (No. 1), p 97-98

Tabor, D., 1969. *Gases, Liquids and Solids*, Penguin Books

Tunca, N., Delamore, G.W., and Smith, R.W., 1990. Corrosion of Mo, Nb, Cr, and Y in Molten Aluminum, *Metall. Trans. A*, Vol 21A (No. 11), p 2919-2928

Watkins, M., Johnson, K.I., and Breyer, N.N., 1975. Effect of Cold Work on Liquid-Metal Embrittlement by Pb Alloys on 4145 Steel, *Proc. 4th Interamerican Conf. Materials Technologies*, Caracas, p 31-38

Weeks, J.R. and Gurinsky, D.H., 1958. *Liquid Metals and Solidification*, American Society for Metals

Yost, F.G. and Romig, A.D., 1988. Thermodynamics of Wetting by Liquid Metals, *Mater. Res. Soc. Symp. Proc.*, Vol 108, p 385-390

Chapter 2
Filler Alloys and Their Metallurgy

2.1 *Introduction*

This chapter presents an overview of common filler alloy systems. Necessarily, the survey must include consideration of the parent material with which the filler is used, because the suitability of a filler for a particular joining process depends largely on its compatibility with the base materials.

Extensive reference will be made to phase diagrams in order to highlight particular points. An introduction to alloy constitution and phase diagrams designed for those with no background in the subject is presented in Chapter 3, which covers interpretation of phase diagrams, the associated terminology, and methods for their determination.

For a braze or a solder to be compatible with a particular parent material, it must exhibit the following characteristics:

- It must have a liquidus temperature below the melting point (solidus temperature) of the parent materials and any surface metallizations. Usually a margin is required between these two temperatures in order to achieve adequate fluidity of the molten filler.
- It must be capable of producing joints at temperatures at which the properties of the base materials are not degraded. For example, many work-hardened and precipitation-hardened alloys cannot withstand brazing temperatures without loss of their beneficial mechanical properties. The first type of hardening involves subjecting the alloy to mechanical deformation, such as rolling or hammering, when reasonably cold. As the temperature is raised, the deformation damage is removed by atomic rearrangement in the metal. Precipitation hardening is accomplished by creating a finely divided phase within the material, which can be thought of as akin to a composite material. The dispersed phase is precipitated by means of an appropriate heating schedule, and its strengthening effect is likewise degraded by exposure to high temperatures.
- It must wet the parent materials or a metallization applied to the parent materials in order to ensure good adhesion through the formation of metallic bonds.
- It must not excessively erode parent metals or metallizations at the joint interface. The associated reactions, which must occur to form a metallic bond, should not result in the formation of either large proportions of brittle phases or significant concentrations of brittle phases along interfaces or other critical regions of the joint. Even ductile phases can have weak interfaces with solidified filler alloys.
- It must not contain constituents or impurities that might embrittle or otherwise weaken the resulting joint. Likewise, the parent material must not contribute constituents or impurities to the filler that will have a similar effect.

Besides being compatible with the parent material, the filler and the joining process used must be mutually suited. For example, filler metals containing zinc, lead, or other volatile constituents are not usually appropriate for furnace joining, especially when reduced pressures are involved.

The degree of temperature uniformity that can be achieved over the joint area will have an influence in determining the minimum temperature difference that can be tolerated between the melting temperature of the filler and the melting or degradation temperature of the parent material. This consideration is particularly relevant to the joining of aluminum alloy components at about 600 °C (1100 °F), with the well-known Al-13Si eutectic braze that melts hardly more than 20 °C (35 °F) below this temperature.

The condition of the surface of the parent material may affect its compatibility with the filler, especially when fluxes are not used. As an obvious example, an oxidized surface will be less readily wetted by a filler than will an atomically clean metal surface. This consideration often determines the acceptable shelf life of components prior to joining.

In order to establish whether a particular parent metal (or nonmetal with a surface metallic coating) is compatible with a given filler, the appraisal must be carried out under conditions representative of those used in any practical implementation of the process. Parameters such as process time and temperature can be critical in this regard. Storage shelf life of the filler and the components is another relevant factor, but is often neglected during transfer of a process from the laboratory to the factory.

The properties of the filler metal and the joint that it is used to make must also be compatible with the service requirements. These are likely to involve a combination of at least some of the following considerations:

- The strength and ductility of the joint must meet certain minimum requirements over the range of service temperatures.
- The design of the joint should not introduce stress concentrations in the assembly that might arise through solidification shrinkage or the formation of intermetallic phases.
- The joint must be appropriate for the service environment, in terms of corrosion and oxidation resistance and compatibility with vacuum, in accordance with functional requirements.
- The filler must comply with statutory needs, such as hallmarking regulations for precious metals and health restrictions on lead and cadmium for certain culinary and medical applications.
- Aesthetic requirements must sometimes be met—for example, color, color matching in jewelry and utensils, and the ability of joints to accept surface finishes such as paints, electroplatings, and so on. Good fillet formation is often demanded for aesthetic reasons and also as a criterion of acceptable joint quality.
- Requisite thermal and electrical properties must be achieved.

The simultaneous attainment of several of these desired characteristics is frequently achievable with common filler metals, provided that simple design guidelines and process conditions are satisfied. Indeed, lead-tin solders and silver-copper-zinc-base brazes account for the largest proportion of joints involving filler metals. It is the more exceptional and demanding service requirements that have given rise to the development of the hundreds of additional filler metal compositions. However, in general, the commercially available filler metals have been designed so that, when used in conjunction with common engineering parent materials, they meet many if not all of the above requirements.

2.2 Survey of Filler Alloy Systems

The survey will begin with consideration of brazing alloys, because they provide an instructive example of how fillers can be progressively tailored to suit particular requirements by modifying the composition. By contrast, solders are less amenable to such manipulation, as they tend to be based on eutectic compositions. The addition of other alloying elements usually destroys the eutectiferous character of these alloys and the desirable soldering properties that stem from it. In any case, there is little incentive to introduce such modifications because there is a ready choice of alternative eutectic solders having similar melting points (see Fig. 1.2).

There are literally hundreds of different brazing alloy compositions available on the market. However, these fall into a relatively small number of alloy families, defined according to the major metal constituent (see Fig. 1.3). Only a few brazes are based on eutectic systems. More commonly, the constituent elements form solid solutions, enabling a large number of brazes with different compositions to be developed from each alloy family. Within each alloy family, the differences in composition usually reflect only slight variations in user requirements, as will be seen. The brazing alloys will be treated on the basis of this classification, with attention devoted to delineating the principal features of each alloy system. For more detailed coverage of the available alloys, the reader should consult reference publications [e.g., Klein Wassink 1989; Manko 1992; Schwartz 1987], as well as data sheets and manuals supplied by manufacturers.

2.2.1 Brazes

Possibly the most widely used brazing alloys are those based on the silver-copper-zinc alloy system. Silver and copper generally constitute 60% by weight or more of these brazing alloys, and both elements are occasionally used on their own as filler metals.

2.2.1.1 *Pure Silver*

Pure silver (melting point = 962 °C, or 1764 °F) is seldom used as a filler metal because of its relatively high cost, although it is perfectly satisfactory in most other respects. It has reasonable mechanical properties, is metallurgically compatible with most parent metals, has a low vapor pressure, and possesses adequate fluidity when heated above its melting point. The mechanical properties of joints made using silver as the braze are often superior to those of pure silver in bulk form, because the joint microstructure is likely to be extensively modified through alloying with parent metals—a consequence of the fact that most metallic elements are soluble in silver. The alloying results in the formation of solid solutions and occasionally dispersed intermetallic compounds, both of which usually serve to strengthen the resulting joint.

Pure silver is mostly used as a braze for bonding metals such as beryllium and titanium. Silver is chosen for this application because silver oxide is not stable above 190 °C (375 °F), even in air. Hence, molten silver will usually form a metallic bond to these highly refractory materials. Where less noble metals, including copper, are used as the filler metal, there is a tendency for a layer of oxide to remain at the joint interface.

2.2.1.2 *Pure Copper*

Copper (melting point = 1085 °C, or 1985 °F) is appreciably cheaper than silver, but this advantage is partly offset by the higher process temperatures required and, more particularly, by the need to pay closer attention to the quality of the atmosphere in order to prevent excessive oxidation of the filler and components.

The largest use of pure copper as a braze is for fluxless joining of mild steel in vacuum and reducing atmospheres. This combination of materials is almost ideal from a metallurgical point of view. Not only does molten copper wet steel extremely well, but there is negligible erosion and molten copper is very fluid. Consequently, joints can be made between components that have an interference fit even when the filler is required to flow relatively long distances (several centimeters). The narrowness of the resulting joints, coupled with the high degree of joint filling obtained, means that the shear strength of joints can easily exceed that of the parent metal [Sloboda 1961].

However, the benefits accruing from the lack of extensive alloying with mild steel mean that the joints effectively comprise pure copper, which is not a particularly strong, hard, or fatigue-resistant metal. For this reason, when the joint is a loose fit, minor additions of elements such as nickel, chromium, or cobalt (0.5 to 5%) are often made to improve the mechanical properties of the braze through solid solution and precipitation strengthening. This is not necessary if the parent metal is stainless steel, which contains one or more of chromium, nickel, or cobalt, because the wetting by molten copper will incorporate these elements into the braze.

Another modification that can be made to improve the characteristics of pure copper as a braze is the addition of either boron with nickel to produce an alloy (Cu-2B-1Ni) melting in the range of 1080 to 1100 °C (1975 to 2010 °F), or phosphorus (Cu-8P; melting point = 714 °C, or 1317 °F). The metalloid elements depress the melting point of copper, due to the formation of copper-rich eutectics, as can be seen from the copper-phosphorus phase diagram (Fig. 2.1). They also tend to flux surface oxides, as described in section 5.3.4.

A number of joining processes use thin copper metallizations and rely on the generation of a molten filler, at temperatures below the melting point of copper, through alloying with the parent metal. Electroplatings about 5 μm (0.2 mil) thick are used as the "braze" for the fluxless joining at about 800 °C (1470 °F) of silver alloys that are free of

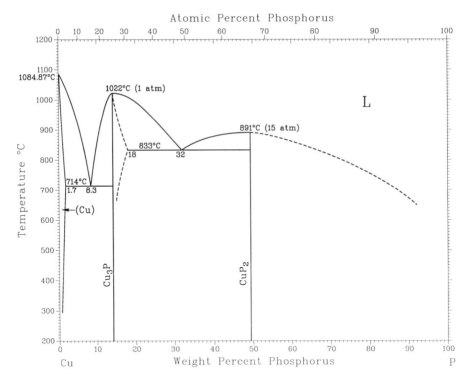

Fig. 2.1 Copper-phosphorus phase diagram

zinc and other volatile metals [Tuah-Poku, Dollar, and Massalski 1988]. The formation of liquid at these temperatures relies on solid-state diffusion to establish a thin layer of the silver-copper eutectic composition alloy at the joint interface, which is molten at the process temperature. Aluminum alloy components can be similarly joined using this approach at a temperature close to 575 °C (1065 °F) [Niemann and Wille 1978]. Such joints tend to be strong, as they are both thin and usually extremely well filled. Because of the exceptional thinness of the layer of braze, extended bonding times can result in the copper diffusing completely into the parent material. This can forestall the problem of galvanic corrosion that might otherwise arise when there is an abrupt interface between two dissimilar metals. Joint remelting during service is also prevented as the zone of low-melting-point filler metal is completely dissipated in this isothermal bonding process. Further details on this type of process are given in section 4.4.2.

2.2.1.3 Silver-Copper Eutectic

Silver and copper enter into eutectic equilibrium at 779 °C (1435 °F), when 28% of copper is added to pure silver. The silver-copper phase diagram is given in Fig. 2.2. The eutectic microstructure is lamellar, with a copper-rich phase interspersed with a silver-rich phase. This alloy system forms the basis of the widely used ternary and quaternary silver-base brazes, although it is sometimes used on its own.

Silver-copper eutectic is a malleable alloy and can be readily worked into a wide variety of preform geometries. It is mostly employed in the fluxless joining of copper-base alloys in vacuum; there is insufficient thermal activation in such an environment to reduce the iron, chromium, and cobalt oxides that are present on the surface of ferrous alloys at the typical temperatures at which the silver-copper braze is used (800 to 850 °C, or 1470 to 1560 °F). For this reason, reducing atmospheres or fluxes are needed for brazing other materials.

A variant of the silver-copper eutectic braze is obtained by adding phosphorus—for example, Ag-92Cu-6P (melting range = 644 to 740 °C, or 1191 to 1365 °F) and Ag-80Cu-5P (melting range = 644 to 700 °C, or 1191 to 1290 °F). These alloys are known as self-fluxing filler metals in that they can be used in air without the addition of flux to the joint region, provided that the parent metal is not too refractory. The self-fluxing mechanism is described in section 5.3.4. A higher silver content confers both enhanced strength and elongation to

Filler Alloys and Their Metallurgy

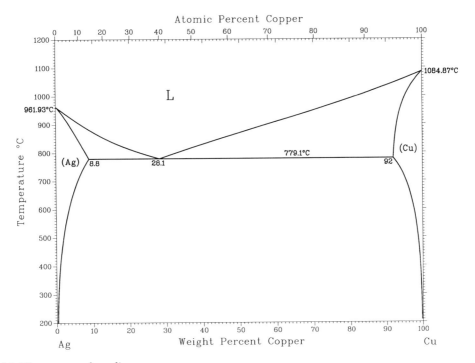

Fig. 2.2 Silver-copper phase diagram

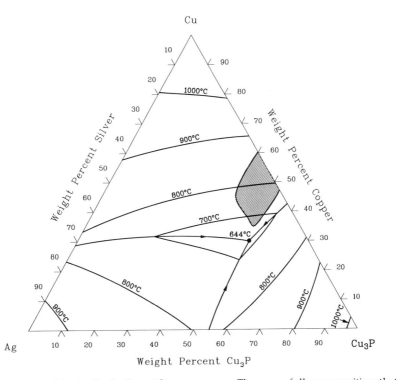

Fig. 2.3 Liquidus surface of the Ag-Cu-Cu₃P partial ternary system. The range of alloy compositions that are widely used as self-fluxing brazes is indicated by the shaded region.

failure to the braze because the Cu_3P phase is brittle, but this must be balanced against the additional cost. The range of alloy compositions that is used as brazes is mapped on the liquidus projection of the Ag-Cu-Cu_3P partial ternary system in Fig. 2.3.

The self-fluxing silver-copper-phosphorus brazes are generally restricted to joining copper-base parent metals and are widely used in plumbing applications. This is because brittle phosphides form at or close to the joint interface if the parent metal contains either iron, nickel, cobalt, or chromium in significant proportions. However, nickel-free steels can be joined by adding minute quantities of nickel to the braze. This converts the normally brittle Fe_3P interfacial phase to Fe_2P, which is considerably more ductile, and moderate joint shear strengths can then be achieved. The maximum continuous service temperature of joints made with the phosphorus-containing alloys is usually restricted to below about 200 °C (390 °F) to avoid selective oxidation of any residual phosphorus and a consequent degradation in joint properties. Long-term exposure of joined assemblies to sulfur-bearing atmospheres should be avoided for similar reasons.

A recent development of the silver-copper eutectic braze has involved the addition of up to 5% titanium [Mizuhara and Mally 1985]. The braze is prepared either as a trifoil, consisting of a titanium sheet that is roll-clad with the silver-copper braze, as a silver-copper alloy wire with a titanium core, or, more recently, as a foil of the ternary alloy. The ternary alloy foil is most readily prepared by rapid solidification directly from the melt; cast ingots of the alloy have poor mechanical properties.

The purpose of adding titanium is to introduce a highly active metal that is capable of directly wetting and bonding to nonmetallic materials, particularly various ceramics and graphite. At elevated temperatures, typically 900 °C (1650 °F) and above, the titanium reacts with the nonmetallic parent material to produce a complex interfacial layer that is wettable by the silver-copper braze. This same principle operates for titanium-containing solders in their application to the joining of nonmetallic materials, provided that a sufficiently high temperature is attained during the process cycle to activate the reaction between titanium and the nonmetallic components [Kapoor and Eager 1989]. For certain applications, other similarly active elements, such as hafnium, lithium, zirconium, and so on, are sometimes used in place of titanium, as titanium does not readily wet all nonmetallic materials. These so-called active filler metals are discussed further in section 5.4.

The relatively high melting point of these silver-copper-base brazes has several disadvantages. These are the high energy costs involved in heating the components to the elevated process temperature, the tendency to thermally degrade the mechanical properties of many parent metals (particularly if they are work or precipitation hardened), and the costs associated with removing the residues of silaceous fluxes that are required for torch brazing in air and which are not soluble in water (see section 5.3.2).

2.2.1.4 Copper-Zinc and Silver-Zinc Alloys

Binary copper-zinc alloys (brasses) have been used for centuries as brazes, but those compositions that have a narrow melting range contain approximately 40 to 50% Zn, as can be seen from the copper-zinc phase diagram given in Fig. 2.4. Their melting points are all above 800 °C (1470 °F), and they possess reasonable ductility. However, the high percentage of volatile zinc is undesirable for furnace brazing operations.

Silver-zinc binary alloys exhibit similar features, but present the added disadvantage of the higher cost of silver in the alloys that have a narrow melting range (60 to 70%). The silver-zinc binary phase diagram is shown in Fig. 2.5. The combination of high silver content, matching color, and resistance to silver cleaning fluids, most of which contain ammonia, do make them attractive for joining certain types of silverware. By comparison, ammonia can induce stress-corrosion cracking in copper-rich alloys.

2.2.1.5 Silver-Copper-Zinc

Silver, copper, and zinc form a eutectic alloy in the proportions 56Ag-20Cu-24Zn at 665 °C (1229 °F). The lowest-melting-point alloys in this system extend from the ternary eutectic point toward the zinc corner of the phase diagram. However, the alloys that contain more than about 40% Zn suffer a drastic drop in their mechanical strength and ductility. The fabrication of these alloys as wrought products is difficult, and the joints formed using them tend to be weak. Hence, they are seldom used industrially.

The range of compositions that lend themselves to brazing applications is further restricted with respect to the copper and silver contents. This is because alloys that contain more than about 60% Cu have a liquidus temperature above 900 °C (1650 °F) and reduced strength, while silver concentrations above 85% are undesirable for similar

Filler Alloys and Their Metallurgy

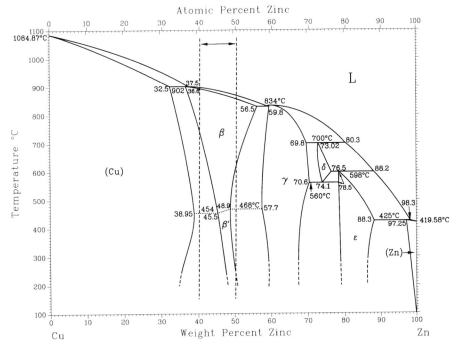

Fig. 2.4 Copper-zinc phase diagram. The range of alloy compositions used as brazes is indicated. These are characterized by having narrow melting ranges.

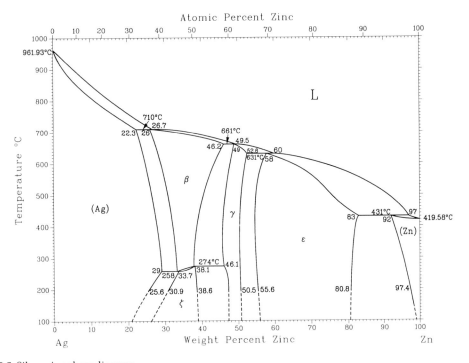

Fig. 2.5 Silver-zinc phase diagram

Principles of Soldering and Brazing

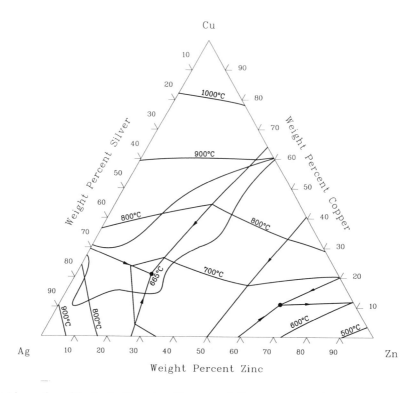

Fig. 2.6 Liquidus surface of the Ag-Cu-Zn system. The range of alloy compositions that is commonly used as brazes is shown shaded.

reasons; high-silver-content alloys are also expensive. The liquidus surface of the silver-copper-zinc ternary phase diagram is given in Fig. 2.6. On this projection, the range of alloy compositions that are widely used as brazes has been mapped.

The silver-rich alloys are frequently used as brazes in the jewelry industry. Although hallmarked sterling silver contains a minimum of 92.5% Ag, it has not been found possible to match this specification with color-matched silver brazing alloys of sufficiently low melting point. Therefore, the hallmarking limits on the silver content for brazes used with silverware have been relaxed to 67.5% Ag. Silver-copper-zinc alloys are then capable of meeting the functional requirements with good color matching.

For the joining of brass components, ternary alloys with a copper:zinc ratio close to 60:40 are used. These possess the color of brass because they have the same majority phases in their microstructure. The silver content depresses the melting range and improves the ductility, fatigue, and corrosion resistance as well as the fluidity of the alloys.

The mechanical strength of the silver-copper-zinc brazes and of joints made with them can be enhanced, particularly for elevated-temperature service, by adding nickel, typically 2 to 4%. Nickel additions also promote wetting of carbides and improve the resistance of joints in steel components to crevice corrosion. There is evidence that the nickel forms interfacial phases, which provide a reactive bond to refractory surfaces [Miller and Schwaneke 1978]. The surfaces of the fillets that form at the edge of the joints are nickel-rich, and thus have superior resistance to corrosion. The nickel-bearing alloys find use in the brazing of tungsten carbide cutting tool tips to steel shanks and for other small components where the reduced fluidity of the braze can be tolerated.

A further refinement is achieved by adding a fifth element, manganese, to form quinary alloys. Manganese has the ability to react with and dissolve carbon, which improves wetting of cemented carbides by the braze. For the same reason, the manganese-containing alloys are capable of wetting many grades of cast iron. The addition of the manganese also further improves the resistance to crevice corrosion of joints made to stain-

Filler Alloys and Their Metallurgy

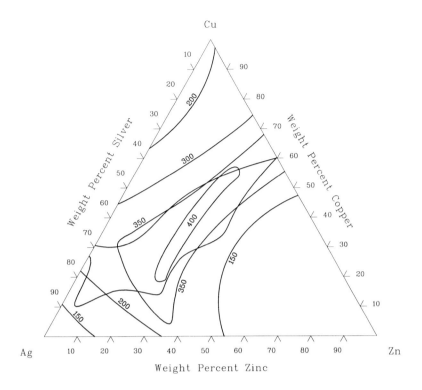

Fig. 2.7 Isostrength (MPa) contours for Ag-Cu-Zn alloys in the cast condition

Table 2.1 Silver-copper-zinc brazing alloys

	Composition, wt%				Melting range		
Ag	Cu	Zn	Ni	Mn	°C	°F	Notes
16.0	50.4	33.6	...	...	780-830	1435-1525	(a)
25.0	38.0	33.0	2.0	...	707-801	1305-1474	(b)
40.0	30.0	28.0	2.0	...	671-779	1246-1434	(b)
45.0	30.0	25.0	...	...	688-774	1270-1425	
49.0	16.0	23.0	4.5	7.5	690-801	1274-1474	(b)
50.0	34.0	16.0	...	...	677-743	1251-1369	
50.0	20.0	28.0	2.0	...	677-727	1251-1341	(b)
56.0	20.0	24.0	...	...	665	1229	(c)
65.0	20.0	15.0	...	...	671-713	1240-1315	
66.7	23.3	10.0	...	...	705-723	1300-1333	
67.5	23.5	9.0	...	...	700-730	1290-1345	(d)
74.0	19.2	6.8	...	...	720-765	1330-1410	(d)
81.0	14.0	5.0	...	...	730-800	1345-1470	(d)

(a) Alloy representative of the brazes used in brassworking. (b) Alloys used for brazing carbide tool tips. (c) Eutectic composition. (d) Alloys chiefly used by silversmiths.

less steel. Both nickel and manganese are cheap in relation to silver and thus help to reduce the intrinsic materials cost of the filler alloy.

A number of silver-copper-zinc brazing alloys used with silverware, brassware, and engineering alloys are listed in Table 2.1. The composition range and selected properties of the usable alloys are mapped out in Fig. 2.7 to 2.9.

2.2.1.6 *Silver-Copper-Zinc-Cadmium*

The addition of cadmium to silver-copper-zinc alloys confers three major benefits. One of these is

Principles of Soldering and Brazing

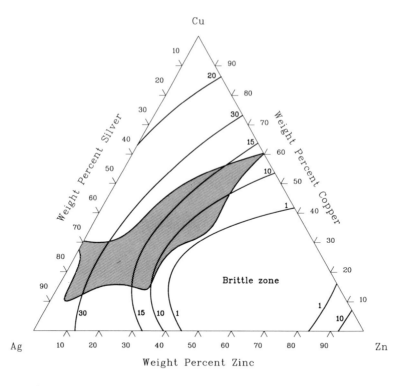

Fig. 2.8 Isoelongation (%) contours of as-cast alloys in the Ag-Cu-Zn system

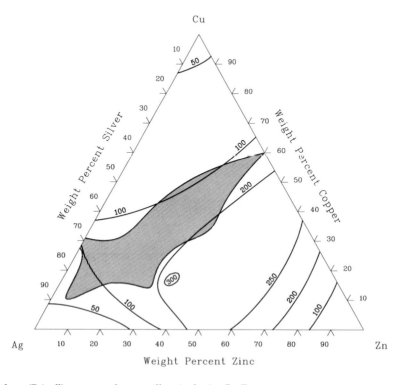

Fig. 2.9 Isohardness (Brinell) contours of as-cast alloys in the Ag-Cu-Zn system

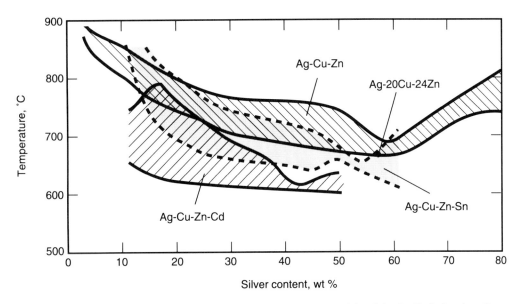

Fig. 2.10 Melting ranges of ternary Ag-Cu-Zn and quaternary Ag-Cu-Zn-Cd and Ag-Cu-Zn-Sn brazing alloys as a function of their silver content. Note that the liquidus and solidus lines do not correspond to fixed ratios of the other constituents.

Table 2.2 Silver-copper-zinc-cadmium brazing alloys

	Composition, wt%					Melting range		
Ag	Cu	Zn	Cd	Ni	Mn	°C	°F	Notes
25.0	40.0	15.0	17.0	0.5	2.5	625-770	1155-1420	
25.0	35.0	26.5	13.5	...	...	607-746	1125-1375	
30.0	27.0	23.0	20.0	...	...	607-710	1125-1310	
35.0	26.0	21.0	18.0	...	...	607-701	1125-1294	
40.0	19.0	21.0	10.0	...	...	595-630	1105-1165	
42.0	17.0	16.0	25.0	...	...	610-620	1130-1150	(a)
45.0	15.0	16.0	24.0	...	...	607-618	1125-1145	
50.0	15.5	16.5	18.0	...	...	627-635	1161-1175	
50.0	15.5	15.5	16.0	3.0	...	632-688	1170-1270	
67.5	11.0	11.5	10.0	...	...	635-720	1175-1330	(b)

(a) Pseudoeutectic composition alloy with the narrowest melting range. (b) Used for joining hallmarked silverware

a reduction in the solidus temperature and the melting range; a second is a reduction in the silver content, as compared with the nearest cadmium-free equivalent alloy (see Fig. 2.10). The liquidus projection of a section through the silver-copper-zinc-cadmium quaternary system at 40% Ag is given in Fig. 2.11, which shows that the effect of cadmium additions to the Ag-Cu-Zn ternary eutectic is to reduce the liquidus and solidus temperatures until a transition reaction occurs at 615 °C (1140 °F). The quaternary alloy that lies at the composition Ag-17Cu-16Zn-25Cd (melting range = 610 to 620 °C, or 1130 to 1150 °F) is commonly referred to as the "pseudoeutectic" composition. Further cadmium additions substantially widen the melting range of alloys. The compositions of the principal brazes in this alloy family are listed in Table 2.2.

The third benefit produced by the introduction of cadmium is an improvement in the fluidity of the alloys. This enhanced fluidity enables the production of narrower and better-filled joints that consequently exhibit improved mechanical properties. Departure from the pseudoeutectic composition widens the melting range of the alloys. The alloy compositions that are richer in silver—namely those containing 67.5% of this element—are needed to satisfy the silver content requirements of the jewelry industry. The alloys with a lower silver concentration are used primarily in

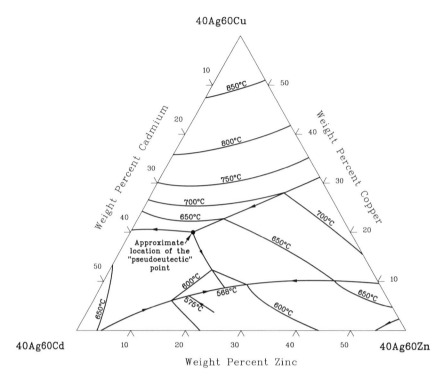

Fig. 2.11 Section through the Ag-Cu-Zn-Cd quaternary system at 40% Ag. The liquidus temperature of alloy compositions is indicated by a series of isotherms. Note: This is not a true ternary phase diagram. Source: After Petzow and Aldinger [1968]

Table 2.3 Silver-copper-zinc-tin brazing alloys

		Composition, wt%				Melting range	
Ag	Cu	Zn	Sn	Ni	Mn	°C	°F
30.0	36.0	32.0	2.0	…	…	655-750	1210-1380
42.0	21.0	27.0	2.0	4.0	4.0	665-750	1230-1380
45.0	25.0	25.5	3.0	1.5	…	645-700	1195-1290
50.0	21.5	27.0	1.0	0.5	…	660-680	1220-1255
55.0	21.0	22.0	2.0	…	…	630-660	1165-1220
56.0	22.0	17.0	5.0	…	…	618-651	1145-1204
60.0	23.0	14.5	2.5	…	…	620-685	1150-1265

engineering applications, where the silver content can represent an important cost item. However, any reduction in the percentage of silver is made at the expense of the mechanical properties of the joints and a widening of the melting range of the braze, which then necessitates increasing the joining temperature.

The silver-copper-zinc-cadmium alloys are also available with nickel and manganese additions. These minor alloying constituents modify the characteristics of these brazes in the same way as they do in the silver-copper-zinc family, described above.

2.2.1.7 Silver-Copper-Zinc-Tin

The cadmium-bearing brazes were originally developed in the 1930s, before the health hazards of cadmium oxide vapor were widely appreciated. In recent years, restrictions have been placed on the use of cadmium-containing brazes, particularly in Sweden, triggering a search for substitutes.

The outcome of this endeavor has been the development of a series of alloys in which tin replaces the cadmium. The principal compositions are listed in Table 2.3. The tin largely preserves the

Table 2.4 Cadmium-free carat gold "solders"

Au	Ag	Cu	Zn	Ni	Ga	°C	°F
95	...	...	...	...	5	415-810	780-1490
88	...	...	...	6	6	550-590	1020-1095
80	...	...	8	12	...	782-871	1440-1600
75	9	6	10	...	...	730-783	1345-1441
66.6	10	6.4	12	5	...	718-810	1324-1490
58.5	11.8	25.7	4	...	...	816-854	1501-1569
50	25	10	9	6	...	724-782	1335-1440
41.7	30	8.3	15	5	...	702-732	1296-1350
33.3	30	16.7	20	...	...	695-704	1285-1299
25	58	...	17	...	...	738-807	1360-1485

Table 2.5 Industrial gold brazing alloys

Au	Ag	Cu	Ni	Pd	°C	°F
92	...	...	...	8	1180-1230	2155-2245
82.5	...	16.5	2	...	899	1650
82	...	...	18	...	955	1750
80	...	20	...	...	910	1670
75	...	...	...	25	1375-1400	2505-2550
62.5	...	37.5	...	...	930-940	1705-1725
60	20	20	...	...	845-855	1555-1570
50	...	...	25	25	1121-1125	2050-2055
37.5	...	62.5	...	...	980-1000	1795-1830
29	45	26	...	...	767	1413
20	...	80	...	...	1018-1040	1864-1905

good flow characteristics conferred by the cadmium and also acts as a melting point depressant, although not to the extent that cadmium does. Moreover, the substitution broadens the melting range of the alloys, as can be seen in Fig. 2.10. Thus, for example, the melting range of the 55Ag-21Cu-22Zn-2Sn alloy is 30 °C (55 °F) (liquidus temperature = 660 °C, or 1220 °F; solidus temperature = 630 °C, or 1165 °F). This compares with only 10 °C (20 °F) for the cadmium-bearing alloy 42Ag-17Cu-16Zn-25Cd (liquidus temperature = 620 °C, or 1150 °F; solidus temperature = 610 °C, or 1130 °F), which also contains significantly less silver. The low solid solubility of tin in zinc restricts the percentage of tin that can be added. Larger additions lead to the formation of embrittling intermetallic phases.

Nickel- and manganese-bearing versions of these alloys are also commercially available, as is a silver-copper-zinc-indium alloy. The substitution of tin by indium does not offer any particular improvement in the properties of the filler, and it marginally increases the cost of the braze. Other alternative alloys to the Ag-Cu-Zn-Cd brazes are mentioned in section 2.2.1.11.

2.2.1.8 Gold-Bearing Filler Metals

For many years, gold-bearing brazing alloys were used almost exclusively in the jewelry industry. These alloys, known as carat gold "solders" by the trade, were developed to provide goldsmiths with a selection of relatively low-melting-point brazes that match the gold content and color of carat jewelry alloys. In general, these brazes have little to offer industrial users that cannot be found in the much cheaper silver and base-metal brazing alloys. Table 2.4 lists a representative range of carat gold brazing alloys. The silver, copper, zinc, nickel, and gallium serve to adjust both the melting temperatures and, more importantly for jewelry applications, the caratage and color of the alloys. The color of an alloy is determined by that of the phase that constitutes the largest volume fraction. Therefore, the Au-3Si solder, which has as its phases gold (85 vol%) and gray silicon (15 vol%), has a golden hue, whereas the Au-20Sn solder, which comprises silvery Au_5Sn (50 vol%) and silvery AuSn (50 vol%) is metallic white in color.

More recently, a new range of gold-bearing brazing alloys has been developed in response to technological demand, particularly from the elec-

Principles of Soldering and Brazing

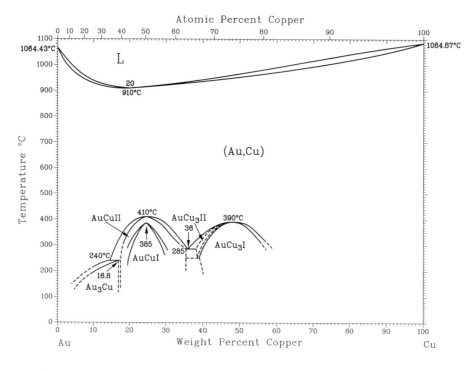

Fig. 2.12 Gold-copper phase diagram

tronics, nuclear power, and aerospace industries. Some examples are listed in Table 2.5. They are particularly suited for use in corrosion-resistant assemblies with enhanced mechanical properties that require joining alloys with matching properties. It is possible to enhance the oxidation resistance of the silver-bearing and nickel-bearing brazing alloy families by adding small percentages of certain elements such as aluminum, nickel, chromium, and manganese. The improved properties derive from the ability of these elements to form relatively stable and inert oxide films. However, in aggressive chemical environments, many of these elements corrode easily. By contrast, noble metals are, by their very definition, largely inert chemically and therefore capable of surviving in harsh environments.

The gold-base brazing alloys can be divided into three principal families—namely, gold-copper, gold-nickel, and gold-palladium. The gold content of some of these alloys is less than 50%. Nevertheless, by convention they are classed as gold brazes, because the presence of the expensive precious metal constituent, rather than its proportion, governs the designation. The same is true of brazes containing the platinum-group metals.

The gold-bearing alloys designed for industrial use are superior to base-alloy brazes in the following respects:

- Improved oxidation resistance
- Enhanced mechanical properties of joints at elevated temperatures
- Relatively low degree of erosion of component metal surfaces during the joining operation
- Good corrosion resistance in most chemical environments
- Usable with only mild fluxes, owing to the nobility of these alloys

The three principal families of gold-bearing alloy brazes will be considered briefly in turn.

Gold-Copper. The gold-copper binary alloy system, which also provides the basis for many of the brazes used in jewelry, is characterized by a continuous solid solution between the two constituent metals, with a liquidus minimum at the Au-20Cu composition where there is a single melting point (910 °C, or 1670 °F). The gold-copper phase diagram is shown in Fig. 2.12. All other alloy compositions in this system have narrow

Filler Alloys and Their Metallurgy

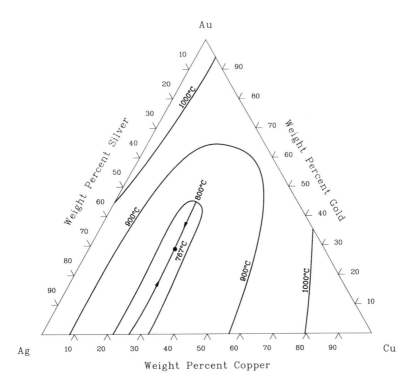

Fig. 2.13 Liquidus surface of the Ag-Au-Cu ternary system

melting ranges, typically less than 20 °C (40 °F). This means that gold-copper brazes possess excellent fluidity and readily form good fillets. Alloys within the range of approximately 40 to 90% Au undergo ordering transformations at low temperature that produce a hardening effect. However, they are sufficiently ductile to be mechanically worked to foil and wire, suitable for making preforms. Gold-copper alloys will readily wet a range of base metals, including many refractory metals.

The addition of nickel to the copper-gold brazes helps to improve their ductility: a typical composition is Au-16.5Cu-2Ni, which has a melting point of 899 °C (1650 °F). Adding silver to copper-gold binary alloys effects a significant reduction in melting point to 767 °C (1413 °F) at the composition 45Ag-29Au-26Cu. The liquidus surface of the phase diagram for this alloy system is given in Fig. 2.13. However, for applications where corrosion resistance is required, the gold concentration must be kept above 60%, which increases the melting point to about 850 °C (1560 °F).

Gold-Nickel. Gold forms a continuous series of solid solutions with nickel in a similar manner to the gold-copper alloys, except that the melting range tends to be wider. The liquidus/solidus minimum occurs at the Au-18Ni composition and 955 °C (1750 °F). The gold-nickel phase diagram is shown in Fig. 2.14.

A nickel content of 35% is generally the maximum used for brazing alloys, owing to the considerable widening of the melting range as the proportion of nickel is increased. The gold-nickel brazes possess many of the advantages of the gold-copper alloys, but provide the additional benefits of superior resistance to oxidation and improved wetting.

The elevated-temperature properties (namely, strength and corrosion resistance) can be further enhanced by introducing alloying elements such as chromium, manganese, and palladium. These additions also promote wetting of refractory materials, notably graphite and carbides. A typical example is the Au-34Cu-16Mn-10Ni-10Pd alloy, which was developed for brazing components in the Space Shuttle main engine at about 1000 °C (1830 °F), which demanded joints with superior oxidation resistance.

The extra constituents enable the proportion of costly gold in the alloy to be significantly reduced without sacrificing essential properties. Thus, the

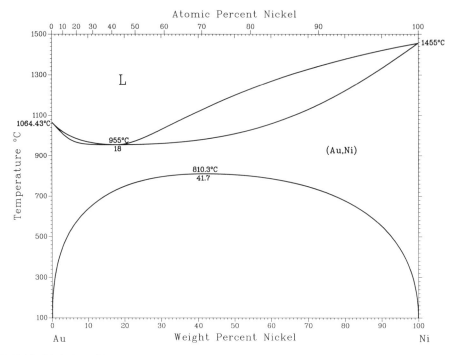

Fig. 2.14 Gold-nickel phase diagram

Au-22Cu-8.9Ni-1Cr-0.1B alloy, containing 68% Au (melting range 960 to 980 °C, or 1760 to 1795 °F) was developed as a cheaper equivalent to the minimum-melting-point Au-18Ni composition [Sloboda 1971]. This substitution provides an instructive example of the considerations that frequently apply in the development of new filler compositions. First, target properties are defined. In this case, the new alloy was required to:

- Readily wet heat-resistant steels, in particular Jethete (Fe-0.45C-0.7Mn-12Cr-2.5Ni-0.3V-2Mo), and produce sound joints to components of these materials
- Completely melt below 1000 °C (1830 °F) in order to be compatible with the special steels, which are heat treated (the beneficial effects of the heat treatment are lost above this temperature)
- Be free of volatile constituents to enable the braze to be used in vacuum joining operations
- Confer high strength and oxidation resistance to the joints
- Be sufficiently ductile to be able to be mechanically worked to foil and wire for preforms

The alloy was designed such that copper substituted for some of the gold and the ensuing loss of oxidation resistance was compensated for by the chromium addition. The deterioration of the wetting characteristics due to the presence of the chromium was made good by the introduction of a small fraction of boron. Finally, the relative proportions of the constituents were then adjusted to minimize the melting range and optimize the mechanical properties.

Gold-Palladium. The addition of palladium to the gold-copper and gold-nickel alloys improves their resistance to oxidation at elevated temperatures. These alloys are therefore used mostly for joining refractory metal components that serve in relatively aggressive environments, such as exist in modern jet engines. Commercially available brazes of this type have melting temperatures that reach approximately 1200 °C (2190 °F). All these alloys are classified simply as gold-palladium alloys, drawing attention to the two precious metals in this series. Some examples are listed in Table 2.5.

2.2.1.9 *Palladium-Bearing Filler Metals*

Palladium is the major constituent of a series of brazing alloys that contain copper, nickel, and

Table 2.6 Palladium-bearing filler metals

	Composition, wt%				Melting range	
Pd	Ag	Cu	Ni	Other	°C	°F
65	...	...	...	35Co	1230-1235	2245-2255
60	...	...	40	...	1237	2259
54	...	...	36	10Cr	1232-1260	2250-2300
25	54	21	...	...	900-950	1650-1740
21	...	...	48	31Mn	1120	2050
15	65	20	...	...	850-900	1560-1650
5	68.5	26.5	...	...	805-810	1480-1490

silver. These alloys possess many of the beneficial properties of the gold-bearing brazes, but are less expensive. They confer on joints:

- Good mechanical integrity and freedom from brittle intermetallics. This is a consequence of the fact that palladium forms solid solutions with most common metals
- Enhanced mechanical strength at elevated temperatures. In this respect, they tend to be superior to the family of gold alloys that do not contain palladium or other platinum-group metals
- High oxidation resistance at elevated temperatures, especially in the case of the palladium-nickel alloys
- Good corrosion resistance, although not as good as the gold-bearing brazing alloys
- Low vapor pressure at typical brazing temperatures, comparable to that of the gold alloys
- A consistently narrow melting range, in most cases no more than 25 to 50 °C (45 to 90 °F)

A representative range of commercial palladium-bearing alloys is listed in Table 2.6. These brazing alloys find application in refractory metal structures where cost is more critical than with certain aerospace uses, which can justify the gold-bearing alloys. Platinum is superior to palladium in terms of its chemical inertness and thus finds occasional use in brazing alloys, but the applications are restricted by the high cost of this metal and by its relatively poor mechanical workability.

2.2.1.10 *Nickel-Bearing Filler Metals*

Nickel-base brazing filler metals are more extensive in composition range and properties than even the silver-base alloys described above. Their primary merit is the ability to endure high-temperature service (above 1100 °C, or 2010 °F), even in moderately aggressive environments. These properties can be largely attributed to the corrosion resistance of elemental nickel and its relatively high melting point (1455 °C, or 2650 °F). Pure nickel is not widely used as a braze because of its high melting point, except for certain specialist applications, such as the joining of molybdenum and tungsten components intended for subsequent operation at elevated temperatures.

The melting point of nickel is depressed by alloying additions of an appreciable number of elements (e.g., boron, carbon, chromium, copper, manganese, phosphorus, sulfur, antimony, silicon, and zinc), many of which are introduced in combination to produce the variety of nickel-bearing filler metals that are commercially available. Of these, phosphorus and boron are particularly effective in low concentrations at promoting the wetting characteristics of nickel-base brazes. The nickel-boron phase diagram is shown in Fig. 2.15. This is similar to the copper-phosphorus diagram in that there is a "low"-melting-point eutectic between nickel and Ni_3B, as there is between copper and Cu_3P. All of these elements are inexpensive in comparison with the constituents of noble brazing alloys. One of the most widely used nickel-base brazes, BNi-1, has the composition Ni-3B-0.7C-14Cr-5Fe-5Si and melts in the range of 977 to 1038 °C (1790 to 1900 °F).

A feature of filler metals containing boron, carbon, silicon, and phosphorus is that these elements will diffuse rapidly into many parent metals, thereby increasing the remelt temperature of the solidified braze. Therefore, the process is commonly referred to as diffusion brazing [Duvall, Owczarski, and Paulonis 1974]. This process is fully described in section 4.4.2. It is exploited in the manufacture of engines with superalloy turbine blades, because it enables the brazed component to be used in service at temperatures that exceed the peak brazing process temperature. However, care must be taken to select an appropriate braze/parent metal combination, because the diffusion of certain elements into structural engi-

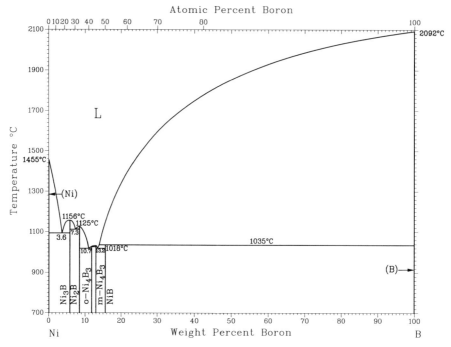

Fig. 2.15 Nickel-boron phase diagram

neering alloys can result in intergranular embrittlement. Conversely, the formation of borides, carbides, phosphides, and so on in the joint gap, through reaction with the constituents of the parent metals, can cause a reduction in joint ductility and impact strength.

Nickel has extensive intersolubility with many engineering base metals, notably iron, chromium, and manganese. Erosion of the parent metal by the braze during the joining process can therefore be severe unless close attention is paid to controlling the parameters of the brazing cycle, particularly the excess temperature above the melting point of the braze.

Many nickel-base brazing alloys themselves contain significant proportions of phosphide and boride phases and thus tend to be brittle. Until recently, this has meant that the compositions of nickel-base brazing alloys that could be fabricated as foils and wire were extremely limited. Instead, the compositions that are brittle were applied in the form of pastes, that is, alloy powder in a fluid binder. Organic binders burn during the brazing cycle, which tends to introduce voids and carbonaceous residues that weaken joints. This restriction has been overcome by the development of rapid solidification casting technology (see section 1.4.2.2). Because of the low melting point of the alloys in relation to that of the constituent elements, particularly when the brazing alloy contains boron and/or phosphorus, and because during rapid solidification the cooling rate typically exceeds 10^5 °C/s (2×10^5 °F/s), these nickel-base alloys generally solidify with an amorphous structure [DeCristofaro and Henschel 1978]. Examples of rapid solidification casting and foils prepared by chill-block melt spinning are shown in Fig. 1.13 and 1.14. A selection of nickel-base alloys that are available commercially as amorphous strip is included in Table 2.7. This family of alloys covers melting temperatures that range from about 700 °C (1290 °F) to more than 1100 °C (2010 °F).

Although the nickel-base brazing alloys have particularly benefited from rapid solidification technology, other filler metals that are normally brittle when cast conventionally have also been upgraded by this preparation technique [DeCristofaro and Bose 1986]. These include brazes based on copper, palladium, and cobalt, with the copper alloys being developed as cheaper alternatives to the nickel-bearing alloys and the palladium- and cobalt-base alloys for more demanding application environments. Solders based on aluminum, gold, bismuth, tin, and lead have also been successfully prepared as ductile foils by this technique [Bowen and Peterson 1987]. The main alloy

Filler Alloys and Their Metallurgy

Table 2.7   Examples of commercially available rapidly solidified filler metals

Composition, wt%	Melting range °C	Melting range °F	Structure	Typical applications
Ag-28Cu	779	1432	Microcrystalline	Most engineering materials
Ag-28Cu-5Ti	775-790	1425-1455	Microcrystalline	Engineering ceramics
Al-13Si	577	1071	Microcrystalline	Aluminum alloys
Au-20Sn	278	535	Amorphous	Sealing electronics packages
Au-3Si	363	685	Microcrystalline	Microelectronic die attach
Bi-43Sn	139	282	Microcrystalline	Hermetic solder seals
Cu-10Mn-30Sn	640-700	1185-1290	Amorphous	Copper alloys and mild steel
Cu-10Ni-4Sn-8P	610-645	1130-1195	Microcrystalline	Copper alloys and mild steel
Cu-8P	714	1317	Amorphous	Copper alloys and mild steel
Cu-20Sn	770-925	1420-1695	Microcrystalline	Copper alloys and mild steel
Co-19Cr-19Ni-8Si-1B	1120-1150	2050-2100	Amorphous	Cobalt-base alloys and superalloys
Ni-10P	880	1615	Amorphous	Steels, stainless steels, superalloys
Ni-32Pd-8Cr-3B-1Fe	940-990	1725-1815	Amorphous	Steels, stainless steels, superalloys
Ni-14Cr-5Si-5Fe-3B	970-1075	1780-1965	Amorphous	Steels, stainless steels, superalloys
Ni-15Cr-3B	1020-1065	1870-1950	Amorphous	Steels, stainless steels, superalloys
Ni-41Pd-9Si	712-745	1314-1375	Amorphous	Steels, stainless steels, superalloys, cemented carbides
Pb-62Sn	183	361	Microcrystalline	Electronics systems fabrication
Pb-5In-2.5Ag	300	570	Microcrystalline	Electronics systems fabrication
Pd-38Ni-8Si	830-875	1525-1605	Amorphous	Stainless steels, superalloys, cemented carbides
Sn-3.5Ag	221	430	Microcrystalline	Electronics systems fabrication
Sn-25Ag-10Sb	240-290	465-555	Microcrystalline	Sealing electronics packages and die attach
Ti-15Cu-15Ni	902-932	1656-1710	Amorphous	Superalloys and engineering ceramics
Ti-20Zr-20Ni	848-856	1558-1573	Amorphous	Superalloys and engineering ceramics
Zr-17Ni	961	1762	Amorphous	Titanium-base alloys
Zr-16Ti-28V	1193-1250	2179-2280	Amorphous	Titanium-base alloys

compositions that have been commercially developed in the form of rapidly solidified foils, wire, and powder are listed in Table 2.7. These are produced with either an amorphous or a microcrystalline microstructure.

The use of this casting technology for producing foil and wire preforms not only allows brazes with new compositions to be manufactured, but also confers a number of associated benefits. First, the brazing alloy is comparatively ductile in the amorphous state, as there are no discrete phases or grain boundaries in the microstructure that might be sources of embrittlement. On heating an amorphous alloy to approximately half its melting point in degrees Kelvin, the microstructure will revert to a crystalline form. After melting and solidification, the braze will have a conventional microstructure. However, there is strong evidence in the published literature that the microstructure of joints made with both solders and brazes prepared by rapid solidification are finer and more uniform than joints made with conventionally prepared filler metals [DeCristofaro and Bose 1986; Bowen and Peterson 1987]. The reasons for this are not fully understood, but are likely to be associated with a degree of atomic ordering in the liquid phase [Johnson 1990]. Evidence for this explanation comes from the gray/white allotropic transformation of tin. Gray tin does not resolidify as purely white tin after a cursory melting excursion as one might expect if the liquid retained no "memory" of its previous state as a solid.

Second, the absence of a dendritic microstructure in both amorphous and microcrystalline metals means that a rapidly solidified alloy is metallurgically homogeneous, as cast, akin to a solid solution. The alloy therefore melts in a highly uniform manner, which helps to minimize local fluctuations in the erosion of the parent metals due to an absence of segregated phases in the filler.

Third, filler metals prepared by rapid solidification tend to be cleaner both on the surface and within the bulk, as there is only a single processing step and that is usually carried out in a protective atmosphere. Metallic contamination from rolling and drawing machinery is eliminated, as is carbonaceous contamination from lubricating fluids emanating from the machinery.

Finally, the homogeneous filler metal preforms have a much narrower melting range compared with the equivalent composition alloy prepared as a trifoil or clad wire, comprising a core of titanium with a cladding of the other constituents. Thus, the Ti-15Cu-15Ni composition braze has a melting range of 902 to 932 °C (1656 to 1710 °F) as a homogeneous alloy and 912 to 1007 °C (1674 to 1845 °F) when prepared as a trifoil in the form of a titanium layer sandwiched between foils of Cu-50Ni and heated at 10 °C/min (20 °F/min).

When using nickel-base fillers containing metalloids, there is a tendency for embrittling nickel-metalloid compounds to precipitate in the joint if the width of the braze is more than about 50 µm (2 mils). Where wide joints are made, this problem may be avoided by fitting into the joint gap a porous shim of a metal that will rapidly soak up the filler and dilute the metalloid [Lugscheider and Schittny 1988]. This technique is particularly useful in repair work.

2.2.1.11 *Other Brazing Alloy Families*

There are many other families of brazing alloys; those based on aluminum will be considered below. For example, families of copper-manganese-tin and copper-nickel-tin-phosphorus alloys have been developed as inexpensive alternatives to silver-containing brazing alloys [Chatterjee, Mingxi, and Chilton 1991; Datta, Rabinkin, and Bose 1984]. These have melting ranges and physical properties in joints that are similar to those of the silver-copper-zinc-base brazes, and the phosphorus-containing alloys are self-fluxing when used to join copper and steel, because the phosphorus is present at a concentration of about 8%. Other brazing alloy families, including alloys of magnesium, cobalt, titanium, zirconium, and tungsten, will not be dealt with here because they are not widely used and their behavior as filler metals is largely similar to that of the alloy compositions which are described in this chapter; in addition, the metallurgical principles on which they were developed are not significantly different. Nevertheless, it should be borne in mind that these brazes are offered on the market because they provide specific property advantages for certain applications. Details can be found in reference works such as Schwartz [1987] and data sheets and manuals supplied by manufacturers.

2.2.2 Aluminum- and Zinc-Bearing Filler Metals

2.2.2.1 *Aluminum-Bearing Brazes*

Aluminum-bearing brazes are primarily used for joining components of aluminum and its alloys. An exceptional application is described in Chapter 7. Joints to aluminum components tend to be more susceptible to corrosion than similar joints between other common metals. This is because aluminum has an electrode potential that is more negative than most other metallic elements, and this can give rise to a galvanic corrosion problem. The necessary emphasis placed on corrosion and its prevention has resulted in the adoption of a few select aluminum- and zinc-base alloys as filler metals for joining aluminum alloy components.

The degree of galvanic corrosion that can occur is a function of the difference between the electrode potentials of two metals when they are electrically coupled by an electrolyte, such as ionized water. The more negatively biased metal is the one that corrodes. Table 2.8 lists the standard electrode potentials of several common metals with reference to hydrogen and shows that aluminum is usually the metal subject to chemical attack. However, it should be borne in mind that in the majority of filler alloys the phases present are not pure metals but solid solutions and compounds of various sorts, the electrode potentials of which are often not known and cannot be predicted from the constituent elements. Even for elemental metals, the data in Table 2.8, which are obtained under closely specified conditions, can serve only as a rough guide, because measured electrode potentials depend on such factors as the nature and strength of the electrolyte and the surface condition of the metal electrodes.

Susceptibility to galvanic corrosion provides one of the main criteria for aluminum being the major constituent of brazes used for joining the same metal. Indeed, all of the commercially available aluminum-bearing fillers are based on the Al-13Si eutectic composition alloy, which melts at 577 °C (1071 °F). The aluminum-silicon phase diagram is shown in Fig. 2.16. Besides depressing the melting point of aluminum, silicon also confers some fluidity, and it is one of the few elements that does not greatly enhance the corrosion of the parent materials.

Aluminum-silicon binary alloy fillers are available as wires, foils, and cladded sheets. The cladded material comprises a sheet of a suitable aluminum engineering alloy, coated on one or

Filler Alloys and Their Metallurgy

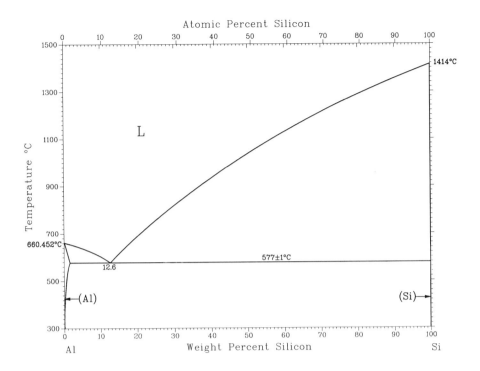

Fig. 2.16 Aluminum-silicon phase diagram

Table 2.8 Electrode potential of selected elements at 25 °C (77 °F)

Element	Electrode potential Volts
Gold	+1.50
Silver	+0.80
Copper	+0.34
Hydrogen	0.00
Lead	−0.13
Tin	−0.14
Nickel	−0.25
Cadmium	−0.40
Iron	−0.44
Zinc	−0.74
Silicon	−1.30
Aluminum	−1.66
Magnesium	−2.37

both sides with filler metal. Each cladding typically constitutes up to 15% by thickness of the total thickness of the sheet. Clad materials are particularly suited for fluxless brazing processes; the joint interface between two mated components that have been roll-clad will liquefy on heating above the solidus temperature, and this helps to displace the alumina layer and other surface films present.

Lower-melting-point filler alloys can be achieved by adding copper, magnesium, and zinc, which further depress the melting point of aluminum-silicon, due to the existence of ternary eutectic points in all three Al-Si-X (X = copper, magnesium, zinc) systems. Magnesium has the additional benefit of aiding in wetting for the reasons discussed in section 5.3.3. Representative aluminum filler metals suitable for brazing aluminum alloys are listed in Table 2.9. Aluminum alloy brazes are perfectly satisfactory for joining metals such as stainless steels, molybdenum, and tungsten, provided that the different electrode potentials of these metals are recognized and appropriate measures are taken to minimize the risk of corrosion of the joint.

Several problems are associated with the use of aluminum-base brazes with aluminum alloy components, which represents their main area of application:

- Their melting points are very close to those of most aluminum engineering alloys. For some of these alloys, the melting points of the available brazes can actually exceed the solidus temperatures of the engineering alloys. For this reason, many of the wrought and casting alloys cannot be joined by brazing.
- Aluminum has high solubility in these brazes, resulting in extensive erosion of the parent

Table 2.9 Typical aluminum-bearing filler metals

	Composition, wt%			Melting range	
Al	Si	Cu	Mg	°C	°F
93	7	...	...	577-612	1071-1134
90	10	...	...	577-593	1071-1099
87	13	...	...	577	1071
86	10	4	...	524-585	975-1085
88	10	...	2	555-585	1030-1085

metal. Consequently, thin-wall components with a thickness of less than about 0.5 mm (0.02 in) cannot be joined easily by this means. The alloying increases the melting point of the filler and tends to impede lateral flow and fillet formation by the molten braze.

- They exhibit poor wetting on aluminum alloys when used without fluxes. This is a consequence of the high reactivity of aluminum with oxygen in the atmosphere and the refractory nature of the alumina that forms.
- Heavy fluxing is required. The fluxing agents react extensively with the aluminum-base parent materials. This results in the formation of large quantities of flux residues that are difficult to remove and which are strongly corrosive.
- Most engineering alloys of aluminum rely on precipitation hardening for their boosted mechanical properties (i.e., hardening through the presence of a finely divided second phase in the material). The temperatures required for brazing with the available filler metals are too high to be compatible with the heat treatment step that precedes precipitation of the second phase. There have been attempts to remedy this situation by developing new alloy compositions with substantially lower melting points. Recently, a series of new multicomponent aluminum brazes containing silicon, copper, nickel, and (optionally) zinc as the principal additions have been developed for fluxless brazing processes in inert gas atmospheres at temperatures as low as 520 °C (970 °F) [Humpston, Jacobson, and Sangha 1992].

Notwithstanding these difficulties, the fluxless brazing of aluminum has been successfully developed for the fabrication of radiators and heat exchangers of this metal and is now routinely employed for mass production in the automotive industry. The process, as established by VAW in Germany, is performed in either high vacuum or in a high-quality nitrogen atmosphere and uses an aluminum-silicon filler alloy containing specified minor additions that promote spreading and fillet formation. Special cleaning procedures applied to component surfaces are an essential part of the process [Schultze and Schoer 1973].

2.2.2.2 Zinc-Bearing Solders

The majority of solders that are used for joining aluminum and its alloys primarily contain zinc [Finch 1985]. Zinc is the only low-melting-point metal that does not grossly enhance the corrosion of aluminum alloys, because its electrode potential is close to that of aluminum, as indicated by the data in Table 2.8. This metal plays an additional role in helping to disrupt alumina surface layers and so helps, to a lesser or greater degree, depending on the proportion of zinc present and the joining atmosphere, in promoting wetting by the solder. A flux-cored zinc solder that has high resistance to corrosion is available for joining aluminum [Rubin 1982].

One of the zinc-bearing solders that is widely used for joining aluminum is the eutectic composition alloy Al-94Zn (melting point = 381 °C, or 718 °F). The aluminum-zinc phase diagram is shown in Fig. 2.17. Often a few parts per million of gallium are added to the solder to aid wetting, as molten gallium is one of the few metals that will not ball-up on alumina. Spreading of the solder is promoted by a reaction between gallium oxide and alumina. Like other solders containing large fractions of volatile constituents, the zinc-bearing alloys used for soldering aluminum are not suitable for joining processes that employ reduced-pressure atmospheres.

All of the zinc-bearing solders contain high percentages of zinc, generally in the range of 75 to 95%, and their solidus temperatures cover the range of 197 to 419 °C (387 to 786 °F). The other elements present in significant percentages are aluminum, cadmium, copper, and tin. A representative list of these solders is given in Table 2.10. Pure zinc is not used as a filler metal because

Filler Alloys and Their Metallurgy

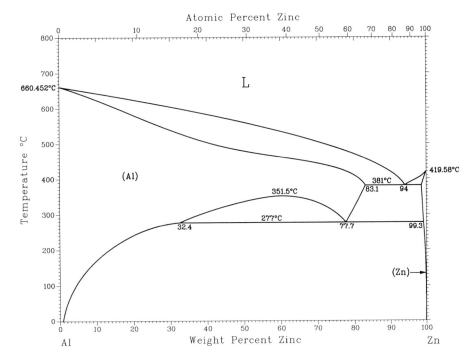

Fig. 2.17 Aluminum-zinc phase diagram

it tends to ball-up on heating when used in normal atmospheres, whereas when small amounts of other elements are added, wetting of surfaces is more readily achieved. The cadmium-containing alloy listed in Table 2.10 is now generally avoided owing to its relative toxicity.

The choice of alloying additions is made on the grounds of reducing the erosion of the substrate metals and, as far as possible, of improving the spreading and flowing characteristics. In general, the additions usually do not reduce the liquidus temperature, but only the solidus temperature.

In recent years, considerable research effort has been devoted to the development of improved new solders based on zinc [Harrison and Knights 1984]. These alloys have the following potential benefits:

- They can be used at sufficiently low temperatures so as not to destroy the work-hardened strength of the copper alloys. Yet, at the same time, the joints that are produced are typically two to four times stronger than those obtained with the common solders.
- The zinc solders are compatible with galvanized steel components.

Table 2.10 Zinc-bearing alloys used as solders

Composition, wt%	Melting range °C	°F
Zn-6Al	381	718
Zn-5Al-2Ag-1Ni	381-387	718-729
Zn-7Al-4Cu	379-390	714-735
Zn-10Cd	266-399	511-750
Zn-25Cd	266-370	511-700
Zn-1Cu	418-424	784-795
Zn-30Cu-3Sb-1Ag	416-424	781-795
Zn-10Sn-1Pb	197-385	387-725
Zn-70Sn	199-311	390-592
Zn-2Ni	419-560	786-1040

- They are inexpensive in relation to most brazes and solders, being approximately one-fourth the cost of the lead-tin alloys, for a given volume of filler metal.
- They are lightweight, having approximately half the density of the Pb-60Sn eutectic solder.
- They possess high thermal conductivity, exceeding 100 W/m · K (60 Btu/h · ft · °F).
- They are not hazardous: unlike cadmium and lead, the toxicity effects of small amounts of zinc are noncumulative and temporary. Zinc is one of the vital trace elements in the human

diet, as are many elements that are toxic in larger concentrations.

The factors that have greatly limited the adoption of the zinc solders include:

- The potential strengths to copper alloys are compromised by the presence in joints of embrittling copper-zinc compounds that form as interfacial phases. For this reason, the zinc-base solders that have been formulated for use with copper alloys often contain numerous minor alloying additions to modify the growth of these phases.
- Zinc alloys in joints to most base metals are susceptible to galvanic corrosion, a problem that they share with aluminum filler alloys.
- The low intrinsic material costs are largely offset by high fabrication costs arising from the low ductility of most zinc-base alloys.
- The relatively high volume contraction on solidification, which is typically 1 to 1.5 vol%, is detrimental to joint filling and can cause stress concentration in joints.
- The zinc alloys generally exhibit poor flow characteristics. The high oxidation rate of zinc, which in air is on the order of 2 to 3 µm/min at 400 °C (750 °F), coupled with high surface tension and viscosity (both double those of lead-tin alloys), are the main factors.
- The high vapor pressure of zinc means that these solders cannot be used in reduced-pressure atmospheres, because the volatility of this element exacerbates the formation of voids in joints and contaminates furnacing equipment.
- The high affinity of zinc for oxygen requires the use of aggressive fluxes when soldering in ambient atmospheres, which in turn leads to cleaning and corrosion problems. The flux efficacy with particular parent metal/solder combinations is strongly dependent on its actual composition. This makes it extremely difficult to achieve tolerant soldering processes using zinc-base filler metals.

Because, for many situations, the disadvantages outweigh the prospective benefits, even these "modern" zinc-bearing solders have not found much favor in industry.

Table 2.11   Gold-bearing solders

Composition, wt%	Eutectic temperature °C	°F
Au-3Si	363	685
Au-25Sb	356	673
Au-12Ge	361	682
Au-20Sn	278	532

2.2.3 Solders

Whereas the brazing alloys comprise families of kindred alloys based on silver, gold, palladium, aluminum, and so on, each solder family tends to be different from the next one, in terms of composition and physical properties (see section 1.2). The principal solder alloy families are reviewed below.

2.2.3.1 *Gold-Bearing Filler Metals*

Gold is unusual in that it is the only element on which both brazes and solders are based, that is, this element is the major constituent of both types of filler. The gold-bearing solders are all gold-rich eutectic composition alloys and have melting points between 278 and 363 °C (532 and 685 °F). They are listed in Table 2.11, and their associated phase diagrams are presented in Fig. 2.18 to 2.21. In view of their high gold content and thus high cost, the applications of these alloys tend to be limited and specialized. One of the chief attractions of these solders is their melting point, which falls within the 300 to 550 °C (570 to 930 °F) gap that separates the upper limit of the lead-base solders from the lower limit of the available aluminum-bearing brazes.

Gold-bearing solders have the advantage of being suitable for joining to gold metallized components, which are employed extensively in high performance products of the microelectronics industry. The compatibility with gold metallizations is due to the high slope of the liquidus surface between the various eutectic compositions and pure gold. Accordingly, the volume of gold that will dissolve in these solders is small at typical excess temperatures above the melting point. This is explained more fully in section 3.2. The high percentage of gold in these solders means that their applications are limited to those where the value of the product is high and, therefore, the cost of the precious metal can be tolerated. The two principal gold-base solders are considered briefly below.

Gold-silicon alloys are primarily used in the form of a foil preform for bonding silicon semiconductor chips to gold-metallized pads in ce-

Filler Alloys and Their Metallurgy

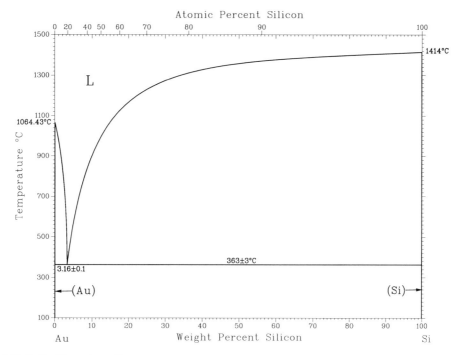

Fig. 2.18 Gold-silicon phase diagram

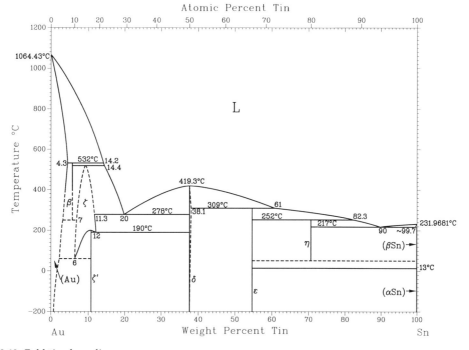

Fig. 2.19 Gold-tin phase diagram

Principles of Soldering and Brazing

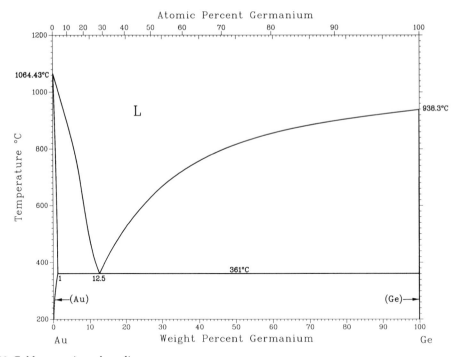

Fig. 2.20 Gold-germanium phase diagram

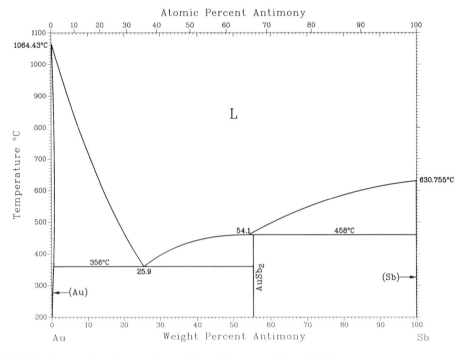

Fig. 2.21 Gold-antimony phase diagram. Source: Prince, Raynor, and Evans [1990]

ramic packages. The alloy compositions used as solders are gold-rich with respect to the eutectic composition. This is deliberate because the eutectic alloy is too hard and brittle even to hot-roll to foil but, by making the alloy gold-rich, the proportion of the ductile gold phase in the microstructure is sufficient to improve the mechanical properties to tolerable limits. Rapid solidification is unable to provide ductile foil, because rapid cooling of molten gold-silicon alloys results in the formation of a large volume fraction of (metastable) gold-silicon intermetallic compounds, principally Au_3Si, that render the foil too brittle to handle [Johnson and Johnson 1983]. An alternative method of applying the solder to silicon components is to coat the back surface of the silicon die with a thin layer of gold, applied by a vapor-phase technique. On heating the gold-metallized silicon to above 363 °C (685 °F), interdiffusion between the gold and silicon results in the formation of the liquid filler metal *in situ*.

The gold-silicon solder is an exception to the rule of eutectic composition alloys possessing favorable characteristics as filler metals. In particular, it suffers from high viscosity when molten. This characteristic is a direct consequence of the low temperature of the eutectic transformation, relative to the high melting points of both constituent phases and the silaceous dross that tends to form on the surface of the alloy. Furthermore, when using gold-silicon solders, precautions must be taken to ensure that the initial cooling rate of solidified joints does not exceed 5 °C/s (9 °F/s). If this condition is not satisfied, Au_3Si is formed and its subsequent decomposition with time to gold and silicon can produce cracks within the joint, due to the associated volume contraction [Johnson and Johnson 1984]. Improved versions of this filler alloy that do not suffer from these deficiencies and that offer other functional benefits are becoming available. Their development is described in Chapter 7.

The gold-germanium eutectic composition alloy is sometimes recommended in place of the gold-silicon solder where cost margins are particularly tight. As can be seen from the phase diagram in Fig. 2.20, it offers the benefit of lower gold concentration with little change in melting point. However, the price of germanium is fairly close to that of gold so that the saving in materials costs is relatively small, and, as solders, gold-germanium alloys are more difficult to use than their silicon equivalents because they are more brittle and more prone to dross formation.

Gold-Tin. The Au-20Sn eutectic solder is hard and moderately brittle. These properties arise from the fact that its constituent phases are two gold-tin intermetallic compounds—namely, $AuSn(\delta)$ and $Au_5Sn(\zeta)$. The Au_5Sn phase is stable over a wide range of composition and consequently has some limited ductility, which is imparted to the solder alloy (approximately 2%). Although difficult, this solder can be hot rolled to foil and preforms stamped from it. By using rapid solidification casting technology, it is possible to produce thin ductile foil, up to about 75 μm (3 mils) thick, and having a fine microstructure. However, this state is somewhat unstable, and within about 30 minutes at room temperature, the rapidly solidified strip is indistinguishable in its mechanical properties from foil prepared by conventional fabrication methods [Mattern 1989]. Nevertheless, the hardening can be suppressed for about a year if the quench-cooled material is stored under liquid nitrogen (–196 °C, or –320 °F), and for close to 1 month at –20 °C (–4 °F), so that it is possible to manufacture shaped foil preforms while the strip is still ductile, which can then be either immediately placed in a jig or returned to cold storage.

An alternative method of introducing the Au-20Sn solder into an assembly is to selectively coat the joint area with a thick layer of gold, overlaid with a thinner layer of tin in the thickness ratio of 3Au:2Sn. On melting, the tin will alloy with the gold to form the AuSn intermetallic compound as a stable barrier layer at the interface between the gold and the molten tin, leaving the solder alloy rich in tin and with a correspondingly low melting point. To ensure that no tin-rich phases remain, it is necessary to heat the coated components to a temperature exceeding the melting point of the AuSn phase (419 °C, or 786 °F) either before or during the bonding cycle. This will melt the barrier layer and enable the reaction between the tin and the gold to continue to completion.

The principal applications of the Au-20Sn solder are die attach of gold-metallized chips and the hermetic sealing of ceramic semiconductor packages. The die attach process that is used when the sealing operation is performed with the Au-20Sn solder is usually the higher-melting-point gold-silicon solder. Thus, the gold-silicon and gold-tin solders are used in the initial joining operations of a step soldering sequence, which ends with the soldering of the packages containing the chips onto printed circuit boards. This final soldering operation is most frequently performed using a

Table 2.12 Low-melting-point eutectic composition alloys used as solders

		Composition, wt%				Eutectic temperature	
Ag	Bi	In	Pb	Sn	Other	°C	°F
...	49.0	21.0	18.0	12.0	...	57	135
...	33.7	66.3	...	...	...	72	162
...	52.0	...	32.0	16.0	...	95	203
...	67.0	33.0	...	...	...	109	228
...	...	51.0	...	49.0	...	120	248
...	55.5	...	44.5	...	...	123	253
...	57.0	...	...	43.0	...	139	282
3.0	...	97.0	...	...	...	144	291
...	...	96.0	...	...	4.0Zn	144	291
5.0	...	80.0	15.0	...	...	149	300
...	...	99.5	...	...	0.5Au	156	313
...	...	75.0	25.0	...	...	165-170(a)	330-340(a)
1.5	...	...	36.0	62.5	...	179	354
...	...	...	38.0	62.0	...	183	361
...	...	50.0	50.0	...	...	190-210(a)	375-410(a)
...	...	...	...	91.0	9.0Zn	198	388
...	...	...	85.0	...	15.0Au	215	419
3.5	...	...	...	96.5	...	221	430
...	...	...	...	99.3	0.7Cu	227	441
...	...	...	...	95.0	5.0Sb	235-245(a)	455-475(a)
...	...	...	85.0	3.5	11.5Sb	240	464
25.0	...	...	...	65.0	10.0Sb	240-290(a)	465-555(a)
...	...	...	88.9	...	11.1Sb	251	484
...	...	25.0	75.0	...	...	255-265(a)	490-510(a)
2.5	...	5.0	92.5	...	...	300	572
2.5	...	...	97.5	...	...	304	579
1.5	...	...	97.5	1.0	...	309-310(a)	588-590(a)
...	...	...	99.5	...	0.5Zn	318	604

(a) These alloys are not eutectic compositions, but have been included because of their narrow melting range and their industrial exploitation.

lead-tin solder at a process temperature of about 220 °C (430 °F).

Of the gold-bearing solders, only the Au-20Sn alloy has a moderate degree of fluidity when molten. In their major field of application, that of semiconductor manufacture, the use of fluxes to promote spreading is not usually permitted. Instead, mechanical agitation is normally applied to the components to displace surface oxides and thereby achieve wetting while the solder is molten. This type of mechanical fluxing process is described in detail in section 5.3.5. Of the other low-melting-point eutectic alloys of gold, gold-germanium and gold-antimony are inferior to gold-silicon and gold-tin alloys with regard to their wetting characteristics and thus are seldom used as solders.

2.2.3.2 Other Solder Alloy Families

All other solder alloys are based on one or more constituents, chosen from seven common elements: antimony, bismuth, cadmium, indium, lead, silver, and tin. Of these, the cadmium-containing alloys have largely been removed from manufacturers' catalogs because they offer no clear advantages over other solders and their use is subject to restrictions arising from the toxicity of cadmium. Therefore, they will not be discussed. In recent years, the use of lead-containing solders has also been questioned on health grounds, especially in applications relating to food and drinking water, which has led to the development of a range of lead-free solders for plumbing applications [Irving 1992]. These solders comprise typically 95% Sn, with the balance being one or more of silver, antimony, bismuth, copper, nickel, and zinc.

A similar concern has been voiced in connection with the use of lead-tin solders in electronic fabrication. This presents not only a health risk to operators, but electronic scrap and spent plating solutions also pose a significant effluent hazard. However, any restrictions placed on lead-containing solders could have serious implications for industries that use solders, because the majority of low-melting-point eutectic alloys contain lead in significant concentrations. Although several lead-

Filler Alloys and Their Metallurgy

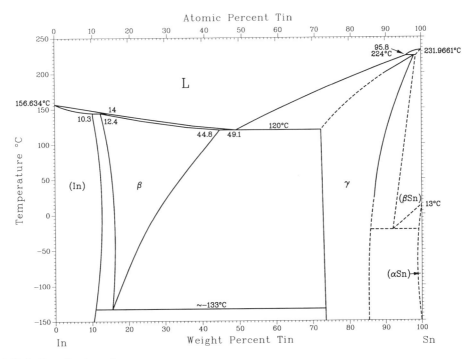

Fig. 2.22 Indium-tin phase diagram

free solders have been developed, they all have significant melting ranges and inferior spreading characteristics and frequently contain a high proportion of tin, which makes them expensive compared with the equivalent lead-tin solders. None has yet found widespread commercial acceptance. Moreover, replacing the lead content entirely would require at least 20,000 tonnes per annum of the substitute element. Metals such as silver, bismuth, and indium are simply not available in this quantity.

A list of the common binary, ternary, and quaternary composition solder alloys is given in Table 2.12, and the associated phase diagrams of those in widest use are shown in Fig. 2.22 to 2.30. All of these alloys are ductile and can be mechanically worked to produce preforms of virtually any desired geometry. The majority of solders are based on eutectic compositions, for reasons explained in section 3.2.1. Many solder manufacturers also offer off-eutectic compositions in these and other alloy systems. They are included in the product range because they are either cheaper, easier to fabricate as wire and foil, or have a melting range. It is usually the case that fluidity and mechanical properties suffer on moving to an off-eutectic composition. Some high-lead-content and noneutectic solders have been included in Table 2.12.

The high-lead solders were formulated with the express purpose of providing an elevated melting point (>300 °C, or 570 °F), the indium-lead solders for soldering to gold metallizations, and the 25Ag-65Sn-10Sb solder (frequently referred to as alloy J [Olsen and Spanjer 1981; Mackay and Levine 1986]) for the fabrication of semiconductor devices.

The majority of solders melting below 300 °C (570 °F) can be classified into three groups, depending on whether the alloy contains tin, indium, or neither element. The criterion for this classification is made on the basis of the likely reaction of the solder with the substrate material which, in the temperature range covered by solders, is necessarily a metal because there is insufficient thermal activation available for wetting nonmetals.

In the majority of cases, wetting of a component by a solder results in the formation of intermetallic compounds, either within the filler or at the interface between the solder and the parent material. These intermetallic phases have a pronounced effect on the mechanical properties of the joint. As constituents of solders, indium and tin have a dominant role in determining the intermetallic compounds that form by reaction with the components: the compounds invariably contain

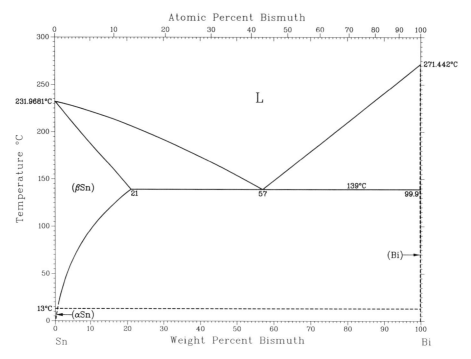

Fig. 2.23 Bismuth-tin phase diagram

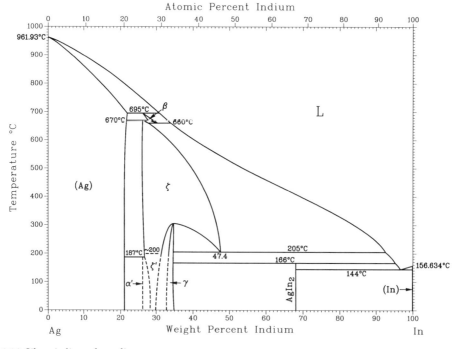

Fig. 2.24 Silver-indium phase diagram

Filler Alloys and Their Metallurgy

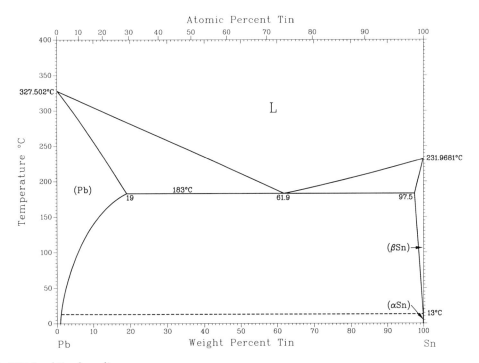

Fig. 2.25 Lead-tin phase diagram

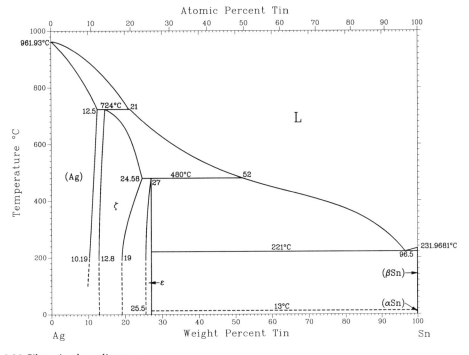

Fig. 2.26 Silver-tin phase diagram

Principles of Soldering and Brazing

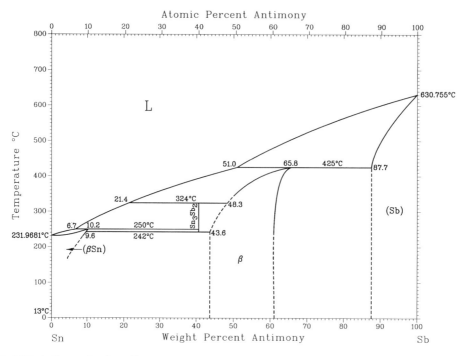

Fig. 2.27 Antimony-tin phase diagram

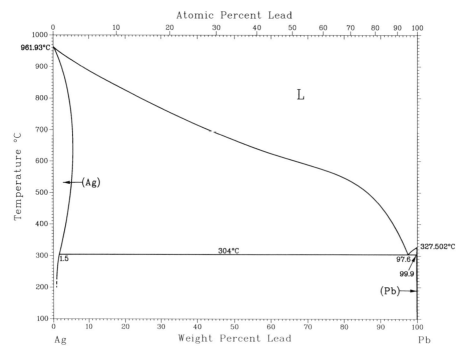

Fig. 2.28 Silver-lead phase diagram

Filler Alloys and Their Metallurgy

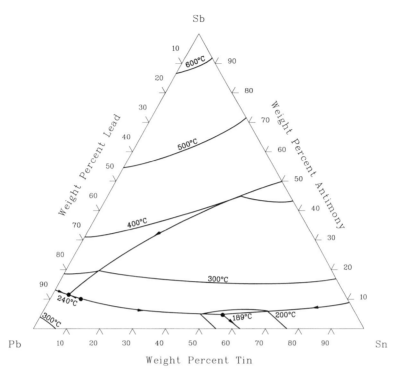

Fig. 2.29 Liquidus surface of the Pb-Sb-Sn system

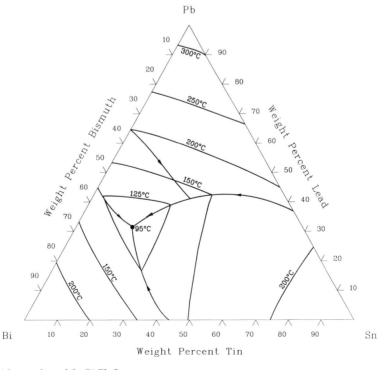

Fig. 2.30 Liquidus surface of the Bi-Pb-Sn system

either indium or tin. When both indium and tin are present, the composition of the parent materials determines which of these elements forms the intermetallic compounds. For solder alloys that contain neither element, intermetallic compounds are not always formed and those that do necessarily depend on the composition of both the solder and of the components. A list of the compounds that form between common solder alloys and a selection of engineering metals is given in Table 2.13. The mechanism by which they form and some of the implications of their presence in joints are discussed in section 3.2.

There are five principal solder alloy families based on the constituents bismuth, indium, lead, silver, and tin. Each of these elements confers different properties on the filler metals, and often in a manner that is not obvious. For example, the fact that solders tend to be based on eutectic composition alloys means that their melting points are determined by the eutectic reaction rather than by the melting points of the individual elements present. There is no simple relationship between the melting temperatures of the constituent elements and the eutectic temperature. Therefore, while the elements fall into the following sequence, Ag > Pb > Bi > Sn > In, when ranked in descending order of melting point, the solder alloys do not fall into this pattern; those containing bismuth have the lowest melting points of all. Despite such complexities, it is possible to make some generalizations about the role of each element in a solder, as discussed in the following paragraphs.

Silver is present only as a relatively minor constituent of solder alloys, because higher concentrations (more than about 5%) result in a sudden increase in the liquidus temperature toward those of the silver-bearing brazes. Small additions of silver are used primarily to enhance mechanical properties of solders and joints and to promote solder fluidity by destabilizing native surface oxides on the molten filler. Silver oxide is not stable in air at temperatures above 190 °C (375 °F). The Ag-96Sn solder combines these advantages and has among the best mechanical and physical properties of any solder alloy [Harada and Satoh 1990]. Silver-containing solders tend to be preferred when joining silver-coated components, because the presence of silver in the solder reduces the rate and extent of scavenging from the metallization, as shown by the data in Fig. 2.31.

Bismuth is the most brittle constituent of the common solders; for this reason, few solders contain more than 50% of this element. Bismuth-bearing alloys comprise the majority of the lowest-melting-point solders, as can be seen from Fig. 1.2. Bismuth exhibits the unusual characteristic of expanding on freezing, enabling solders to be tailored to have essentially zero liquid-to-solid vol-

Table 2.13 Principal intermetallic phases formed by alloying between the common binary solders and engineering parent metals and metallizations

Solder alloy	Cu	Fe	Parent metal/metallization Ni	Ag	Au
Ag-Bi	(a)	(a)	$NiBi_3$	(a)	Au_2Bi
Ag-In	$CuIn_2$ Cu_3In_2	(a)	Ni_3In_7	$AgIn_2$ $AgIn2$	$AuIn_2$
Ag-Sn	Cu_3Sn Cu_6Sn_5	$FeSn_2$	Ni_3Sn_4 Ni_3Sn_2	Ag_3Sn	$AuSn_4$ $AuSn_2$
Ag-Pb	(a)	(a)	(a)	(a)	$AuPb_3$ $AuPb_2$
Bi-In	(a)	(a)	(b)	$AgIn_2$ Ag_2In	$AuIn_2$
Bi-Sn	Cu_3Sn Cu_6Sn_5	$FeSn_2$	Ni_3Sn_4 Ni_3Sn_2	Ag_3Sn	$AuSn_4$ $AuSn_2$
Bi-Pb	(a)	(a)	$NiBi_3$	(a)	Au_2Bi
In-Sn	Cu_3Sn Cu_6Sn_5	$FeSn_2$	(b)	$AgIn_2$ Ag_2In	$AuIn_2$
In-Pb	$CuIn_2$ Cu_3In_2	(a)	Ni_3In_7	$AgIn_2$ Ag_2In	$AuIn_2$
Sn-Pb	Cu_3Sn Cu_6Sn_5	$FeSn_2$	Ni_3Sn_4 Ni_3Sn_2	Ag_3Sn	$AuSn_4$ $AuSn_2$

(a) No intermetallic compounds formed. (b) No data available. Note: Not all of the phases that can form are listed, only those commonly observed following typical soldering process temperatures and times. Also, not all of the phases listed have exactly the listed stoichiometry, as many are stable over a range of composition.

ume contraction by appropriately adjusting the bismuth concentration. This property can confer benefits in certain applications—for example, in making hermetic soldered joints [Dogra 1985], as discussed in section 4.4.1.2. The low melting point and inferior fluidity of the bismuth-bearing solders impose constraints on joint design and processing conditions. For example, their relatively low melting temperature means that there is little thermal activation available and aggressive inorganic fluxes are needed to chemically clean the surfaces of the parent materials.

Indium and lead are the two softest and most ductile constituents of solder alloys. Despite their inferior mechanical properties, solders with high lead concentrations find wide application because they are the cheapest and the easiest to use of the high-melting-point solder alloys available. The indium-bearing solders are particularly attractive for use with gold metallizations because these are not readily dissolved, thereby preventing dewetting, and the interfacial phases that form are comparatively ductile, so joints are not embrittled by their presence. The low level of gold erosion stems from a combination of the steep slope of the liquidus line on the phase diagram between indium and gold (see Fig. 2.32) and the formation of a thin, continuous intermetallic compound ($AuIn_2$) between the molten solder and the gold metallization. This layer of compound then acts as a barrier against further gold dissolution and alloying, resulting in the profile of the erosion curves shown in Fig. 2.33. Thus, indium-contain-

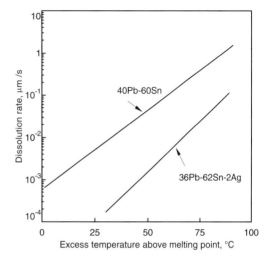

Fig. 2.31 Substantial reduction of the dissolution rate of silver in tin/lead eutectic composition solder obtained by small silver additions to the alloy. Source: After Bulwith and Mackay [1985]

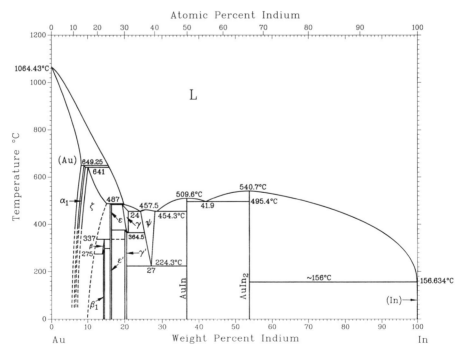

Fig. 2.32 Gold-indium phase diagram

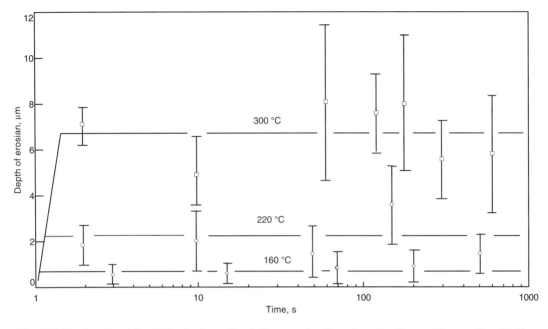

Fig. 2.33 Erosion of a gold metallization by molten indium as a function of reaction time and temperature. Similar results are obtained for indium-base solders, including gold-indium, silver-indium, indium-lead, and indium-tin.

ing solders can be reliably used in conjunction with very thin gold metallizations.

Tin is a preferred constituent of solder alloys because it confers fluidity, benefits wetting, enhances mechanical and physical properties, and possesses exceptionally low vapor pressure. The effect of tin additions on the promotion of solder spread is discussed in section 1.3.4. After lead, cadmium, and zinc, tin is the cheapest ingredient of solder alloys. Nevertheless, the price of tin is still some 30 times that of lead, which accounts for the popularity of lead-rich tin-lead solders. By way of comparison, bismuth and antimony are about the same price as tin, with indium and silver being more than an order of magnitude more expensive still. However, tin-bearing solders tend to form brittle compounds on reaction with many parent materials and metallizations used in engineering, particularly copper and gold. Silver is one of the few exceptions, with silver-tin intermetallic phases being comparatively ductile. Therefore, considerable care must be taken in designing the joining process and specifying the service environment so as to restrict the formation of intermetallic compounds to concentrations below those which would otherwise weaken and embrittle the joints. This point is discussed in connection with the phase diagrams of the relevant alloy systems in section 3.2.

2.2.3.3 Lead-Tin Solders

Lead-tin solders need to be singled out and discussed in additional detail because they account for about 95% of all soldered joints rather than because their metallurgical properties are significantly different from those of other solders. More than 50,000 tonnes per annum of lead-tin solder are used in the electronics assembly industry alone. The beneficial characteristics of lead-tin alloys have been appreciated since at least Roman times, and the elder Pliny, writing in the first century A.D. in his *Historia Naturalis* [Rackham 1952, p 242], specifically mentions an alloy containing two parts of (black) lead and one part of white lead (i.e., tin) used for soldering pipes. He also remarks that the price of this alloy is 20 denari per pound. It is interesting to note that this works out at $110 per kilogram, assuming that gold has maintained its purchasing power since Pliny's day. The current price for the same solder is lower by an order of magnitude.

Tin-lead alloys offer the following advantages compared with other solders:

- Superior wetting and spreading characteristics, compared with most other solders, especially those containing bismuth, antimony, and indium. The difference is quantified in section 1.3.4.

- Relatively inexpensive to produce and use. Lead, in particular, is cheap compared with other solder metals, notably bismuth, indium, and silver. Moreover, the fabrication costs are low in comparison with those of most solders, as lead-tin alloys are readily cold worked and do not suffer the environmental restrictions associated with the zinc- and cadmium-containing solders, although these are set to grow as of the hazards associated with lead become better understood.

- Greater versatility. Lead-tin solders readily wet a wide range of metals to produce sound joints with minimal substrate erosion. Soldering to thick gold metallizations represents one of the few situations in which restrictions need to be observed in order to prevent embrittlement through the formation of unfavorable intermetallic phases. The reaction between lead-tin solders and gold metallizations is discussed in detail in section 3.2.2.

- Lead-tin solders can be applied readily as coatings on components by electroplating. This alloy system is the only solder for which electroplating technology has been developed successfully.

- Satisfactory mechanical properties for many applications. In terms of mechanical strength, stiffness, and fatigue resistance, only silver-tin solders are superior and even then by less than an order of magnitude, and generally by a factor of less than three.

In most other respects, such as corrosion resistance and electrical and thermal conductivity, there is little to distinguish lead-tin alloys from the majority of other low-melting-point solders.

Lead-tin solders are seldom pure, but often contain other elements, either as incidental impurities or as deliberate additions designed to modify specific properties. In most cases, these extra elements impair the wetting and spreading characteristics. Three major additions are routinely encountered:

- *Antimony* is frequently present to a maximum concentration of about 1%. It has a beneficial role in improving the mechanical properties of joints though a solid-solution strengthening mechanism. Antimony can substitute for double the proportion of tin, without greatly widening the melting range of the solder and thus is often favored by solder manufacturers as it decreases materials costs. At higher concentrations of antimony, there are solder alloys based on the ternary system antimony-lead-tin. The liquidus projection of this alloy system is given in Fig. 2.29, from which it can be seen that there is a ternary eutectic at 240 °C (465 °F) containing 11.5% Sb.

- *Bismuth* is added for similar functional reasons as antimony. The maximum concentration is usually limited to about 3%, as higher levels result in a significant widening of the melting range and a noticeable impairment to wetting and spreading behavior. In addition, there is a range of low-melting-point ternary solders based on bismuth-lead-tin alloys in which bismuth is present in high concentrations. The liquidus surface of this alloy system is given in Fig. 2.30. Small quantities of bismuth and also antimony are added to lead-tin solders to prevent degradation at low temperatures through a mechanism known as a "tin pest," which is described in section 7.2.1.

- *Silver*. Silver-lead-tin solders find application for soldering silver-coated surfaces. The addition of 2% Ag to the lead-tin eutectic alloy has a marked effect in reducing the erosion of silver coatings, as shown by the data in Fig. 2.31. The presence of silver in lead-tin solders results in the formation of a fine dispersion of the relatively ductile intermetallic compound Ag_3Sn, on solidification, which boosts the mechanical properties of joints. Small quantities of copper and gold have a similar beneficial effect, as described in section 3.2.2. However, these elements are not usually present as deliberate alloying additions, but rather are incorporated automatically when soldering copper- and gold-coated components, respectively.

All of the above additions to lead-tin solder, in small quantities, enhance the strength of joints to copper, because they alloy preferentially with some of the tin and thereby serve to reduce the thickness of the copper-tin intermetallic phases that form at copper/solder interfaces and which are a source of weakness (see section 3.2.2) [Quan et al. 1987].

Although solders are relatively weak compared with brazes, soldered joints are seldom used in structural applications and therefore do not usually fail due to inadequate tensile or shear strength.

Soldered joints are found primarily in electronics assemblies; their failure in this area of application is often associated with the thermal cycling that inevitably occurs during the operation of electronic circuits. Because the joined components tend to have different thermal expansivities, the joints will suffer creep and fatigue on thermal cycling, which can lead to failure.

A further complication of soldered joints in electronics assemblies arises from the high temperatures at which lead-tin solder is used relative to its melting point, typically up to 0.9 of the melting point expressed in degrees Kelvin. This means that the microstructure of solidified lead-tin joints tends to be unstable under typical service conditions and grain growth frequently will occur, to the detriment of the creep and fatigue resistance of joints. In addition to this mode of degradation, lead-tin solders are susceptible to stress-induced microstructural changes. The superposition of these microstructural changes on a cyclic strain regime results in concentrations of shear bands developing in the alloy microstructure. Such features are prone to initiating fatigue cracks, and hence soldered joints differ from brazed joints in that fatigue failure is initiated internally, whereas in the engineering structures where brazes are commonly used, fatigue cracks almost always start at an exterior surface. It is precisely because the failure mode of soldered joints in electronics systems is so complex and depends on many interrelated variables that the test methods used in the electronics industry to assess the integrity and reliability of soldered joints are specific to that industry, with the testing usually carried out on fabricated assemblies. For further details, see Section 6.5.5 and Lund 1985; Coombs 1988.

The key to conferring fatigue resistance to joints made with lead-tin solder is the development of a thermally stable and fine-grained microstructure. Two approaches are being pursued in current research in an attempt to address this need. The first approach involves grain refinement. Minor additions (0.01 to 0.1%) of reactive elements such as lithium and beryllium have been explored as grain refiners for lead-tin alloys. A summary of published results is given by Klein Wassink [1989]. They operate by creating a fine dispersion of either oxide/nitride particles or stable intermetallic phases when the solder is molten, which act as sites from which the solid phases can grow on solidification (heterogeneous nucleation). The presence of a large number of such particles when the alloy is molten is reflected in a fine alloy microstructure. Although grain refinement is beneficial to the mechanical properties of metals, the fine microstructure is not thermally stable and will gradually coarsen by solid-state diffusion when thermal energy is provided by exposure to elevated temperature (typically above 75 °C, or 165 °F).

The second approach is dispersion strengthening, which involves the generation or mechanical incorporation of fine insoluble particles into the solder, typically 1 μm (40 μin.) in diameter and frequently less. These provide some grain refinement on solidification of the solder by heterogeneous nucleation, but their principal role in this instance is twofold. First, they inhibit coarsening of the microstructure because, as they are insoluble in the solder, it is thermodynamically favorable for the particles to reside at regions in the alloy matrix where there are natural departures from a regular atomic lattice, such as the boundaries between the tin- and lead-rich phases in lead-tin solders. Second, provided the particles are sufficiently fine and numerous, they will, for similar reasons, impede the movement of dislocations through the solder matrix. Dislocations are faults in the atomic lattice that can be induced to move through it by the application of mechanical stress; the movement of many such dislocations results in plastic flow of the material. Hence, by restricting the movement of dislocations, the mechanical properties of the solder are enhanced. Because the solid particles are insoluble in the solder, dispersion strengthening remains effective even when the alloy is heated up to its solidus temperature.

A new area of development in solder technology concerns composite solders [NEPCON West 1992; Betrabet et al. 1991; Betrabet, McGee, and McKinlay 1991; Marshall et al. 1991]. The objective is to improve the mechanical and thermomechanical properties of existing solders by incorporating particles, which are generally of intermetallic compounds. The reinforcing particles need to be wetted by the solder and must be of similar density so that gravitational separation does not occur and the mixture remains homogeneous. The published literature reveals a number of features of this activity, discussed in the following paragraphs.

One type of composite solder contains fine particles (<5 μm, or 0.2 mil, diam) at low volume fractions (0.5 to 5 vol%) to produce grain refinement. All reported work has been on lead-tin solders. The particulate content is primarily Cu_6Sn_5, Cu_3Sn, Ni_3Sn_4, or Cu_9NiSn_3. These compounds are soluble in the solder and the composites behave differently from the dispersion strengthened materials. Tests have shown that the grain refinement is thermally stable to temperatures approaching the melting point of the solder matrix, so that the thermal fatigue resistance of joints is significantly improved. However, the improvements in solder strength are marginal, because at ambient temperatures and above, the yield stress of solders decreases with reducing grain size so that the strengthening effect produced by the hard particulate phase is almost totally offset by the plastic behavior of the soft matrix.

The second type of composite solder is filled with larger particles (typically 10 to 100 µm, or 0.4 to 4 mils, diam) at volume fractions of 10 to 40%. These are akin to conventional metal-matrix composite materials and benefit from enhanced tensile strength, resistance to creep, and improvements to other mechanical properties at the expense of a small reduction in ductility. The combination of higher volume and larger particle size is effective in strengthening the solder by restraining the yield behavior of the soft matrix. The material combinations that have been evaluated are largely the same as those described above for the first type of composite. To date, both types of composite solders have been difficult to produce without a substantial level of micropores, which offset the gains in properties discussed above.

Neither grain refinement nor dispersion strengthening has yet found significant commercial use by industry, although there is clearly scope for further research in this area.

References

Betrabet, H. *et al.*, 1991. "Dispersion Strengthened Lead-Tin Alloy Solder," European Patent Application No. 91202114.4

Betrabet, H.S., McGee, S.M., and McKinlay, J.K., 1991. Processing Dispersion-Strengthened Sn-Pb Solders To Achieve Microstructural Refinement and Stability, *Scr. Metall. Mater.*, Vol 25, p 2323-2328

Bowen, R.C. and Peterson, D.M., 1987. A Comparison of Rapid Solidification Cast Versus Conventional Die Attach Soft Solders, *IEEE Trans. Components Hybrids Manuf. Technol.*, Vol 10 (No. 3), p 341-345

Bulwith, R.A. and Mackay, C.A., 1985. Silver Scavenging Inhibition of Some Silver Loaded Solders, *Weld. J.*, Vol 64 (No. 3), p 86s-90s

Chatterjee, S.K., Mingxi, Z., and Chilton, C., 1991. A Study of Some Cu-Mn-Sn Brazing Alloys, *Weld. J.*, Vol 70 (No. 5), p 118s-122s

Coombs, C.F., Jr., 1988. *Printed Circuits Handbook*, 3rd ed., McGraw-Hill

Datta, A., Rabinkin, A., and Bose, D., 1984. Rapidly Solidified Copper-Phosphorus Base Brazing Foils, *Weld. J.*, Vol 63 (No. 10), p 14-21

DeCristofaro, N. and Henschel, C., 1978. Metglass Brazing Foil, *Weld. J.*, Vol 57 (No. 7), p 33-38

DeCristofaro, N. and Bose, D., 1986. Brazing and Soldering With Rapidly Solidified Filler Metals, *Proc. Conf. Rapidly Solidified Materials*, ASM International, San Diego, p 415-424

Dogra, K.S., 1985. A Bismuth Tin Alloy for Hermetic Seals, *Brazing Soldering*, Vol 9 (No. 3), p 28-30

Duvall, D.S., Owczarski, W.A., and Paulonis, D.F., 1974. TLP Bonding: A New Method for Joining Heat Resistant Alloys, *Weld. J.*, Vol 53 (No. 4), p 203-214

Finch, N.J., 1985. Soft Soldering Aluminum and its Alloys, *Brazing Soldering*, Vol 9 (No. 59), 46-49

Harada, M. and Satoh, R., 1990. Mechanical Characteristics of 96.5Sn/3.5Ag Solder in Microbonding, *IEEE Trans. Components Hybrids Manuf. Technol.*, Vol 13 (No. 4), p 736-742

Harrison, K.T. and Knights, C.F., 1984. Development of Zinc Based Solders, *Brazing Soldering*, Vol 6 (No. 1), p 5-9

Humpston, G., Jacobson, D.M., and Sangha, S.P.S., 1992. "Low Melting Point Aluminium Alloy Braze," U.K. Patent Application No. 9218404.3

Irving, B., 1992. Host of New Lead-Free Solders Introduced, *Weld. J.*, Vol 71 (No. 10), p 47-49

Johnson, A. 1990. Evidence for Au-Si Bonding in Liquid Gold-Silicon Alloys From Electrical Resistivity Measurements, *Solid State Commun.*, Vol 76 (No. 6), p 733-775

Johnson, A.A. and Johnson, D.N., 1983. The Room Temperature Dissociation of Au$_3$Si in Hypoeutectic Au-Si Alloys, *Mater. Sci. Eng.*, Vol 61, p 231-235

Johnson, D.N. and Johnson, A.A., 1984. Surface Cracking in Gold-Silicon Alloys, *Solid State Electron.*, Vol 27 (No. 12), p 1107-1109

Kapoor, R.R. and Eagar, T.W., 1989. Tin-Based Reactive Solders for Ceramic/Metal Joints, *Metall. Trans. B*, Vol 20B (No. 6), p 919-924

Lugscheider, E. and Schitty, Th., 1988. Wide Gap Brazing of Stainless and Carbon Steel, *Brazing Soldering*, Vol 14 (No. 1), p 27-29

Lund, P., 1985. "Quality Assessment of Printed Circuit Boards," Bishop Graphics, Inc.

Mackay, C.A. and Levine, S.W., 1986, Solder Sealing Semiconductor Packages, *IEEE Trans. Components Hybrids Manuf. Technol.*, Vol 9 (No. 2), p 195-201

Marshall, J.L. *et al.*, Composite Solders, *IEEE Trans. Components Hybrids, Manuf. Technol.*, Vol 14 (No. 4), p 698-702

Mattern, N., 1989. Dynamical X-ray Diffraction Study on the Phase Transformation in Rapidly Quenched

Au71Sn29 Alloy, *Proc. Conf. Advanced Methods of X-ray and Neutron Structural Analysis of Materials*, p 73-76

Miller, V.R. and Schwaneke, A.E., 1978. Interfacial Compositions of Silver Filler Metals on Copper, Brass and Steel, *Weld J.*, Vol 57 (No. 10), p 303s-310s

Mizuhara, H. and Mally, K., 1985. Ceramic-to-Metal Joining With Active Brazing Filler Metal, *Weld. J.*, Vol 64 (No. 10), p 27-32

NEPCON West 1992. *Proc. Conf. National Electronic Packaging and Production*, Technical Session 34, "Composite Solders," National Electronic Packaging and Production Conference, Anaheim, 23-27 Feb, p 1245-1284 (6 papers)

Niemann, J.T. and Wille, G.W., 1978. Fluxless Diffusion Brazing of Aluminum Castings, *Weld. J.*, Vol 57 (No. 10), p 285s-291s

Olsen, D.R. and Spanjer, K.G., 1981. Improved Cost Effectiveness and Product Reliability Through Solder Alloy Development, *Solid State Technol.*, Vol 9, p 121-126

Petzow, G. and Aldinger, F., 1968. Nonvariante Gleichgewichte und Schmelzraume im System Silber-Kupfer-Zink-Kadmium, *Z. Metallkd.*, Vol 59 (No. 2), p 145-153; Vol 59 (No. 5), p 390-395; Vol 59 (No. 7), p 583-589

Prince, A., Raynor, G.V., and Evans, D.S., 1990. *Phase Diagrams of Ternary Gold Alloys*, The Institute of Metals, London

Quan, L. *et al.*, 1987. Tensile Behavior of Pb-Sn Solder/Cu Joints, *J. Electron. Mater.*, Vol 16 (No. 3), p 203-208

Rackham, H., 1952. *Natural History*, Vol 9, Cambridge; transl. of Pliny. *Historia Naturalis XXXIV.161*

Rubin, W., 1982. Some Recent Advances in Flux Technology, *Brazing Soldering*, Vol 2 (No. 1), p 24-28

Schultze, W. and Schoer, H., 1973. Fluxless Brazing of Aluminum Using Protective Gas, *Weld. J.*, Vol 52 (No. 9), p 644-651

Sloboda, M.H., 1961. "Design and Strength of Brazed Joints," *Welding and Metal Fabrication*, Vol 7, 291-296

Sloboda, M.H., 1971. Industrial Gold Brazing Alloys, *Gold Bull.*, Vol 4 (No. 1), p 2-8

Tuah-Poku, I., Dollar, M., and Massalski, T.B., 1988. A Study of the Transient Liquid Phase Bonding Process Applied to a Ag/Cu/Ag Sandwich Joint, *Metall. Trans.*, Vol 19A (No. 3), p 675-686

Chapter 3

Application of Phase Diagrams to Soldering and Brazing

3.1 Introduction

The selection of a solder or braze for a particular application is frequently based exclusively on the melting point and mechanical properties of the filler and its ability to wet the parent materials. The filler is regarded as a uniform layer of metal that simply bridges the gap between the components and binds them together. If only life were that simple! In reality, the formation of the desired metallic bond between a filler and a component requires a degree of alloying. The ensuing metallurgical reactions usually lead to a heterogeneity of phases in the joint. To further complicate matters, kinetic factors tend to accentuate the development of this nonuniformity. Such inhomogeneities often determine the quality and overall characteristics of joints, such as their mechanical properties, the ease and extent of filler spreading, the nature of any fillets formed, and so on.

Metallurgical reactions do not cease once the joint has been made, but continue to proceed, to a greater or lesser extent, during the service life of the assembly. The rate-controlling step for reaction between two solid metals is the diffusion of atoms between the reacting phases. The relative position of the product of the reaction to that of the reacting phases will be governed largely by the diffusion coefficients of the participating metals. For individual metals, it has been established empirically that the rate of diffusion, R, increases rapidly with absolute temperature, T, following an exponential relationship:

$$R = A \cdot \exp\left(\frac{-Q}{kT}\right) \qquad \text{(Eq 3.1)}$$

where k and A are constants, k being the Boltzmann constant and A an experimentally determined factor for each combination of reacting phases and which may vary with concentration. Q is the activation energy for diffusion, which, to a first approximation, is proportional to the melting point, T_m, of the particular metal [Birchenall 1959, p 200].

The rate of reaction will therefore be dependent on the homologous temperature defined as the ratio of T/T_m and will be more pronounced for soldered than for brazed joints that see service at or close to normal ambient temperatures. The reactions produce perceptible metallurgical changes in the constitution and microstructure of soldered joints [Frear, Jones, and Kinsman 1991, Chapter 2]. However, with the increasing use of brazed ceramic assemblies at elevated temperatures, in some cases to within 0.9 T_m, these joints are subject to metallurgical changes that are comparable to those observed in soldered joints.

For a proper understanding of filler metal reactions with parent materials, it is essential to have some grasp of the subject of alloy constitution.

The "constitution" of an alloy refers to features such as its composition, melting range, range of phase stability, solubility limits, and related parameters that can be deduced from the phase diagram of the system in which the alloy appears.

Soldering and brazing technologists are prone to evade consideration of alloy phase diagrams, which accounts for the superficial analysis of joint microstructures that frequently characterizes published studies on soldering and brazing. Part of the problem lies with the highly academic framework in which alloy constitution studies are pursued, which has resulted in a concentration of effort on the finer points of phase equilibria. Although perfectly reasonable for a scientific endeavor that aims to furnish definitive data, this aspect tends to be emphasized at the expense of practical considerations directed toward optimizing the use of materials. The net result, as in other fields such as nondestructive testing, is the emergence of a gulf that separates the scientists from the applied technologists. In the following sections, some attention will be given to highlighting the value of phase diagrams to soldering and brazing technologists and suggesting how this valuable source of information might be tapped.

Frequently, the appropriate phase diagram for elucidating specific filler/substrate reactions and joint microstructures is not available in the literature. Surprisingly as it may seem, no complete phase diagram has been published that fully describes the reaction between lead-tin solders and copper substrates. As a means of showing how a lack of this information can be made good, experimental techniques and a methodology for elucidating gaps in phase diagrams are described in section 3.3. Note that all of the phase diagrams in this book are defined in weight percentages of the constituent elements, as this is more appropriate to soldering and brazing technology than the atomic percentage scale. However, some of the calculations cited for constructing phase diagrams from limited empirical data involve atomic percentages. General equations for converting atomic to weight percent of constituents in alloys, and vice versa, are given in Appendix 3.1.

3.2 *Application of Phase Diagrams to Soldering and Brazing*

The fundamentals of alloy phase diagrams are covered in many metallurgical textbooks and will not be repeated here. Readers without a background in this field are referred to the publications listed in the Preface under the appropriate heading, which provide an excellent introduction to the subject. Here it will suffice to state that a phase diagram is a representation of the thermodynamic stability of phases as a function of composition with respect to particular thermodynamic variables such as temperature or, less commonly, pressure.

What is important to remember is that the information given by the diagrams relates to essentially equilibrium conditions. The phase diagram tells us about the ultimate balance of phases within the joint and those that are likely to be encountered during the progression toward equilibrium. A joined assembly in which the filler and abutting components are different materials are never in true compositional equilibrium, as long as the joint remains distinct. In most practical contexts, the composition of a joint will be tending toward equilibrium over most of its width, and therefore phase diagrams are applicable to an assessment of its constitution. However, at the edges of the joint, marked compositional gradients may exist, causing a significant deviation from equilibrium. These will be exacerbated by any temperature gradients that develop during the process cycle and are manifested as the appearance of different phases in those regions. Even here, phase diagrams can assist in the elucidation of the metallurgical reactions and the resulting phases, as shown in the following section.

Phase diagrams can provide the following practical information:

- The melting temperatures of the "virgin" filler metal and of the abutting components
- The probable freezing range of the filler metal following alloying with the components and hence the remelt temperature of the joint
- Whether the filler in the joint remains homogeneous after reaction with the components and, if not homogeneous, the phases that are likely to be present, or which may form subsequently, with their elemental compositions and melting temperatures

Phase diagrams do not reveal:

- The rate of reactions that might occur between the filler and the components and their variation with time and temperature. This applies both when the filler is molten and when it is solid during service.

- The spatial distribution and morphology of phases in the joint, although frequently it is possible to deduce whether intermetallic phases are likely to form as interfacial layers or will be dispersed throughout the solidified filler. This is explained in section 3.2.2

- The wetting characteristics of a particular filler/parent materials combination, even of perfectly clean surfaces. In practice, wetting is likely to be heavily influenced by the oxides, impurities, and residues that are inevitably present on component and filler surfaces, but which are extraneous to the alloy phase diagram.

- Physical properties of joints, in particular the mechanical and corrosion characteristics. However, it is often possible to predict the likely range of certain physical properties by comparison with other known alloy systems.

The simplest diagrams that are encountered in a joining context are those relating to binary alloys where, for example, the filler is a pure metal being used to join components of another metal. This situation is represented by the copper brazing of iron or mild steel. A compilation of all known binary alloy phase diagrams is given in Massalski [1991].

More commonly encountered are ternary systems, exemplified by a binary alloy filler used with components of a third metal. Ternary phase diagrams tend to be less completely established, and the information provided in the technical literature often must be complemented by experimental "gap filling" of essential constitutional data. A compendium of authoritative ternary alloy phase diagrams is being prepared under the auspices of the Alloy Phase Diagram International Commission (APDIC), and the first volumes have already appeared [Prince, Raynor, and Evans 1990; Petzow and Effenberg 1988; Raghavan 1986; ASM International 1993].

Higher-order alloy systems are naturally more complex and are less well documented. However, for a given joining process only a very limited portion of the phase diagram is required, and it is often possible to experimentally determine the necessary data using the methods outlined in section 3.3.

The value of alloy phase diagrams for understanding and optimizing soldering and brazing processes can best be appreciated by describing a few specific examples.

3.2.1 Examples Drawn From Binary Alloy Systems

A binary filler alloy with complete solubility in the solid state, used to join components of one of the constituent metals. Consider a gold-nickel braze used with nickel components. The gold-nickel phase diagram is given in Fig. 3.1, which shows this binary alloy system to possess a minimum melting point of 955 °C (1751 °F), at 18% nickel. On either side of this composition, the liquidus and solidus phase boundaries separate, with the alloys melting over a range of temperatures. Within the melting range, the alloy is partly liquid and partly solid. On cooling below the solidus temperature, an alloy in this system exists as a single-phase solid but, as the temperature is slowly lowered and equilibrium maintained, this phase separates into two—one gold-rich and the other nickel-rich. The temperature of the decomposition varies with composition, reaching a maximum of 810.3 °C (1490.5 °F), which corresponds to an alloy containing 41.7% nickel.

The 18% nickel composition corresponding to the minimum melting point is normally used for brazing alloys, because it completely melts at a unique temperature. The benefits of this property are shared by eutectic alloys and are explained in the next example, which relates to a eutectic alloy braze. On solidification of an alloy with the minimum melting point composition, the molten alloy, L, transforms to a single solid phase, S, at a unique temperature. This may be written as:

$$L \leftrightarrow S$$

The formation of a single solid phase distinguishes this type of transformation from a eutectic, which is referred to below.

For the purposes of this example, it will be assumed that an alloy of composition Au-50Ni is used as the braze. The gold-nickel phase diagram shows that the Au-50Ni alloy has a solidus temperature of 1000 °C (1830 °F) and a liquidus temperature of 1200 °C (2190 °F). Thus, the melting range of an alloy of this composition is 1000 to 1200 °C (1830 to 2190 °F), and the brazing operation must be carried out at a temperature above 1200 °C (2190 °F) in order to obtain any significant flow by the filler metal.

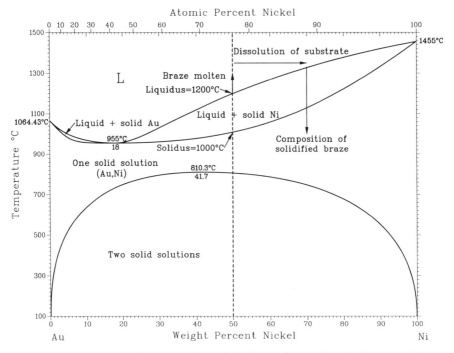

Fig. 3.1 Gold-nickel phase diagram. The erosion of a nickel substrate by a gold-nickel braze and the associated change to the composition of the filler metal are indicated.

On heating the braze above its liquidus temperature, the braze will wet and simultaneously alloy with the nickel substrate. If the reaction is allowed to proceed further, the braze will dissolve nickel up to an "equilibrium" value, determined by the intersection of a line drawn on the phase diagram at the process temperature with the liquidus curve, as indicated in Fig. 3.1. Thus, at 1350 °C (2460 °F) the dissolution of nickel by the Au-50Ni braze changes its composition to approximately Au-70Ni

Because some of the surface region of the nickel component is dissolved by the braze, this process is commonly referred to as dissolution or erosion. As a consequence of this alloying process, the liquidus and solidus temperatures of the filler will change, as is evident from Fig. 3.1, and the microstructure of the solidified braze will be modified accordingly. The practical manifestation of the increase in the melting point of the braze is an inevitable reduction in its fluidity and also in the driving force for spreading (see section 1.3.4). Heavy erosion of the parent materials is often undesirable. If erosion is to be minimized, then the filler metal must be chosen such that alloying with the parent metal causes the liquidus to rise in temperature rapidly as the composition changes toward that of the parent metal. At the same time, the joining operation should be carried out at the lowest practicable process temperature and within the shortest possible time in order that the concentration of the parent material in the filler alloy remain low.

The phase diagram represented in Fig. 3.1 shows that, at equilibrium, above the liquidus temperature the braze exists as a single homogeneous liquid phase, and immediately below the solidus temperature it should comprise a homogeneous solid phase. Between these two temperatures, the braze is a pasty, two-phase mixture of solid and liquid, the proportions of which are given by the lever rule. Referring to the enlarged portion of the phase diagram given in Fig. 3.2, at 1100 °C (2010 °F) the weight fraction of braze that does not melt under equilibrium conditions is:

$$\% \text{ solid} = \left(\frac{OX}{XY}\right) \cdot 100$$

The remainder of the braze will have melted, that is:

$$\% \text{ liquid} = \left(\frac{OY}{XY}\right) \cdot 100$$

Table 3.1 Application of the lever rule to solidification of a Au-70Ni alloy

Temperature		Liquid		Solid	
°C	°F	Fraction, %	Composition	Fraction, %	Composition
1320	2410	100	Au-70Ni	0	—
1275	2325	62	Au-61Ni	38	Au-84Ni
1230	2245	36	Au-54Ni	64	Au-78Ni
1185	2165	17	Au-48Ni	83	Au-74Ni
1140	2085	0	—	100	Au-70Ni

where X is the composition of the liquid phase, Y is the composition of the solid phase, and O is the composition of the alloy. By way of an example, for an alloy of composition Au-50Ni, at 1100 °C (2010 °F), X = Au-38Ni and Y = Au-66Ni. Therefore, at this temperature, the percentage of solid will be $(38 - 50)/(38 - 66) \cdot 100 = 43\%$; the percentage of liquid will be 57%.

Although the alloy remains close to compositional equilibrium during melting, in practice it does not do so during solidification. The resolidified braze will tend to develop local concentration gradients of a type known as "coring." Coring occurs because the compositions of the liquid and solid phases change as the alloy cools through the two-phase region.

Using the lever rule as above, the proportions and composition of liquid and solid can be calculated at various temperatures between the liquidus temperature, at which the alloy is completely liquid, and the solidus, when it has completely solidified. Data determined at five temperatures, corresponding to an aggregate composition of Au-70Ni, are presented in Table 3.1. In this case, it can be seen that the first liquid to solidify is rich in nickel, with respect to the nominal composition of the alloy, while the last liquid to solidify is of that composition. If cooling can be controlled to a suitably slow rate, the solid fraction at each stage will continually and uniformly adjust by diffusion to the composition that is indicated on the phase diagram. Under these conditions, when solidification is complete the alloy will be homogeneous. However, in most practical situations, the rate of solidification will be faster than diffusion can act to homogenize the alloy, and coring will result. The cored microstructure will normally persist on cooling to room temperature.

Being the result of nonequilibrium cooling, coring tends to broaden the temperature range over which an alloy melts and is usually undesirable in virgin filler metals. It can be removed by isothermally annealing the stock alloy at elevated temperature but below the solidus temperature.

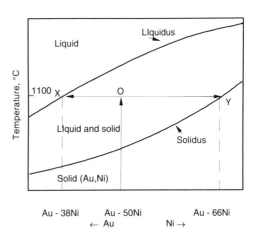

Fig. 3.2 Application of the lever rule to the gold-nickel system

This treatment will furnish the necessary activation energy for diffusion to bring the alloy toward its equilibrium state, and hence toward compositional uniformity, as explained in section 3.3.5.1 below.

A binary eutectic composition braze used with components of one of the constituent metals, with no intermetallic compound formation. A representative example of this type of reaction is a silver-copper braze used to join copper components. The silver-copper phase diagram represented in Fig. 3.3 shows that there is a single composition, that of the eutectic, e (Ag-28Cu), at which the alloy transforms between a liquid (L) and two solid phases (S1 and S2) at a unique temperature, 779 °C (1436 °F), according to the reaction:

$$L \leftrightarrow S1 + S2$$

At the eutectic composition, solid alloys form as a mixture of two finely divided phases, one silver-rich and the other copper-rich. For all other com-

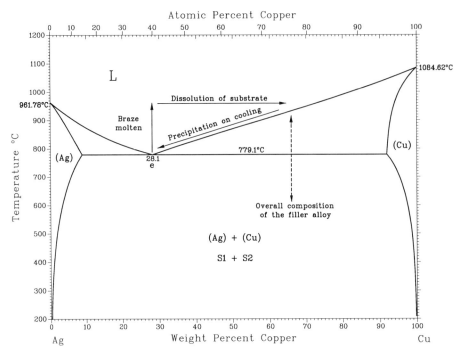

Fig. 3.3 Silver-copper phase diagram. Brazing of copper components with Ag-28Cu eutectic braze results in dissolution of copper into the molten filler metal, followed by precipitation of the excess copper as the assembly cools from the peak process temperature. These processes can be readily quantified by referring to the phase diagram.

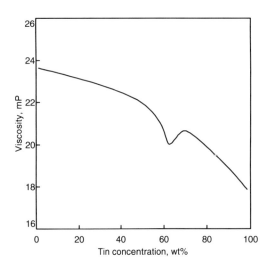

Fig. 3.4 Viscosity of lead-tin alloys at 50 °C (90 °F) above the liquidus temperature. The Pb-62Sn eutectic composition alloy has an abnormally low viscosity compared with adjacent compositions. Source: After Fisher and Phillips [1954]

positions, except those of the pure metals, there is a separation between the liquidus and solidus temperatures.

It can be seen from the silver-copper phase diagram (Fig. 3.3) that copper is soluble in the Ag-28Cu braze when molten. Thus, at the joining process temperature, the braze will dissolve copper from the components until the "equilibrium" concentration of copper is attained, as described above for the gold-nickel braze with nickel components. However, in the case of the eutectic silver-copper braze, the dissolution of copper increases the liquidus temperature of the filler metal in the joint but not its solidus temperature, because eutectic transformations are isothermal.

At the commencement of the cooling stage of the process cycle, the molten braze no longer corresponds to the eutectic composition, but is rich in copper and consequently now possesses a freezing (i.e., melting) range. On cooling below the liquidus temperature, the excess copper will solidify first, as indicated by the phase diagram. This precipitation tends to occur preferentially at the interface between the components and the braze, because this interface tends to be slightly cooler than the volume of the molten braze because heat is lost via the extremeties of the assembly. Precipitation continues until the temperature and composition of the remaining liquid reach the eu-

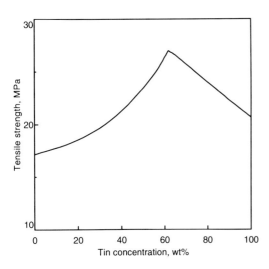

Fig. 3.5 Tensile strength of cast bars of lead-tin alloys. Optimum mechanical properties are coincident with the eutectic composition (Pb-62Sn). Source: After Inoue, Kurihara, and Hachino [1986]

tectic point, so that final solidification by the molten filler results in the formation of a small volume fraction of finely divided eutectic.

Alloys of eutectic composition are preferred as filler metals because of the following characteristics:

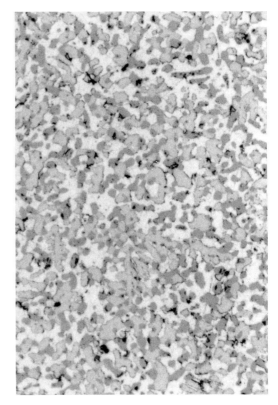

Fig. 3.6 An alloy microstructure characteristic of a (ternary) eutectic transformation. The alloy composition is 35.8Ag-34.2Al-30.0Ge. 100×

- Superior spreading behavior when molten. This feature is an immediate consequence of there being no temperature range over which the alloys coexist as solid and liquid, which also applies to the Au-18Ni alloy described in the previous example. Where a pasty mixture can occur, the partly molten alloy is made sluggish by its high viscosity. In addition, the viscosity of molten eutectic alloys is lower than for that of near-eutectic compositions when these are totally liquid. This effect is well documented for lead-tin alloys and is illustrated in Fig. 3.4.
- Superior mechanical properties, arising from the interspersed or duplex character of the eutectic microstructure and the fine grain size. Grain refinement is the only metallurgical process that enhances both the strength and ductility of a metal. The superior mechanical properties of the lead-tin eutectic solder, with its duplex microstructure, over the adjacent alloy compositions is illustrated in Fig. 3.5. The fine triplex microstructure of a ternary eutectic in an aluminum alloy can be seen in Fig. 3.6.
- Joining process temperatures can be chosen to be only slightly above the melting point of the alloy, precisely because eutectic composition alloys melt completely at a single temperature.
- A reduced risk of disturbing located components, which can easily occur when the filler appears to be solid but is actually in a pasty state. A rapid liquid-to-solid transformation on cooling, without an intervening pasty stage, minimizes the chance of such an interruption. However, this assumes that alloying of the filler with the component materials does not greatly shift the composition of the filler from its eutectic point. A disturbed joint generally has inferior mechanical properties, and the fillets will acquire a rough surface with a frosty appearance.

For these reasons, most solders and brazes are either eutectic compositions or have many of the

Principles of Soldering and Brazing

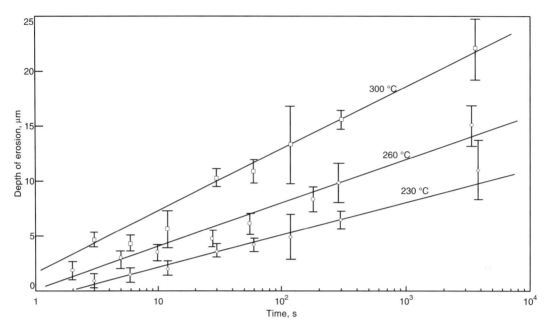

Fig. 3.7 Erosion of silver by molten tin as a function of reaction time and temperature. Source: After Evans and Denner [1978], together with the authors' own data

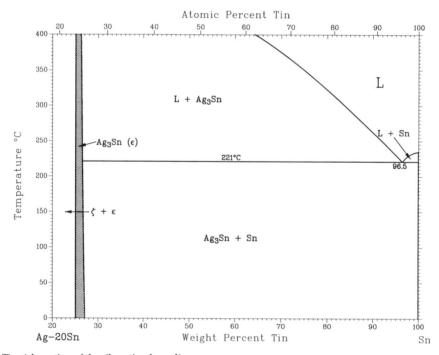

Fig. 3.8 Tin-rich portion of the silver-tin phase diagram

features of eutectic alloys. This is true of the commonly used lead-tin solders and also of the silver-copper-zinc-cadmium brazes.

A binary eutectic composition solder used with components of one of the constituent metals, with intermetallic compound formation.

Application of Phase Diagrams to Soldering and Brazing

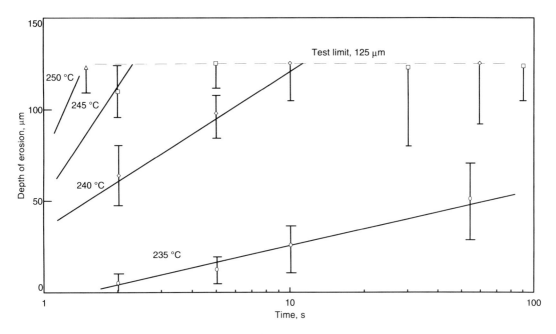

Fig. 3.9 Erosion of gold by molten tin as a function of reaction time and temperature

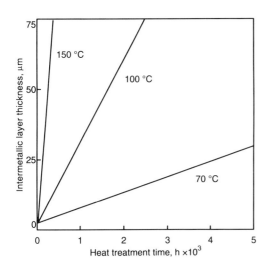

Fig. 3.10 Continued growth of gold-indium intermetallic phases at the interface between a gold metallization and In-Pb solder at elevated temperature but below the solidus temperature of the filler metal. Source: After Frear, Jones, and Kingsman [1991]

The silver-tin eutectic solder (Ag-96Sn) is widely used to join silver-coated components. One of the main reasons for this choice is that the erosion of the metallization on such components is low and is highly predictable. The restricted erosion is a consequence of the formation of the Ag_3Sn intermetallic compound as a layer that separates the molten solder from the remaining silver. This interfacial layer restricts further reaction to a degree determined by temperature and the duration of the soldering process, as indicated by the data given in Fig. 3.7. Using such data, it is possible to determine the minimum thickness of silver that is required for a given volume of solder, without erosion causing encroachment on the underlying material. This feature can be used to advantage in soldering to nonmetals by ensuring that the application of a silver metallization of defined thickness will prevent erosion through to the nonmetallic base material in a prescribed soldering cycle and thereby avoid catastrophic dewetting.

The entire silver-tin phase diagram is shown in Fig. 2.26, but only a portion of this is important in soldering applications, which are carried out below 300 °C (570 °F). This part of the diagram is illustrated in Fig. 3.8. It contains a single eutectic reaction between tin and the Ag_3Sn intermetallic compound (ε) at composition Ag-96.5Sn. Because the melting point of the Ag_3Sn phase is higher than normal soldering temperatures, the phase diagram indicates that it should form interfacially between the molten solder and the solid silver. This is indeed observed.

However, although the phase diagram can provide guidance about where new phases will form,

Principles of Soldering and Brazing

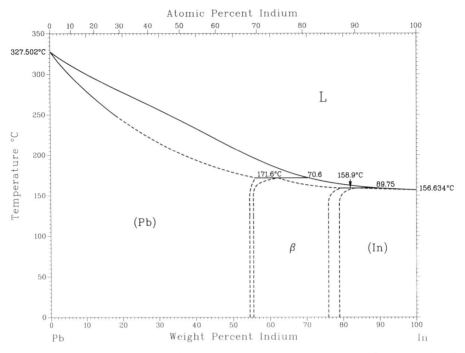

Fig. 3.11 Indium-lead phase diagram. The phase diagram contains two peritectic transformations of the form: L + (Pb) ↔ β, and L + β ↔ (In).

Fig. 3.12 An alloy microstructure characteristic of a peritectic transformation. The alloy contains four constituents: aluminum, copper, nickel, and silicon. The primary phase is totally surrounded by a rim of a second phase as a result of the peritectic reaction failing to maintain equilibrium conditions during solidification.

it cannot be used to determine their ultimate distribution within the material. In an analogous situation—namely, tungsten wetted by the Al-13Si eutectic braze—the high-melting-point WSi_2 intermetallic phase initially forms at the interface but promptly spalls off, due to intrinsic stress in the layer, and is redistributed as needles within the body of the braze.

Another piece of important information that cannot be ascertained from equilibrium phase diagrams is the growth rate of phases. The different rates at which solders can react with substrate metals to form intermetallic phases are vividly illustrated by comparing the rate of erosion of gold by molten indium in Fig. 2.33 with that by molten tin in Fig. 3.9.

Although it would appear from Fig. 2.33 that the indium-gold reaction is self-limiting, this is only true in relation to the short timescales of typical soldering operations (<10 min). Over longer periods of time, the reaction will proceed to a significant extent. Even when the solder is solid, the extent of the reaction can be appreciable, as shown in Fig. 3.10, despite the fact that the reaction rate may be one or two orders of magnitude slower than when the filler is molten (the exact ratio depending on the respective temperatures). This point must be borne in mind when considering the stability of a joint over the service life of the associated assembly. The design life may be as long as 25 years, which is longer than the duration of the joining cycle by a factor of up to six orders of magnitude.

A binary peritectic solder, illustrating problems associated with using a filler in this category. The second common type of phase transformation is the peritectic reaction, where a liquid, L, on cooling, partly solidifies to form a solid phase, S1, and at the peritectic temperature the remaining liquid reacts with S1 to form a new solid phase, S2. This may be written as:

$$L + S1 \leftrightarrow S2$$

Indium-lead solders are frequently employed in situations where the standard tin-base solders cannot be used—for example, for soldering to thick gold metallizations. The indium-lead phase diagram, shown in Fig. 3.11, contains two peritectic reactions:

$$L + Pb = \beta \text{ at } 171.6 \text{ °C } (340.9 \text{ °F})$$
$$L + \beta = In \text{ at } 158.9 \text{ °C } (318.0 \text{ °F})$$

Alloys exhibiting this type of transformation are generally undesirable as fillers because, during a peritectic reaction, it is almost impossible to maintain equilibrium conditions. This is because diffusion rates in solids are about two orders of magnitude slower than in liquids, so that a nonequilibrium microstructure develops consisting of islands of the primary solid phase, S1, completely surrounded by a rim of the second solid phase, S2.

A quaternary aluminum alloy microstructure exhibiting a peritectic transformation is shown in Fig. 3.12. In such an alloy, liquid that is rich in the lower-melting-point elements will be retained below the peritectic transformation temperature. Thus, the melting and freezing range of the alloy is widened, and the remelt temperature cannot be reliably predicted. Furthermore, its microstructure will be grossly inhomogeneous and relatively coarse, to the detriment of the mechanical properties of joints made with this alloy.

In higher-order alloys, a number of other types of phase transformations can occur; these are generally referred to as transition reactions. The majority of these possess features akin to a peritectic transformation, and they tend to be avoided in the selection of alloy compositions as filler metals on the same grounds.

3.2.2 Examples Drawn From Ternary Alloy Systems

It is rare for a soldered or brazed joint to be limited to a combination of just two elements, forming a binary alloy system. Usually the filler is an alloy of at least two metals, while engineering alloy substrates are frequently multicomponent. In certain cases, alloying does not result in the formation of new phases, but more commonly intermetallic compounds form between the constituents. The volume, distribution, and morphology of these intermetallic phases in a joint can have a pronounced effect on mechanical properties in particular. From the relevant phase diagram it is possible to predict whether the intermetallic compound will tend to form as a continuous interfacial layer against the parent materials or is more likely to be dispersed throughout the joint. If the intermetallic compound has poor mechanical properties, then a dispersion is preferred. Examples of these different situations are described below.

Complete intersolubility between a eutectic braze and the metal on the joint surfaces. A typical example of this type of reaction is provided by a Ag-28Cu eutectic braze used in the fluxless brazing of gold-plated molybdenum components. For the three-component system of silver-gold-copper, a ternary phase diagram is required to follow the reaction between the molten filler and the solid components and also to obtain quantitative data on the phases that form. A ternary system is most usually represented by an equilateral triangle, with each of the vertices corresponding to the three elements. A grid is normally drawn on the triangle to provide a linear scale of composition. Temperatures are then represented by a series of isotherms, so that the position of the liquidus on the diagram is mapped as a topographical surface viewed in plan.

The liquidus surface of the silver-gold-copper system is shown in Fig. 2.13. This surface contains a valley that runs from approximately the center of the diagram, at the 800 °C (1470 °F) isotherm, to the eutectic point on the silver-copper binary axis. The minimum in the valley is 767 °C (1413 °F) at a composition of 45Ag-29Au-26Cu, which represents the lowest temperature at which liquid can exist in the silver-gold-copper system at atmospheric pressure. Figure 3.13 depicts an isothermal section at 700 °C (1290 °F), where all the phases are solid. These comprise two solid solutions, one combining silver and copper and the other rich in gold, separated in the triangle by a parabolic boundary.

Phase stability as a function of temperature is commonly represented by a diagram resembling a binary alloy phase diagram, where either one of the constituents or the ratio of two constituents is

Principles of Soldering and Brazing

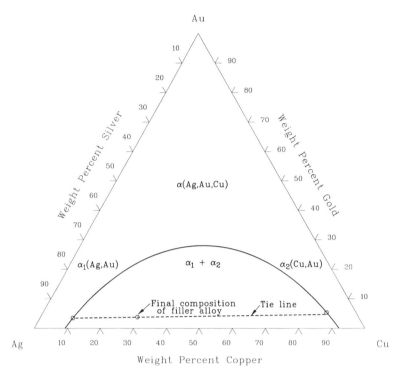

Fig. 3.13 Isothermal projection of the Ag-Au-Cu phase diagram at 700 °C (1290 °F). A hypothetical tie line is shown linking the compositions of the two conjugate phases formed by reaction of the Ag-28Cu braze with a thin gold metallization.

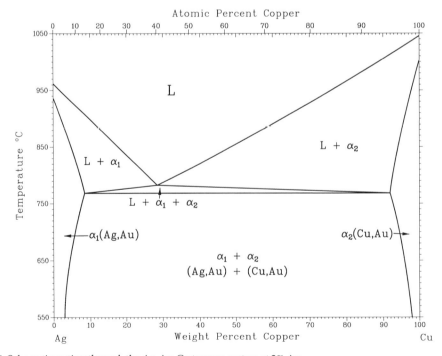

Fig. 3.14 Schematic section through the Ag-Au-Cu ternary system at 2% Au

Application of Phase Diagrams to Soldering and Brazing

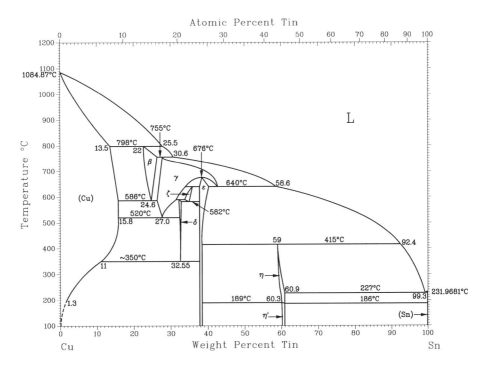

Fig. 3.15 Copper-tin phase diagram

held fixed. Such a section of the silver-gold-copper phase diagram at 2% gold is shown in Fig. 3.14. A single diagram cannot be used to track the solidification sequence, because the ensuing composition changes can extend outside the plane of the diagram. For a similar reason, the lever rule cannot be applied to this representation in order to calculate the proportions of phases that exist in equilibrium. However, the lever rule can be used in conjunction with a series of isothermal sections.

When gold coatings are used with the Ag-28Cu braze, the thickness of the gold layer will normally represent a maximum of about 2% of the volume of the metal in the joint, for reasons of economy. In the joining operation, the molten Ag-28Cu braze fuses with the gold layer to form a single melt. The underlying molybdenum will be wetted by the molten alloy, but does not effectively dissolve. On cooling, the liquid filler solidifies to produce a eutectic microstructure comprising silver- and copper-rich phases, each containing some gold, as indicated by the section through the silver-gold-copper phase diagram at 2% gold shown in Fig. 3.14.

The proportion of gold in each phase cannot be estimated directly from the phase diagram, because the orientation of the relevant tie line join-

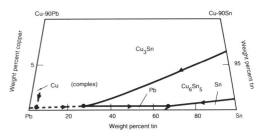

Fig. 3.16 Liquidus surface of the lead-tin-rich portion of the Cu-Pb-Sn ternary system. The primary phase to form on solidification is indicated for each phase field.

ing the two conjugate phases is usually not known. A tie line is a line drawn in a phase diagram; it is always straight and ends at points that correspond to two separate phases. Compositions lying along a tie line will comprise these two phases in a proportion that is given by the lever rule (see section 3.2.1). A typical tie line, drawn on an isothermal

Principles of Soldering and Brazing

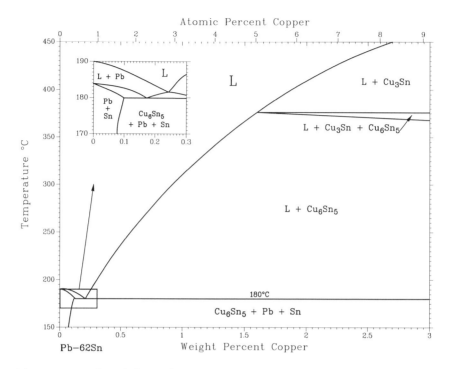

Fig. 3.17 Schematic section through the Cu-Pb-Sn phase diagram between lead-tin eutectic solder and copper, at low percentages of copper

section of the silver-gold-copper phase diagram at 700 °C (1290 °F), is indicated in Fig. 3.13. Using this construction, it is possible to predict the precise composition of one of the constituent phases of a silver-gold-copper alloy, provided that the composition of the other and also their weight ratio are known.

Interfacial compound formation between a eutectic braze and the component metals. As an example of interfacial compound formation, it is well known that lead-tin solders, when used with copper components, form one or two copper-tin intermetallic compounds, predominantly at the copper/solder interface. The phase diagram of the copper-tin system is shown in Fig. 3.15.

Despite the industrial importance of the copper-lead-tin system, as noted in the introduction to this chapter, there is no complete published phase diagram that describes the constitution of lead-tin solders containing small (<10%) concentrations of copper. The tentative liquidus projection of the copper-lead-tin system and a section between the Pb-60Sn solder alloy and copper shown in Fig. 3.16 and 3.17, respectively, are those derived by the authors. These partial phase diagrams are not precise, but they do represent the principal phase relationships in this alloy system. The vertical section through the phase diagram furnishes the following information:

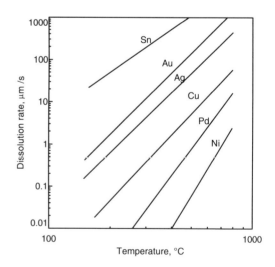

Fig. 3.18 Rate of dissolution of a range of engineering parent metals and metallizations in lead-tin solder as a function of temperature. Source: After Klein Wassink [1989]

Application of Phase Diagrams to Soldering and Brazing

Fig. 3.19 Phases formed by reaction between a tin-lead solder and a copper substrate, following extended heat treatment

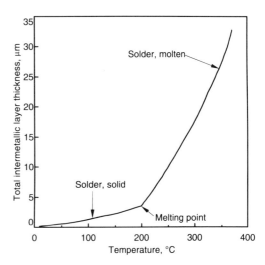

Fig. 3.21 Growth of copper-tin intermetallic compounds on a copper substrate in contact with lead-tin solder for 100 s at different process temperatures

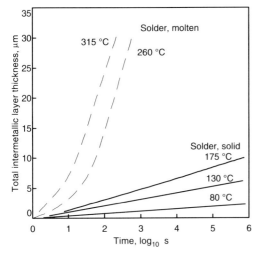

Fig. 3.20 Growth of copper-tin intermetallic compounds on a copper substrate wetted by lead-tin eutectic solder as a function of reaction time and temperature

- Copper is soluble in molten lead-tin solders.
- The dissolution of copper by the solder marginally depresses the melting point of the lead-tin eutectic alloy.
- The dissolution of copper also results in the formation of either one or two intermetallic compounds. The presence of these phases—namely, Cu_3Sn and Cu_6Sn_5, depends on the temperature and duration of the soldering cycle and the overall composition of the filler metal prior to solidification. This is explained below.

The rate of dissolution of copper in molten Pb-60Sn solder is rapid, as is the erosion of most other engineering metals and metallizations by this alloy, as shown in Fig. 3.18. Consequently, despite the short process cycle times normally associated with soldering practice, the molten solder will dissolve sufficient copper to reach the saturation concentration, dictated by the process temperature and the composition of the liquidus surface in the copper-lead-tin phase diagram. At a typical soldering temperature of 230 °C (445 °F), it can be deduced that the solder will dissolve about 0.5% Cu (see Figs. 3.17 and 3.18).

On cooling this ternary filler metal, it would be expected from consideration of the phase diagram that copper-tin intermetallic phases will precipitate and become distributed within the bulk of the joint. If the peak process temperature is kept below 375 °C (705 °F), the phase diagram shows that the primary phase that precipitates on cooling will be Cu_6Sn_5. This has been confirmed by experiment. The precipitates are typically on the order of 5 to 20 nm in size and therefore can only be resolved using high-resolution imaging techniques, such as transmission electron microscopy [Felton et al. 1991]. However, if the peak process temperature exceeds 375 °C (705 °F) and remains so long enough for more than 1.6% Cu to dissolve, then the first phase to precipitate will be Cu_3Sn.

In practice, and not immediately apparent from the liquidus surface of the ternary phase diagram, the copper-tin intermetallic compounds form at the interface between the solid copper and molten solder. Owing to the prevailing concentra-

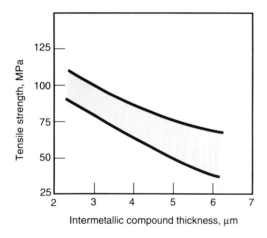

Fig. 3.22 Effect of thickness of copper-tin intermetallic compounds in soldered joints on tensile strength at room temperature

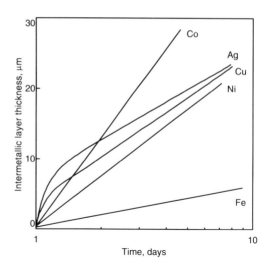

Fig. 3.23 Growth of tin-base intermetallic phases by solid-state diffusion at 170 °C (340 °F) on various substrate materials. Source: After Kay and Mackay [1979]

tion gradient, the copper-rich Cu_3Sn phase forms adjacent to the copper component, with the more tin-rich Cu_6Sn_5 phase between it and the solder, as shown in Fig. 3.19.

The rate of formation of interfacial copper-tin intermetallic phases is governed by the rate of interdiffusion of copper and tin through them and is well characterized. A distillation of the data that are available in the literature is presented in Fig. 3.20. The Cu_3Sn phase layer does not grow to any significant thickness during typical industrial soldering processes, as there is insufficient time for the required concentration of tin to diffuse through to the copper [Dirnfeld and Ramon 1990].

The formation of continuous layers of intermetallic compounds between the copper and the molten solder confers the benefit of greatly restricting the erosion of the copper substrate. Hence, during a typical soldering cycle, there is only slight dissolution of copper by the solder, which therefore remains largely free from copper-tin intermetallic precipitates [Felton et al. 1991].

Copper-tin intermetallic compounds are, however, fairly brittle. Due to the relatively high rate of diffusion of copper and tin through them, the thickness of these phases will continue to increase during elevated-temperature service of the assembly, albeit much more slowly than when the solder is molten, as shown in Fig. 3.21. The presence of thick layers of copper-tin intermetallic phases is widely reported as being detrimental to the mechanical properties of the joints, and this has prompted many studies of the effect of their presence (e.g., Dirnfeld and Ramon [1990]). Figure 3.22 shows the relationship between the tensile strength of joints and the thickness of the copper-tin intermetallic layer as determined in these studies. Notwithstanding the general concern, there are very few reports of joint failures that can definitely be pinpointed to the copper-tin intermetallic phases. This is possibly because tin forms hard intermetallic phases, of comparable thickness, by solid-state diffusion at elevated temperature with virtually all the metals commonly used in engineering [Kay and Mackay 1976]. Some relevant data are reproduced in Fig. 3.23. Not surprisingly, then, joints made to copper testpieces using lead-tin solder are not significantly weaker than joints to other common substrate materials, in terms of their strength, ductility, and fatigue resistance.

Not all the binary lead-tin solders form two intermetallic phases on reaction with copper. If the tin level is less than 25%, the phase diagram in Fig. 3.16 indicates that only the Cu_3Sn intermetallic will form under near-equilibrium conditions [Grivas et al. 1986].

Distributed compound formation between a eutectic solder and the component metals. Another industrially important alloy system comprises gold, lead, and tin. Its importance lies in the widespread use of gold as a solderable metallization for fluxless joining in the electronics industry.

The liquidus surface of the gold-lead-tin ternary phase diagram is given in Fig. 3.24, and a section from the eutectic Pb-62Sn composition toward gold is shown in Fig. 3.25. The principal

Application of Phase Diagrams to Soldering and Brazing

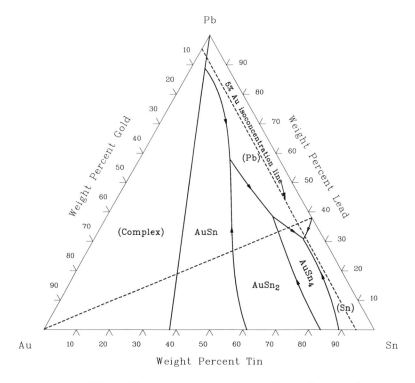

Fig. 3.24 Liquidus projection of the Au-Pb-Sn ternary system. The first phase to form on solidification is labeled for each phase field. Source: After Humpston and Davies [1984, 1985]; Humpston and Evans [1987]

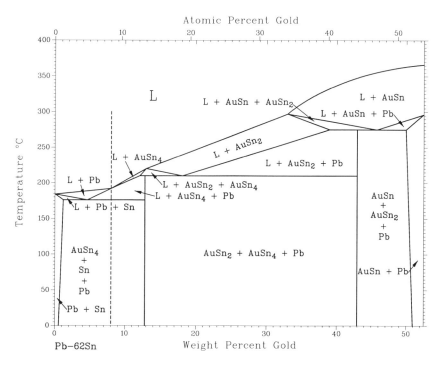

Fig. 3.25 Vertical section through the Au-Pb-Sn ternary system between eutectic tin-lead (Pb-62Sn) composition and gold. The plan view of this section is marked by a dashed line on the liquidus projection of the gold-lead-tin ternary system shown in Fig. 3.24.

features of these diagrams are closely related to those of the copper-lead-tin system discussed above, with the difference that, in the gold-base system, at least three embrittling intermetallic compounds are able to form directly from the liquid. Therefore, it might be expected that, when tin-lead eutectic solder is used to join solid gold components, there is little difference in microstructure of the joints.

Gold metallizations are seldom more than a few microns thick, largely because of the cost of this metal, so there is a limit to the concentration of gold that will accumulate in the molten filler. The precise value is determined by both the volume of solder alloy and the thickness of the metallization, but is usually well below the saturation concentration. Consequently, due to the high rate of dissolution of gold in molten tin-lead solders, the metallization will be completely dissolved during the heating cycle.

On cooling a gold-lead-tin ternary alloy formed by reaction of the lead-tin eutectic solder with a gold metallization, gold-tin phases will be precipitated, as indicated by the phase diagram. The precipitate will form as a dispersed phase throughout the volume of the filler. In joints of typical geometry, the most common intermetallic compound is usually $AuSn_4$, which forms either as large conglomerates or as a finely divided phase, depending whether this compound comprises the primary phase or is a minority phase (secondary or tertiary precipitate).

When gold-tin compounds form as the primary or major phases, they embrittle the soldered joints because of the weak interface between them

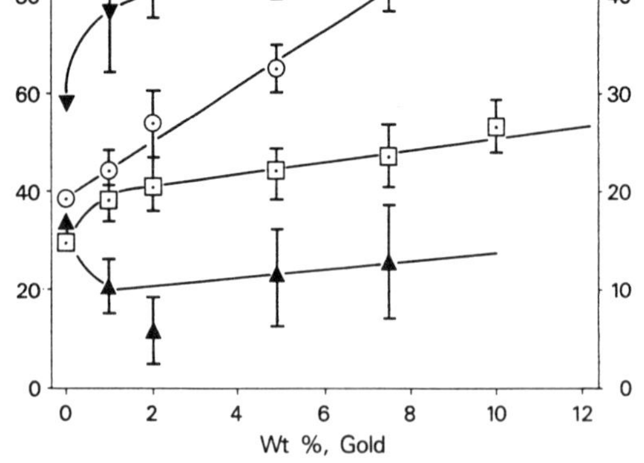

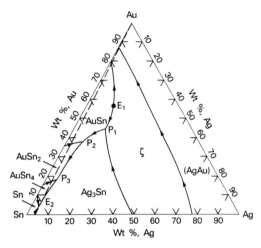

Fig. 3.27 Liquidus projection of the Ag-Au-Sn ternary system. The first phase to form on solidification is labeled for each phase field.

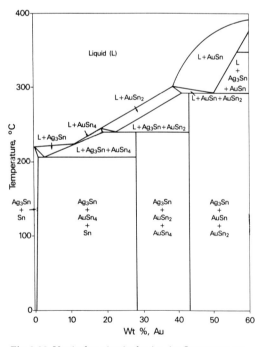

Fig. 3.28 Vertical section in the Ag-Au-Sn ternary system between eutectic tin-silver (Ag-96.5Sn) alloy and gold. The plan view of this section is marked by a dashed line on the liquidus projection of the silver-gold-tin ternary system shown in Fig. 3.27.

Fig. 3.26 A selection of mechanical properties of Ag-96Sn solder as a function of gold addition

and the lead-tin eutectic [Harding and Pressly 1963]. However, provided that the concentration of gold in the solder does not exceed 8%, then the $AuSn_4$ compound will be present only as a minor phase, as indicated by the 8% Au concentration line (dashed) on the phase diagram in Fig. 3.25. A maximum gold concentration of 5% is usually considered a safe working limit in industry. By adjusting the volume fraction of gold to solder to below this critical value, it is possible to prevent embrittlement of the solder in joints. As can be seen from Fig. 3.24, high-lead-content tin-lead solders can accommodate higher levels of gold before gold-tin intermetallic compounds become dominant (i.e., constitute the primary phase) and cause joint embrittlement. Likewise, from the appropriate sections of the phase diagrams of other solder alloys in combination with gold, it is possible to derive safe working limits for them in a similar manner.

Low concentrations of the $AuSn_4$ phase actually enhance the mechanical properties of many tin-containing solders, including lead-tin [Wild 1968]. This strengthening is illustrated in Fig. 3.26 for the Ag-96Sn solder. Up to a concentration of about 8% gold, the solder microstructure is characterized by a fine dispersion of the $AuSn_4$ phase. However, when the level of gold in the solder rises toward 10%, there is a sudden change in properties, and ingots of the solder become completely unworkable. This change corresponds to the $AuSn_4$ intermetallic compound becoming the primary phase in the alloy, in accordance with the ternary silver-gold-tin phase diagram shown in Fig. 3.27 and 3.28. It is noteworthy that the Ag-96Sn solder is tolerant to approximately twice the volume fraction of gold that the Pb-60Sn solder can accommodate before the intermetallic becomes the dominant phase and the alloy is embrittled. The critical levels of gold that give rise to $AuSn_4$ as the primary phase are listed in Table 7.2 for a selection of solders. Fluxless soldering processes that involve gold metallizations are discussed in section 7.3.

A point to be aware of is that if the soldering cycle time is exceedingly short, then it is possible for some of the gold metallization to survive. The composition of the joint can then range from the solder composition, at the centerline, through to the pure gold of the metallization. An example is shown in Fig. 3.29. Because this is a nonequilibrium situation, the morphology and distribution of the phases that form cannot be predicted from the phase diagram. The micrographs in Fig. 3.29 show the presence of a layer of $AuSn_4$ between the solder and a region of residual gold coating (toward the circumference of the area of solder spread where the solder thickness tails off). Here, the gold-tin intermetallics form interfacially against the gold metallization, exactly as in the copper-lead-tin system.

The mechanical properties of joints containing intermetallic phases can be inferred from a phase diagram according to whether they are compounds of exact stoichiometric composition (i.e., they are in integral atomic ratios of their constituents) or exist over a range of compositions. Exact stoichiometric compounds tend to form when one of the two elements is strongly metallic in character and the other significantly less so, in terms of the density of the free electrons that bind the atoms of the metal together. Cu_3P, Cu_3Sn, and Cu_6Sn_5 are typical examples of this type of compound. These compounds tend to be hard and brittle because their crystal structures are frequently of low symmetry (i.e., they deviate from simple cubic or hexagonal structures), and this limits the directions in which volume-conserving plastic flow can occur. This point is illustrated for the intermetallic compound Co_3V, which has a complex hexagonal

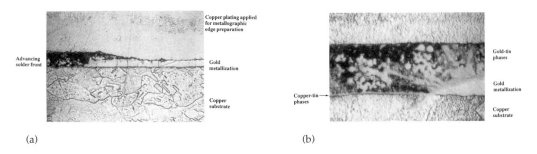

Fig. 3.29 Dissolution of a gold metallization by lead-tin solder. Gold-tin intermetallic phases initially form against the gold, which then dissolves in the molten filler. (a) 17.5×. (b) 175×

structure and is brittle when pure (<1% elongation to failure at room temperature). However, if iron is substituted for part of the cobalt so that the compound is modified to $(Co,Fe)_3V$, the crystal structure will transform to simple face-centered cubic (fcc) and this change enhances the elongation to failure to more than 40% [Baker and George 1992]. Another consequence of the low symmetry of the crystal structures of most exact stoichiometric compounds is that the interfaces between these compounds and other phases tend to be weak. These characteristics are transferred to the joint unless the compounds form as a fine dispersion within the filler; therefore, their occurrence should be minimized or, even better, avoided wherever possible.

Compounds that are stable over a range of compositions tend to be moderately ductile and have crystal structures exhibiting high symmetry, as do most elemental metals. Therefore, they tend to have a benign effect on joint properties. An example of such a compound is Ag_3Sn, which is stable over the composition range from 13 to 20% silver at room temperature and has a close-packed hexagonal crystal structure. This compound forms as an interfacial layer when silver-tin solder is used to join silver-coated components in a manner analogous to the reaction between lead-tin eutectic and copper. An example is shown in Fig. 3.30, with the associated binary phase diagram given in Fig. 2.26. The growth rate of the Ag_3Sn layer decreases exponentially with time at a fixed temperature (Fig. 3.31). The growth of this intermetallic layer correlates with the measured rate of erosion of silver by tin (Fig. 3.7). The growth of the Ag_3Sn layer progressively reduces the dissolution of silver. For this reason, the silver-tin eutectic solder used in conjunction with silver-coated substrates forms the basis of a soldering process that is highly tolerant to the processing time and temperature inasmuch as the risk of totally dissolving a thin silver metallization can be minimized.

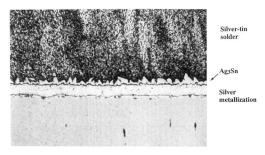

Fig. 3.30 A continuous interfacial layer of the intermetallic compound Ag_3Sn forms on reaction between silver-tin solder and silver. 20.5×

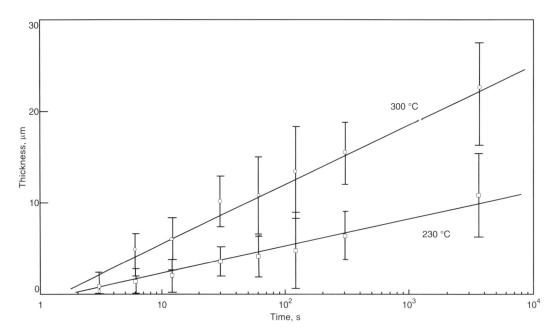

Fig. 3.31 Thickness of the Ag_3Sn intermetallic layer formed by reaction between Ag-96Sn solder and silver as a function of reaction time and temperature. Source: After Evans and Denner [1978], with authors' own data

3.2.3 Complexities Presented by Higher-Order and Nonmetallic Systems

More often than not, higher-order alloy systems are encountered in joining, because both the filler and the parent materials usually each have a minimum of two constituents. Combinations involving five or even larger numbers of elements are not uncommon.

The definition of the plethora of phases that can exist in these higher-order systems represents a daunting task. In order to make the problem more tractable, a reductive approach is often employed. This method usually involves partitioning the multicomponent system into a series of quasi-binary or quasi-ternary alloy systems, each containing different but fixed proportions of the other components, and ascertaining these sections of the relevant phase diagrams. Much care should be exercised in extracting quantitative information from these partial phase diagrams, because the tie lines, triangles, quadrilaterals, and so on that are used with the lever rule to determine the relative proportions of phases present often do not lie in the plane of the selected sections.

Complete or even partial phase diagrams are available in the published literature for only a few higher-order metal systems. Therefore, it is usually necessary to determine the phases that form and establish the phase relationships over the limited range of compositions that pertain to the joining process under consideration by empirical means. A methodology for achieving this is described in section 3.3.

Joining to nonmetallic materials presents additional problems that may not be immediately apparent from the phase diagram, assuming that one is available. A common method for joining to materials that contain glass or ceramic phases is to incorporate into the filler a highly active metal such as titanium, zirconium, or hafnium that will wet and bond to these materials. Further details are given in section 5.4. The bonding process relies on the formation of a compound between the active constituent of the filler metal and one of the elements of the nonmetallic material while liberating others.

For example, a gold-nickel braze containing titanium will wet the engineering ceramic silicon nitride, Si_3N_4, by forming a surface reaction layer of titanium nitride, leaving some free silicon. However, when nickel-containing brazes are used with silicon nitride, the nickel silicide that forms is brittle and thus weakens the joint, while the gold present in the filler forms a eutectic with silicon that melts at 363 °C (685 °F). This results in a joint that melts well below the temperature expected for a brazed joint. In the case of Si_3N_4 brazed with an aluminum-silicon filler alloy, the braze wets the ceramic, producing bonds with the silicon and leaving free nitrogen. The nitrogen gas will produce voids in the joint unless provision is made for it to escape from the joint gap [Lugscheider and Tillmann, 1991].

The various examples discussed in this section underline the care that must be taken in choosing a filler for use with a particular material and again demonstrate the importance of considering the combination of filler and components as an integral materials system.

3.3 A Methodology for Determining Phase Diagrams

A detailed determination of an alloy containing even just three constituents is an arduous task, in terms of both technical effort and time, even when all the binary phase diagrams are known. Where there is a lack of phase diagram information on an alloy system that is represented in a soldered or brazed joint, knowledge of the phase relationships over a limited range of compositions may be quite adequate for the purposes of understanding and controlling the metallurgical reactions that occur. The minimum information that is required is the temperature of the solidus and liquidus phase boundaries as a function of alloy composition, the nature of the solidification (particularly whether it is eutectiferous or peritectiferous), and whether any intermetallic compounds are present. If further resources are available, then intermetallic phases that might be present should be identified, and the slope of the liquidus surface between the composition of the filler alloy and that of the substrate material, for the reasons described in the previous section, should be determined.

There are two main approaches for determining the phase relationships in an alloy system. One is by direct experiment, involving measurement of a property of the material that changes discontinuously at a phase transition. The other involves calculating and comparing the Gibbs free energy of alloys at different temperatures using thermochemical data for the constituent elements.

The computation of phase diagrams from thermochemical data requires access to sophisticated computer programs and reliable input data. For most alloy systems, the necessary information is not available in the open literature and, even if this step can be successfully negotiated, taxing mathematical manipulation is still required to convert these data into a phase diagram.

On the other hand, sensitive experimental techniques for determining phase diagrams from property measurements are widely available and readily accessible to the trained technologist. For these reasons, only experimental techniques will be described here, and reference will be limited to a few widely available methods that are appropriate to designing soldering and brazing processes. A comprehensive description of methods for determining alloy phase diagrams by experiment is given by Hume-Rothery, Christian, and Pearson [1952].

The key to determining the essential features of an alloy system in a cost-effective manner is to use a clearly focused strategy for the experimental work and to make extensive use of data reductive procedures. A methodology that has proved effective in furnishing data on previously unknown phase relationships is described below and illustrated by the following practical example.

The germanium window of an infrared sensor was required to be joined by brazing to an aluminum alloy housing. Fluxes could not be used because of the sensitive electronics that were to be installed within the housing. Accordingly, metallizations of a noble metal were applied to the mating surfaces of the germanium and aluminum components to ensure wetting by the filler. Silver was selected for this function in preference to gold because of the incompatibility of aluminum with gold in joints through the formation of brittle phases. The development of this type of fluxless joining processes is described in section 4.2.1. It was envisaged that, on heating, the aluminum and silver would interdiffuse, so that when the temperature reached 567 °C (1053 °F), the aluminum-silver eutectic phase would melt and fill the joint gap. What would happen when the molten aluminum-silver alloy dissolved the silver coating off the germanium window and reacted with the semiconductor material was unknown, as was the remelt temperature of the joint. Accordingly, efforts were directed at obtaining information on the phases that form between silver, aluminum, and germanium.

3.3.1 Literature Search

A worthwhile first step in obtaining the equilibrium phase diagram of a system of interest is to consult the technical literature. If relevant information is available, it should be critically assessed in terms of reliability. A methodology for critically assessing phase diagrams is given by Evans and Prince [1985].

For the silver-aluminum-germanium system, the published data, at the time of this exercise, was insufficient to elucidate the proposed joining process [Schmidt and Weiss 1978]. Therefore, a practical study of the alloy system was necessary. The following strategy was pursued:

- The microstructures of aluminum-germanium alloys containing small percentages of silver were examined in order to obtain an indication of the phases and the associated metallurgical reactions that produce them during solidification. This information was used to identify eutectic reactions that are associated with that portion of the phase diagram and the likely constituent phases.

- Having obtained a rough indication of the phases of potential interest, the next step was to establish the combination that is attributable to the specific eutectic transformation of interest to the joining application.

- Next, the composition of the eutectic point was estimated by a rule-of-thumb procedure, known as the method of equal liquidus slope, in order to narrow the phase field in which to conduct a more detailed search.

- Thermal analysis was then used to establish accurately the temperature of the eutectic transformation and, with the aid of a modeling technique, the topography of the liquidus surface in the defined phase field.

- Finally, a more refined estimation of the eutectic composition was made using quantitative metallography. The accuracy of the result was confirmed by metallographic examination of an alloy of that composition, prepared by cooling from above the liquidus temperature.

The rationale for this procedure is that each successive step helps to establish more closely the eutectic temperature and to narrow the composition field that contains the eutectic transformation. By this means, the time and effort required for each subsequent step, which become progressively more laborious and involved, are kept to a minimum. Each of these steps is explained in more detail below.

Several of the mathematical procedures used during the investigation require the composition of alloys to be specified in atomic percentage. In order to avoid repeated conversion between atomic and weight ratios, all compositions given in the following section (including Figs. 3.34, 3.37 and Table 3.4) are given in atomic percentage, unless clearly indicated otherwise.

3.3.2 Metallographic Examination

Basic information on the nature of the equilibria in an alloy system can be obtained from metallographic examination of alloys of known constituents that have been slowly cooled from above the liquidus temperature. Eutectic and peritectic phase transformations are usually easy to recognize along with other features such as the primary and secondary phases that precipitate, immiscibility in the molten state, and intermetallic compounds.

Such a preliminary assessment is often useful in identifying appropriate experimental methods for a more detailed analysis of the alloy phase diagram. For example, if the microstructure shows evidence of primary peritectic-type solidification or immiscibility in the molten state, thermal analysis of such alloys using cooling curves is likely to give spurious results because the alloy composition will not be homogeneous following solidification.

Energy-dispersive x-ray analysis (EDAX) or a similar analytical technique can be used to identify positively the phases present. However, this type of quantitative investigation requires access to sophisticated analytical equipment and is often unnecessary; the required information frequently can be deduced by other routes, as will be explained.

In the silver-aluminum-germanium system under consideration, only alloy compositions close to the aluminum-germanium binary system were of interest, because the materials present in the joint comprised relatively massive quantities of aluminum and germanium together with two thin layers of silver. Accordingly, it was decided to prepare a molten alloy containing equal parts of aluminum and germanium by weight with 5% silver, and to cool it slowly. On metallographic examination, the alloy microstructure was found to comprise a small fraction of primary aluminum in a ternary eutectic matrix. From these data, it could be deduced that this alloy system indeed contained a ternary eutectic and that its constituents included aluminum and germanium in roughly equal proportions together with a much smaller fraction of silver. Thereby, the region of the phase diagram to search was identified. By noting thermal effects during cooling it was also established that the temperature of the ternary eutectic transformation was almost certainly below 420 °C (788 °F), which is the melting point of the aluminum-germanium binary eutectic reaction.

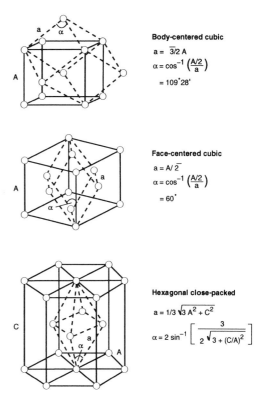

Fig. 3.32 The primitive rhombohedral cells corresponding to the three most common unit cell types adopted by metals

3.3.3 Subdivision of High-Order Systems

Before endeavoring to determine the phase diagram of a ternary or higher-order system, it is advantageous to establish whether the phase diagram subdivides through formation of two-phase regions in the solid state. This is because reaction between a molten filler and a substrate will usually encompass only a limited range of alloy compositions. The most efficient use of resources, then, is

Principles of Soldering and Brazing

Table 3.2 Crystal structures and primitive rhombohedral cells of selected phases in the gold-lead-tin system

Phase	Structure	Lattice parameters, Å		Primitive rhombus parameters	
		A	C	α	a Å
Au	fcc	4.07		60°00′	2.88
Au$_2$Pb	fcc	7.92		60°00′	5.60
AuPb$_2$	fcc	6.17		60°00′	4.36
AuSn	hcp	4.32	5.52	88°20′	3.09
Pb	fcc	4.95		60°00′	3.50
Sn	Tetragonal	5.83	3.18	90°00′	4.64

Note: Refer to Fig. 3.32 for a description of the variables a, c, and α.

to confine the experimental study to the region of the phase diagram that is just large enough to embrace the relevant phase transformations. Identifying how a phase diagram subdivides provides knowledge of the phases participating in each solidification reaction.

An empirical method of predicting the elements and compounds that bound the phase fields of high-order transformations is based on the general rule that the closer that two of these components are to each other in terms of crystal structure, the more readily they will tend to mix to form solid solutions across the phase diagram. The criterion used for this comparison is the size and shape of the rhombohedral primitive cells. The primitive cells of body-centered cubic (bcc), face-centered cubic (fcc), and hexagonal close-packed (hcp) crystal lattice structures are depicted in Fig. 3.32.

The silver-aluminum-germanium system can be triangulated in only one way, as there are no intermetallic compounds in either the aluminum-germanium or silver-germanium binary systems. Consequently, there must be a tie line between the Ag$_2$Al phase and germanium, assuming an absence of ternary intermetallic compounds. Thus, the eutectic transformation must lie in the region bounded by Ag$_2$Al, aluminum, and germanium, and these must be the three phases that participate in the transformation. In fact, they were the phases observed as the characteristic products of a eutectic reaction in the metallographic sample.

By way of a more elaborate example of application of the method, consider the gold-lead-tin system, which is important because of the frequent use of gold metallizations with tin-lead solders. The gold-lead-tin phase diagram can be triangulated in three principal ways, as shown by the tie lines in Fig. 3.33. These triangulations are simplified in that the phases AuSn$_4$, AuSn$_2$,

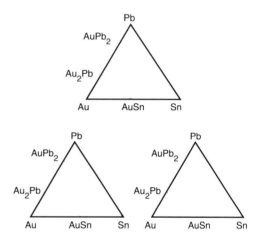

Fig. 3.33 Gold-lead-tin ternary system, triangulated (subdivided) in three different permutations

AuPb$_3$, and Au$_5$Sn are ignored. This is permissible because the AuSn phase is a congruently melting compound (i.e., it behaves like a pure metal in that it transforms directly between the liquid and solid states) and is thermodynamically more stable. Therefore, AuSn is more likely to enter into two-phase equilibrium than are the other incongruently melting gold-tin phases. AuPb$_3$ is also omitted because its tie line is defined by the position of the AuPb$_2$ tie line. Thus, to correctly triangulate the system it is only necessary to establish whether AuSn enters into two-phase equilibrium in the solid state with Au$_2$Pb, AuPb$_2$, or lead. The structural parameters of the primitive cells relating to gold, lead, tin, AuSn, Au$_2$Pb, and AuPb$_2$ are listed in Table 3.2. The primitive cells of AuSn and lead are the closest, implying that the AuSn-Pb

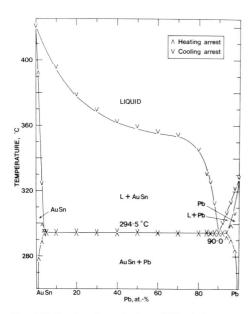

Fig. 3.34 Section through the gold-lead-tin ternary system between the AuSn intermetallic compound and lead. All tie lines between conjugate phases lie within the plane of the section. Source: After Humpston and Davies [1984]

join is a two-phase region. Experimental investigation involving thermal analysis and metallography confirmed that this is indeed the case, as shown in Fig. 3.34. Subdivision of the remainder of the ternary system is then relatively straightforward, as only one permutation is possible. The result is given in Figs. 3.24 and 3.25.

A more scientific approach to identifying the boundaries of the phase fields in the gold-lead-tin system is to compare the Gibbs free energy change for the reactions:

$$AuPb_2 + Sn \leftrightarrow AuSn + 2Pb$$
$$Au_2Pb + 2Sn \leftrightarrow 2AuSn + Pb$$

Thermodynamic calculations show that the combination of AuSn with lead is more stable (i.e., gives rise to a lower Gibbs free energy) than the combination of tin with either of the gold-lead compounds. This indicates that the tie line between AuSn and lead will form one of the phase-field boundaries in this alloy system under equilibrium conditions. These calculations, however, cannot always be made with adequate reliability, because there is often insufficient thermodynamic data on the various intermetallic compounds, in which case there is little alternative but to use primitive cell calculations for partitioning the phase diagram.

3.3.4 Predicting the Composition of Eutectic Points in High-Order Systems

The next step is to predict the likely composition of the eutectic point. Knowledge of the approximate location of the eutectic point fixes the region of the phase diagram that needs to be investigated in detail using thermal analysis and thus limits the number of alloy compositions and the effort involved.

The method of equal liquidus slope allows estimation of the position of a eutectic point. It is based on the empirical fact that the slopes of all liquidus lines, valleys, or surfaces in a given alloy system to a eutectic point are often similar. The magnitude of the slope is then proportional to the temperature difference between the melting points of the participating elements or compounds and the eutectic temperature, with the composition axis scaled in atomic percentage. Thus, in a binary system, the eutectic point will tend toward the 50/50 composition when either the melting points of the two elements are similar and/or the eutectic temperature is well below the melting points of both elements. When the alloy system contains an intermetallic compound that behaves as a pure metal in its melting characteristics (i.e., it completely melts at a single temperature and is thus known as a congruently melting compound), the phase diagram can be broken down into separate regions. Each of these regions is bounded by pure metals or congruently melting compounds and for the purposes of the predictive model can be treated as a complete phase diagram, in terms of composition.

This rule-of-thumb approach allows the composition of high-order eutectic transformations to be predicted with a moderate degree of accuracy from the constituent systems, but with the benefit that there is no need to know which phases participate in the eutectic reaction. However, it is necessary to know the approximate temperature of the eutectic reaction, which can be established either by carrying out a thermal analysis measurement or by plotting a cooling curve. In the case of the silver-aluminum-germanium system, the cooling curve determination showed that the ternary eutectic transformation occurred at about 410 °C (770 °F). It is possible to have some confidence in this first temperature measurement, as it is below the temperature of the lowest-melting-point binary transformation, that of the aluminum-germanium eutectic at 420 °C (790 °F).

The application of the equal liquidus slope method to the silver-aluminum-germanium system involves first collating data on eutectic and peritectic points in the constituent binary phase diagrams and using this information to calculate the likely position of the eutectic point. The available phase diagrams for the constituent binary systems provided the following data:

- The aluminum-germanium system contains a single eutectic reaction that is located at 28 at.% Ge, 420 °C (790 °F).
- There are several peritectic transformations in the aluminum-silver system, but between the Ag_2Al compound and pure aluminum, which has already been established as the only portion of the phase diagram of immediate interest, there is a single eutectic transformation at 61 at.% Al, 567 °C (1053 °F).
- The silver-germanium system contains a single eutectic transformation at 25 at.% Ge, 651 °C (1204 °F).

The equal liquidus slope calculation can be represented as follows:

	Silver	Aluminum	Germanium
		$\dfrac{72}{420-410}$	$\dfrac{28}{420-410}$
+			
	$\dfrac{39}{567-410}$	$\dfrac{61}{567-410}$	
+			
	$\dfrac{75}{651-410}$		$\dfrac{25}{651-410}$
=	0.56	7.59	2.90

Converting to percentages gives 5.1Ag-68.7Al-26.2Ge (at.%) for the ternary eutectic point. This approximate value enables a selected region on the composition triangle to be defined for a more detailed search to home in on the exact eutectic point. Obviously, a composition region centered on 5.1Ag-68.7Al-26.2Ge needed to be considered, but the appropriate search field is determined by a number of considerations, including the actual procedures that are to be used. A suitable and relatively quick experimental method for determining the phase diagram in the vicinity of the eutectic point is thermal analysis, and this was duly chosen. The criteria used to select the most appropriate alloy compositions for such a study are described in the following section.

3.3.5 Thermal Analysis

Thermal analysis is probably the most widely used technique for determining the melting ranges of alloys [Shull and Joshi 1993]. It is familiar to most materials scientists and engineers; therefore, treatment of this subject is limited to a brief overview of the technique, which is presented in Appendix 3.2 of this chapter. This outlines the underlying principles and the various modes of thermal analysis that are employed. Special emphasis is given to a lesser known measurement method, which the authors refer to as calorimetric thermal analysis. This method offers certain advantages for the determination of metallurgical phase diagrams.

The thermal analysis of a number of alloys can establish the ternary reaction temperature and composition and also enable the liquidus and solidus surfaces of the silver-aluminum-germanium alloy phase diagram to be mapped out. The ternary eutectic point will be conspicuous by its zero melting range, that is, the convergence of the liquidus and solidus temperatures.

For the thermal analysis to furnish reliable data, it is necessary to ensure that the alloys of interest are homogeneous, which in turn relies on the method of their preparation. Furthermore, in order to limit the number of alloy compositions that are analyzed, a judicious selection procedure must be employed. These points are addressed below, with reference to the example of silver-aluminum-germanium alloys.

3.3.5.1 Preparation of Alloys for Thermal Analysis

Reliable thermal analysis requires that the alloy under examination be homogeneous. A procedure favored by the authors for preparing alloys involves melting together the constituent elements in a sealed silica crucible and, after stirring or shaking vigorously, quench cooling by immersing it in water. This procedure yields alloys with microstructures having a grain size of a few tens of microns. Prior to thermal analysis, the alloy is subjected to a heat treatment below its melting point. The objectives of the heat treatment stage are to eliminate nonequilibrium segregation in the sample, to allow the dissolution of metastable phases, and to encourage the nucleation and growth of any minor equilibrium phases that might otherwise escape detection.

The homogenization treatment becomes progressively more effective as the temperature is raised, as the diffusion rate increases in accordance with Eq 3.1 above. Because the diffusion rate approximates to an inverse-exponential function of the homologous temperature, T/T_m, a 10% increase in this ratio results in a reduction in the requisite homogenization period by about an order of magnitude.

Furthermore, the time required to achieve homogenization diminishes rapidly with reduction in the grain size (varying roughly with the square of the grain size), as this dimension determines the distance over which atoms must diffuse. For a detailed treatment of diffusion in solids, the reader is referred to standard references such as Darken and Gurry [1953] and Shewmon [1963].

From the foregoing, it is essential that the alloy be prepared with the finest possible grain size. The grain size in cast alloys is determined primarily by the cooling rate, as shown in Fig. 3.35. Hence, a quench-cooling procedure followed by a homogenization heat treatment just below the solidus temperature is suitable for preparing alloys for thermal analysis. For the silver-aluminum-germanium alloys, the appropriate homogenization conditions were a heat treatment temperature of 400 °C (750 °F), maintained for 24 h.

3.3.5.2 Selection of Alloys for Thermal Analysis

The number of different alloy compositions in the silver-aluminum-germanium system that might be examined using thermal analysis is effectively infinite. Some means is required to select a manageable number of samples for study. A statistical approach for specifying the alloy compositions that need to be thermally analyzed is the simplex lattice design method [Scheffe 1958].

The simplex lattice design method is a mathematical technique of interpolating between experimental measurements of intensive parameters such as temperature, which does not require a knowledge of relationships between the value of the intensive property and the composition of the mixture. An intensive property of interest, such as the liquidus temperature of a particular composition, is given by a specific value that is independent of the quantity of material in a sample. When two materials are mixed, the value of the resulting intensive property of the mixture will vary

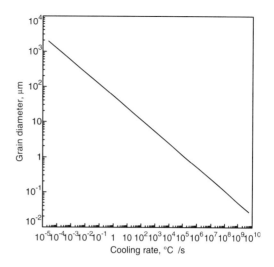

Fig. 3.35 Relationship between grain size and cooling rate. Source: After Jones [1982]

according to the proportions of the two materials present, but not necessarily in a linear fashion.

The application of this method involves obtaining experimental data on, for example, the liquidus temperature of a limited number of alloy compositions and using these values to produce an exactly fitting polynomial equation. This equation is unique, and from it values corresponding to intermediate compositions can be interpolated. A feature of the simplex lattice design method is that the mathematics are considerably simplified if the experimental data are obtained at a set of regular composition intervals.

Suitable composition grids for use with ternary phase diagrams are given in Fig. 3.36. The indexing system used is as follows: The intensive property value of a pure component, p, is denoted Y_p, that of a binary alloy comprising constituents p and q by Y_{pq}, that of a ternary alloy comprising constituents p, q, and r by Y_{pqr}, and so on. $Y_{p=50,q=50}$ (or conveniently abbreviated as $Y_{50,50}$) relates to a 50:50 binary alloy of constituents p and q, while $Y_{50,50,0}$ relates to the same alloy in a ternary phase diagram.

Different orders of polynomial equation can be used to relate the liquidus temperature $Y_{p=i,q=j,r=k}$ to a ternary alloy composition $X_{p=i,q=j,r=k}$, where i, j, and k, respectively, represent the percentages of the constituents p, q, and r in the alloy. The simplest modeling equation that is applicable to phase diagrams is a quadratic polynomial, which is able to model either one maximum or one minimum in a surface and is written out for a ternary system in

Principles of Soldering and Brazing

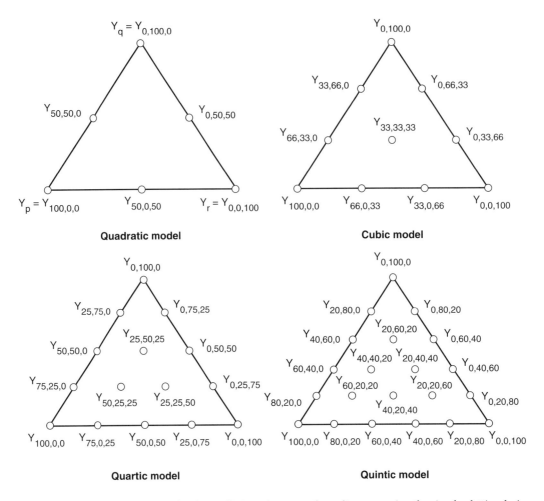

Fig. 3.36 Data point compositions for the prediction of ternary phase diagrams using the simplex lattice design method

Table 3.3 Quadratic polynomial equation for interpolation between experimental data points in ternary systems using the simplex lattice design method (as derived by Scheffe [1958])

Y is the intensive property (here, liquidus temperature) at the composition X_{pqr}.

The expression relating liquidus temperature Y_{pqr} of alloy X_{pqr}, in which X1, X2, and X3 are the percentages of components p, q, and r, is:

$$Y_{pqr} = Y_{100,0,0} \cdot X1 + Y_{0,100,0} \cdot X2 + Y_{0,0,100} \cdot X3 + (4 \cdot Y_{50,50,0} - 2 \cdot Y_{100,0,0} - 2 \cdot Y_{0,100,0}) \cdot X1 \cdot X2$$
$$+ (4 \cdot Y_{50,0,50} - 2 \cdot Y_{100,0,0} - 2 \cdot Y_{0,0,100}) \cdot X1 \cdot X3 + (4 \cdot Y_{0,50,50} - 2 \cdot Y_{0,100,0} - 2 \cdot Y3_{0,0,100}) \cdot X2 \cdot X3$$

where $Y_{100,0,0}$ is the liquidus of pure p, $Y_{50,50,0}$ is the liquidus of an alloy of 50% p and 50% q, and so on.

This expression should be read in conjunction with Fig. 3.36.

Table 3.3. For the general polynomial equations suitable for any number of components and order of polynomial, reference should be made to the paper by Scheffe [1958]. As with any curve-fitting

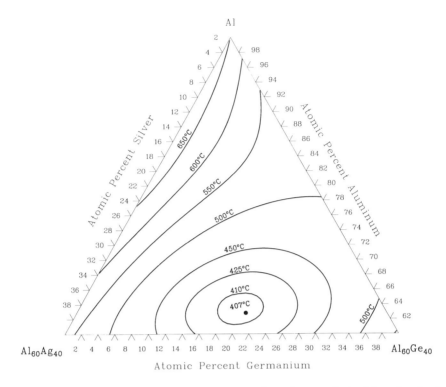

Fig. 3.37 Liquidus surface of the silver-aluminum-germanium phase diagram obtained from the simplex lattice model. Shown in atomic percent

technique, higher-order polynomial equations can reproduce more and sharper inflexions in the liquidus surface, but they require additional empirical data to define their coefficients.

The main limitation of the simplex lattice design method is that the prediction process is smooth and is thus unable to model accurately sharp discontinuities in slope, such as the floor of eutectic valleys, and similar features that appear in complex phase diagrams. Nevertheless, the location of features such as minima corresponding to eutectic points and the general topography of phase boundaries can be determined with sufficient accuracy to facilitate the design of most soldering and brazing processes.

Returning to the example, the region of the silver-aluminum-germanium phase diagram of interest contains only a single minimum, that of the eutectic point, so a quadratic simplex lattice model could be used. However, by using the procedure described below, for only a limited amount of additional work a quartic model can be used, which will permit closer modeling of the contours of the liquidus surface. A considerable reduction in the number of alloy compositions that need to be examined, using thermal analysis, in order to determine the liquidus surface is made possible by fixing the simplex lattice point Y_q coincident with the aluminum corner of the ternary phase diagram. Then, the liquidus values required of nine of the alloy compositions can be obtained from published data on the silver-aluminum and aluminum-germanium binary phase diagrams, as these bound two sides of the composition grid. Rather than attempting to model the liquidus surface of the entire ternary phase diagram, the simplex lattice grid was restricted to a minimum aluminum concentration of 60 at.%. Not only is this the region of specific interest, but the accuracy of the model is improved as the interpolation between data points is reduced. Experimental data are then required only for six ternary compositions.

The liquidus surface of the aluminum-rich portion of the silver-aluminum-germanium system calculated by this method from the experimental results is reproduced in Fig. 3.37. The model indicates that the liquidus surface contains a minimum of 406.8 °C (764.2 °F) at the composition 17.0Ag-62.4Al-20.6Ge (at.%). The predicted temperature of the eutectic reaction agrees closely

Principles of Soldering and Brazing

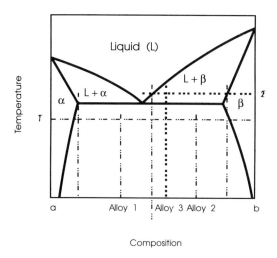

Fig. 3.38 Application of a data reductive method for computing the composition of the liquid solid phases participating in a eutectic equilibrium

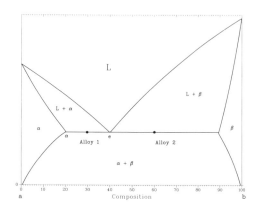

Alloy 1 = 70a + 30b
Alloy 2 = 40a + 60b
Solid solution of a, α = 80a + 20b = A(a,b)
Solid solution of b, β = 10a + 90b = B(a,b)
Eutectic, e = 60a + 40b at Te

Immediately below Te: alloy 1 contains 85% α, 15% β
Immediately below Te: alloy 2 contains 43% α, 57% β

Immediately above Te: alloy 1 contains 50% α, 50% e
Immediately above Te: alloy 2 contains 40% β, 60% e

Fig. 3.39 Data used to calculate the eutectic point in a binary system derived by quantitative image analysis

with the thermal analysis measurements, which showed that the true ternary eutectic transformation temperature is 414.5 ± 1 °C (778.1 ± 2 °F). The composition of the eutectic point is similar to that obtained from the equal liquidus slope method, in that the ratio of aluminum to germanium is essentially the same, but the true eutectic probably contains a higher proportion of silver. Identification of the exact composition of the eutectic reaction was achieved subsequently using the more accurate method of quantitative metallography.

3.3.6 Quantitative Metallography

The methods of equal liquidus slope and simplex lattice used above to predict the composition of the ternary eutectic point in the silver-aluminum-germanium system are, at best, capable only of providing an approximate value. Furthermore, because all of the alloys examined using thermal analysis had a measurable melting range, the ternary eutectic composition could not be verified from the liquidus data. It is thus necessary to resort to an independent method to establish the true ternary eutectic composition.

The aggregate composition of an alloy can be computed using a procedure described by Aldinger [1962]. For an N-component system, it is sufficient to carry out metallographic measurements on $N + 1$ samples. Ideally, these should be evenly distributed over the composition field under consideration, as this improves the reliability of the result. N of these samples can be a selection of the alloys used previously for thermal analysis, reprepared by quench cooling from above the liquidus temperature, followed by a heat treatment just below the solidus temperature. The elemental compositions of these alloys must be known.

The procedure involves using quantitative metallography to measure the relative proportions of the phases present in the N samples. This is readily achieved using an image analysis system, an apparatus that is described more fully in section 6.4.1. The proportions of the constituent phases can then be calculated by solving the simultaneous equations that describe the relationships among these variables. These equations are obtained from the lever rule.

The first step is to calculate the composition of the terminal solid phases that form on solidification, assuming that equilibrium is reached. For the binary eutectic system given in Fig. 3.38, at temperature T_1:

Table 3.4 Summary of the procedure used to establish the ternary eutectic point in the silver-aluminum-germanium ternary system and the results obtained

Step	Method	Temperature °C	Temperature °F	Composition, at.%
1	Cooling curve	410	770	...
2	Equal liquidus slope	...	...	5.1Ag-68.7-1Al-26.2Ge
3	Thermal analysis	414.5	778.1	...
4	Simplex lattice design	406.8	764.2	16.0Ag-62.4Al-20.6Ge
5	Quantitative metallography	...	...	16.5Ag-62.0Al-21.5Ge
Preferred values		414.5	778.1	16.5Ag-63.0Al-20.5Ge
Published data: Schmidt and Weiss [1978]		418	784	17.2Ag-61.8Al-21.0Ge

Alloy 1: $(u\% a + v\% b)$
$= m\% A(a,b) + n\% B(a,b)$
Alloy 2: $(w\% a + x\% b)$
$= p\% A(a,b) + q\% B(a,b)$

where a and b are the elemental constituents and A(a,b) and B(a,b) are the conjugate phases at temperature T_1.

The atomic percentages of the elements in the two alloys (u, v, w, and x) are known, while values for m, n, p, and q are obtained from the quantitative metallographic assessment. By substitution of these values in the above equations, values of A(a,b) and B(a,b) can be deduced.

For example, consider the binary eutectic system shown in Fig. 3.39. Immediately below Te, at temperature T_1, alloy 1 = (70a + 30b), for which the quantitative metallographic assessment showed that it contains 85% A(a,b) and 15% B(a,b). Likewise, alloy 2 = (40a + 60b) and contained the same phases in the proportion 43% A(a,b), 57% B(a,b). Therefore:

Alloy 1: 70a + 30b
$= 85A(a,b) + 15B(a,b)$
Alloy 2: 40a + 60b
$= 43A(a,b) + 57B(a,b)$

Substituting for B(a,b) in these simultaneous equations gives:

70a + 30b = 85A(a,b)
+ 15[40a + 60b − 43A(a,b)]

that is, 100 A(a,b) = 80a + 20b. Likewise, 100 B(a,b) = 10a + 90b. Because $N = 2$, alloy $N + 1$ is number three. The third alloy was prepared by remelting an appropriate alloy that had been used for thermal analysis, but this sample was cooled slowly from above the liquidus temperature to accentuate the primary phase. The assessment in this case was also different, in that measurement was made of the relative proportion of the primary phases in relation to those formed from the eutectic transformation. Again referring to the binary system in Fig. 3.38, at temperature T_2, immediately above Te:

Alloy 3: $(y\% a + z\% b)$
$= r\% L(a,b) + s\% B(a,b)$

Here y, z, and B(a,b) are known, while r and s are obtained from the quantitative metallographic assessment. The composition of the only unknown quantity in this equation, that of the liquid phase, L(a,b), which is the aggregate composition at the eutectic point, can therefore be deduced. Again referring to Fig. 3.39, and using the procedure given above, it was calculated that the composition of the eutectic point is 60a + 40b.

Application of this method to three alloy compositions in the silver-aluminum-germanium system indicated that the ternary point is at the composition 16.5Ag-62.0Al-21.5Ge (at.%). A comparison of this result with data empirically derived using the methods of equal liquidus slope and simplex lattice (see Table 3.4) indicates a close correspondence.

Metallographic examination of an alloy of the estimated ternary eutectic composition revealed a microstructure comprised mostly of the eutectic, but also a small proportion of primary germanium. Two further alloys were then prepared, each with 1 at.% less germanium and the balance made up of silver or aluminum. These were examined using thermal analysis and metallography. The alloy containing the boosted level of silver was also not wholly eutectiferous, whereas the other alloy with the composition 16.5Ag-63.0Al-20.5Ge (at.%) was found to have a solidus temperature of 414.5 °C (778.1 °F) and no detectable melting range. Examination of this alloy after slow cooling from above the liquidus temperature revealed a uniform triplex microstructure with no evidence of a primary phase. A photomicrograph of it is shown in Fig. 3.6. It is therefore reasonable to assume that this composition is close to the true ternary eutectic point and that the calculated isotherms of the liquidus surface generated by the simplex lattice model are sufficiently accurate to serve as the basis for the design of a brazing process.

The partial phase diagram that was determined in this exercise was used to develop a fluxless joining process for attaching germanium windows to aluminum alloy housings using silver interlayers. It involved coating the abutting surfaces of the components with silver to provide wettable surfaces in an inert furnace atmosphere. The original idea was to heat the assembly to about 575 °C (1065 °F), at which temperature the silver-aluminum eutectic phase would melt and fill the joint gap. However, the phase diagram shows that once the silver on the germanium window had dissolved, the molten filler would then react with the germanium of the window, substantially eroding it and generating an appreciable volume of liquid in the process.

The solution that was adopted was to minimize the thickness of the silver layers so as to enable aluminum to diffuse through it and react with germanium to form a molten silver-aluminum-germanium alloy at a much reduced temperature (about 425 °C, or 795 °F). By this means, the erosion of the germanium wafer was reduced to an acceptable level, with the added benefit of a lower brazing temperature. The germanium window was metallized with 0.3 μm (12 μin) of silver by a vapor phase deposition process, and the aluminum body was coated with a similar amount of silver by wet plating. Further details on the main coating methods and their relative merits can be found in the appendix to Chapter 4.

Appendix 3.1: Conversion Between Weight and Atomic Fraction of Constituents of Alloys

In an alloy containing N constituents, conversion from weight to atomic fraction of constituent n may be made using the equation:

$$\% \, n = \frac{P_n \cdot A_n}{\sum_{i=1}^{N} P_i \cdot A_i} \cdot 100$$

where P is the weight percentage of constituent n, A is the atomic weight of constituent n, and i refers to each constituent in turn.

Similarly, in an alloy containing N constituents, conversion from atomic to weight fraction of constituent n may be made using the equation:

$$\% \, n = \frac{P_n / A_n}{\sum_{i=1}^{N} P_i / A_i} \cdot 100$$

Appendix 3.2: Methods of Thermal Analysis

There are three principal types of thermal analysis, known as differential thermal analysis (DTA), differential scanning calorimetry (DSC), and calorimetric thermal analysis (CTA). All of these modes are capable of providing quantitative data relating to the heat capacity of a specimen as its temperature is varied. When the phases in an alloy change, there is a stepwise change in the measured heat capacity, due to the enthalpy of transformation, either exothermic or endothermic, which can be detected. Two other thermal analysis techniques, thermogravimetric analysis (TGA) and thermomechanical analysis (TMA) are also particularly useful for mapping phase boundaries.

Differential Thermal Analysis (DTA)

Differential thermal analysis involves heating or cooling a specimen and a thermally inert reference sample (e.g., alumina powder) at a linear rate, typically 1 °C/min (2 °F/min), through the temperature range of interest. A differential thermocouple measures the temperature difference between the reference material and the specimen, as shown schematically in Fig. 3.A1. In the absence of any phase changes, this difference is negligible. When a phase transformation takes place within the specimen, energy in the form of heat is either evolved or absorbed, and the temperature of the specimen will then differ from that of the reference sample. This enables the phase change that is taking place at that temperature to be detected. Typical heating curves for a thermally inert specimen and one that undergoes an isothermal phase

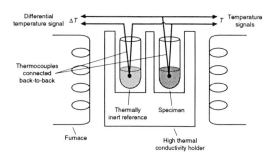

Fig. 3.A1 Differential thermal analysis. The specimen and reference sample are in good thermal contact with the holder. Any departure by the specimen from thermal equilibrium is detected by the differential thermocouple.

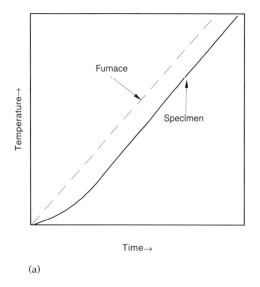
(a)

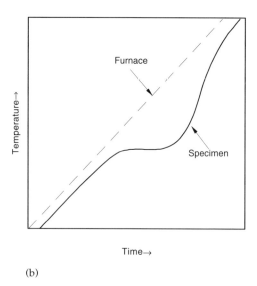
(b)

Fig. 3.A2 DTA time-temperature heating curves. (a) Thermally inert specimen. (b) Specimen that undergoes an isothermal phase transformation

transformation during DTA are shown in Fig. 3.A2.

Differential Scanning Calorimetry (DSC)

Differential scanning calorimetry similarly involves heating or cooling a specimen and a reference sample at a fixed rate, with a differential thermocouple used to detect phase changes. When a specimen undergoes a phase change, a miniature heating/cooling element immediately adjacent to the specimen is used to compensate for any thermal imbalance between the specimen and the thermally inert reference material, which is often the specimen holder itself. A schematic illustration of a representative DSC arrangement is given in Fig. 3.A3, and an energy profile for a specimen undergoing an endothermic phase transition followed by a small exothermic reaction is shown in Fig. 3.A4. The heater/cooler is calibrated in order that the latent heat associated with the phase transformation can be directly measured together with the onset temperature of the phase change.

Both DTA and DSC are prone to error when used to map out metallurgical phase diagrams. The cause of the problem is that the rate of heat transfer to or from the specimen is highly variable over the duration of an individual phase transformation. This arises because a metallurgical phase change involves diffusion processes and thus takes a finite time to occur, but the temperature of the furnace and reference sample meanwhile continue to change at a linear rate. Consequently, the specimen is forced through the phase change relatively quickly and thereafter is unlikely to be metallurgically homogeneous. Any attempt to continue an analysis with a heterogeneous alloy will lead to spurious thermal indications that are diffi-

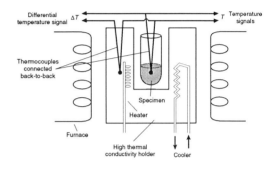

Fig. 3.A3 Differential scanning calorimetry. Any departure from thermal equilibrium by the specimen is detected by the differential thermocouple and compensated for by the calibrated heater/cooler.

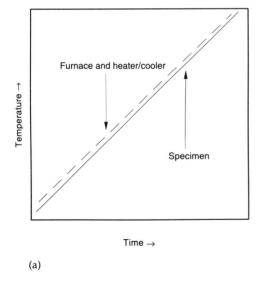

(a)

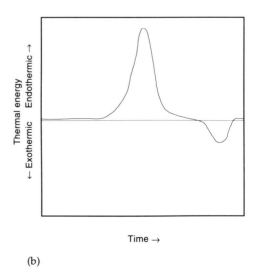

(b)

Fig. 3.A4 (a) DSC time/temperature heating curves. (b) Typical energy input from the heater/cooler. The heater/cooler compensates for any thermal imbalance between the specimen and its holder.

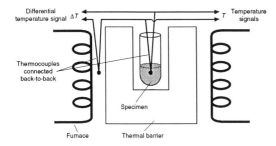

Fig. 3.A5 Calorimetric thermal analysis. The specimen is separated from the furnace by a barrier of moderate thermal conductivity and a differential thermocouple is used to maintain a constant temperature gradient across the barrier accross the barrier.

cult to distinguish from genuine phase transformations. Calorimetric thermal analysis offers the means for overcoming this problem.

Calorimetric Thermal Analysis (CTA)

Calorimetric thermal analysis is a quantitative method of thermal analysis that can preserve the homogeneity of specimens during measurement [Smith 1940; Hagel, Pound, and Mehl 1956; Humpston and Evans 1987]. This benefit is obtained at the expense of speed of analysis. A typical apparatus is illustrated schematically in Fig. 3.A5. A differential thermocouple is used to maintain a constant temperature gradient across a thermal barrier, of known thermal conductivity, which separates the specimen from the furnace. The rate of heat transfer to and from the specimen is thereby fixed in value. Over a temperature interval in which the specimen does not experience a change of phase or state, the temperature of the furnace and of the specimen will vary linearly with time. The rate of temperature change is governed by the heat flux through the barrier and the specific heat capacity of the specimen and its holder.

When a phase transformation occurs, the heat flux reaching the specimen also has to satisfy the energy requirements of the phase change. The rate of temperature change then alters to a new value that is dictated by the rate of material conversion from one phase to the next. Controlling the rate of the phase transformation in this way helps to minimize the tendency of the specimen to become heterogeneous and detect phase changes with only small temperature increments between them [Legendre et al. 1987]. Figure 3.A6 illustrates the response of the temperature sensor to heating and cooling excursions by an alloy through different types of phase transformations.

After calibration, a CTA system is able to provide a quantitative measure of the heat capacity of the specimen as a function of temperature and the enthalpy of any phase transformations that occur, as can DTA and DSC. An additional feature of CTA is that it can provide an indication of the slope of any phase boundaries traversed, thereby offering considerable scope for reducing the number of analyses required for the determination of a phase diagram. Its capability in this regard derives from the fact that the slope of a phase boundary is proportional to the rate at which the phase transformation occurs during a controlled thermal excursion. This slope therefore governs the measured rate of temperature change by the specimen during the phase transformation. The relationship is illustrated schematically in Fig. 3.A7. Knowledge of the slope of a phase boundary helps to establish other features, such as whether there is an invariant transformation associated with the phase change and whether there is phase continuity between two alloy compositions being analyzed.

Thermal analysis is suitable for determining the liquidus and solidus temperatures of most alloys. However, in some alloy systems the temperature of these phase transformations can vary substantially with only minor increments in alloy composition. One practical manifestation of phase boundaries that slope steeply is that the rate of conversion of liquid to solid, or vice versa, during thermal analysis is low. Therefore, accurate detection of the liquidus and solidus temperatures is difficult, and alternative techniques should be used. This point has been highlighted in a study in which the solidus phase boundary at the tin-rich end of the indium-tin binary system was redetermined [Evans and Prince 1983]. By using a more appropriate experimental technique, an apparent kink in the solidus phase boundary over a composition range was proved to be spurious.

Two techniques that can be recommended for analyzing steep liquidus and solidus phase boundaries, respectively, are discussed in the following sections.

Thermogravimetric Analysis (TGA)

Thermogravimetric analysis (TGA), as applied to the determination of steep liquidus boundaries, involves measuring the weight of a

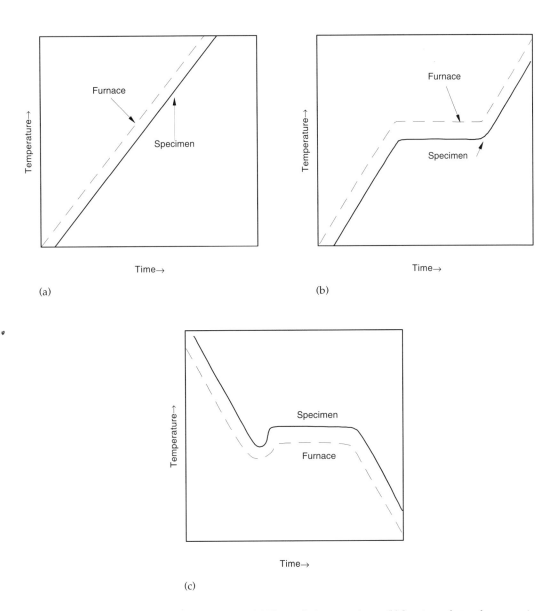

Fig. 3.A6 CTA time/temperature heating curves. (a) Thermally inert specimen. (b) Specimen that undergoes an isothermal phase transformation. (c) When recalescence occurs

thin platinum wire suspended in the melt of the alloy under investigation as this cools through the liquidus temperature. A schematic illustration of a typical measurement configuration is given in Fig. 3.A8. The primary solid phase will tend to crystallize on the wire by heterogeneous nucleation, thereby increasing its weight and so the onset of solidification can be detected. The sensor wire must be selected so that it is wetted, but not significantly dissolved, by the molten alloy.

Thermomechanical analysis (TMA)

Thermomechanical analysis (TMA) is highly effective in charting steep solidus boundaries, because it is extremely sensitive to the formation of a liquid phase in the specimen. The technique relies on the fact that the strength of a metal plummets to virtually zero when the first minute quantities of liquid form as it is heated to above its solidus temperature. Thus, the transition from solid to solid

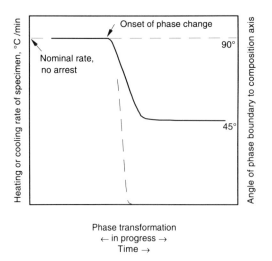

Fig. 3.A7 Rate of heating or cooling of a specimen while it is undergoing a phase change during CTA, as a qualitative indicator of the slope of the phase boundary being traversed

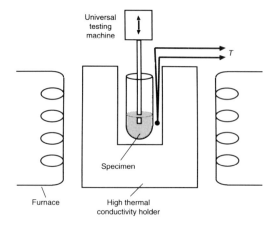

Fig. 3.A9 Thermomechanical analysis. The solidus temperature of an alloy may be detected by assessing its mechanical strength as a function of temperature. The strength decreases to essentially zero if there is liquid present.

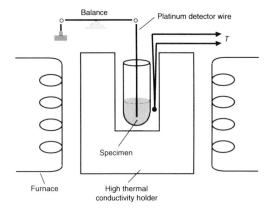

Fig. 3.A8 Thermogravimetric analysis. The liquidus temperature of an alloy can be detected if the primary phase precipitates on the platinum wire by heterogeneous nucleation.

plus liquid can be readily detected by observing the abrupt yield in a mechanically loaded specimen when this change occurs. The measurement can be performed with either a tensile or compressive stress applied to the specimen. A suitable apparatus is illustrated schematically in Fig. 3.A9.

References

Aldinger, F., 1962. A Method of Computing the Composition of Phases, *Metallography*, Vol 2, p 363-374

ASM International, 1993. Alloy phase diagram publications catalogue, Materials Park, Ohio

Baker, I. and George, E.P., 1992. Intermetallic Compounds: An Update, *Met. Mater.*, Vol 8 (No. 6), p 318-323

Birchenall, C.E., 1959. *Physical Metallurgy*, Metallurgy and Metallurgical Engineering Series, McGraw-Hill

Darken, L.S. and Gurry, R.W., 1953. *Physical Chemistry of Metals*, Metallurgy and Mechanical Engineering Series, McGraw-Hill

Dirnfeld, S.F. and Ramon, J.J., 1990. Microstructural Investigations of Copper-Tin Intermetallics and the Influence of Layer Thickness on Shear Strength, *Weld. J.*, Vol 69 (No. 10), p 373s-377s

Evans, D.S. and Denner, S.G., 1978. An Apparatus for the Determination of Solid/Liquid Metal Interactions Under Controlled Conditions, *Pract. Metallogr.*, Vol 15, p 486-493

Evans, D.S. and Prince, A., 1974. Thermal Analysis of the Ag-Au-Sn Ternary System in the Region from 20 to 100%Sn, *Met. Sci.*, Vol 8, p 286-290

Evans, D.S. and Prince, A., 1983. The In-Sn Phase Diagram, *Proc. Symp. Mater. Res. Soc.*, Vol 19, p 389-394

Evans, D.S. and Prince, A., 1985. The Critical Assessment of Ternary Alloy Phase Diagram Data, *J. Less Common Met.*, Vol 114, p 225-239

Felton, L.E. *et al.*, 1991. η-Cu_6Sn_5 Precipitates in Cu/Pb-Sn Solder Joints, *Scr. Metall. Mater.*, Vol 25 (No. 10), p 2329-2333

Fisher, H.J. and Phillips, A., 1954. Viscosity and Density of Liquid Lead-Tin and Antimony-Cadmium Alloys, *J. Inst. Met.*, Vol 11, p 1060-1070

Grivas, D. *et al.*, 1986. The Formation of Cu_3Sn Intermetallic on the Reaction of Cu With 95Pb-5Sn Solder, *J. Electron. Mater.*, Vol 15 (No. 6), p 355-359

Hagel, W.C., Pound, G.M., and Mehl, P., 1956. Calorimetric Study of the Austenite:Pearlite Transformation, *Acta Metall.*, Vol 4 (No. 1), p 37-46

Harding, W.B. and Pressly, H.B., 1963. Soldering to Gold Plating, *Proc. 50th Ann. Conf. Amer. Electroplaters Soc.*, p 90-106

Hume-Rothery, W., Christian, J.W., and Pearson, W.B., 1952. *Metallurgical Equilibrium Diagrams*, The Institute of Physics, London

Humpston, G. and Davies, B.L., 1984. Thermal Analysis of the AuSn-Pb Quasibinary Section, *Met. Sci.*, Vol 18 (No. 6), p 329-331

Humpston, G. and Davies, B.L., 1985. Constitution of the AuSn-Pb-Sn Partial Ternary System, *Mater. Sci. Technol.*, Vol 1 (No. 6), p 433-441

Humpston, G. and Evans, D.S., 1987. Constitution of the Au-AuSn-Pb Partial Ternary System, *Mater. Sci. Technol.*, Vol 3 (No. 8), p 621-627

Inoue, H., Kurihara, Y., and Hachino, H., 1986. Pb-Sn Solder for Die Bonding of Silicon Chips, *IEEE Trans. Components, Hybrids Manuf. Technol.*, Vol 9 (No. 2), p 190-194

Jones, H., 1982. *Rapid Solidification of Metals and Alloys*, Monograph No. 8, Institute of Metals, London, p 42

Kay, P.J. and Mackay, C.A., 1976. The Growth of Intermetallic Compounds on Common Base Materials Coated with Tin and Tin-Lead Alloys, *Trans. Inst. Met. Finish.*, Vol 54, p 68-74

Kay, P.J. and Mackay, C.A., 1979. Barrier Layers Against Diffusion, Paper 4, *Proc. 3rd Int. Brazing Soldering Conf.*, London

Legendre, B. *et al.*, 1987. Contribution Towards Clarification of Au-Sn Phase Diagram for Sn Contents <25at%, *Mater. Sci. Technol.*, Vol 3 (No. 10), p 875

Lugscheider, E. and Tillmann, W., 1991. Development of New Active Filler Metals for Joining Silicon-Carbide and -Nitride, Paper 11, *Proc. 6th Int. Brazing Soldering Conf.*, Stratford-upon-Avon

Massalski, T.B., 1991. *Binary Alloy Phase Diagrams*, ASM International (multivolume set), Ohio

Petzow, G. and Effenberg, G., 1988. *Ternary Alloys*, VCH, Germany (multivolume set)

Prince, A., Raynor, G.V., and Evans, D.S., 1990. *Phase Diagrams of Binary Gold Alloys*, The Institute of Metals, London

Raghavan, V., 1986. *Phase Diagrams of Ternary Iron Alloys*, ASM International (multivolume set), Ohio

Scheffe, H., 1958. Experiments With Mixtures, *J. Roy. Statist. Soc.*, Vol 20B, p 344-360

Schmidt, P.C. and Weiss, A., 1978, Schmelzverhalten von Ag-Al-Ge-Legierungen nahe der eutektischen Mischungen, *Metallwissenschaft Technik*, Vol 32 (No. 9), p 911-913

Shewmon, P., 1963. *Diffusion in Solids*, McGraw-Hill

Smith, C.S., 1940. A Simple Method of Thermal Analysis Permitting Quantitative Measurements of Specific and Latent Heats, *Trans. AIME*, Vol 137, p 236-245

Shull, R.D. and Joshi, A., (Ed.) 1993. *Thermal Analysis in Metallurgy*, TMS, Warrendale, Pennsylvania

Wild, R.N., 1968. "Effects of Gold on the Properties of Solders," Report No. 67-825-2157, IBM Federal Systems Division

Chapter 4

The Role of Materials in Defining Process Constraints

4.1 Introduction

The implementation of soldering and brazing in industrial fabrication has been briefly considered in section 1.4. More detailed attention will be given now to the materials and processing aspects and the manner in which these interrelate in the development of joining processes.

The starting point in the development of any practical joining process is a need to fabricate a unitary assembly or product from a set of components. Often the components are of different materials in order to maximize the performance of the product for a given cost. The product itself will have been designed to satisfy certain functional requirements and, for some items, these can be very diverse. The joint properties must also be consistent with the specified purpose and application of the product or system.

To identify and develop an appropriate joining process, it is essential to first give consideration to all possible aspects associated with the product. This will reveal an array of constraints, some of which are not immediately obvious, that govern the feasibility of the joining route. A flow chart outlining the decision-making steps and the constraints at each stage of the design and manufacture of assemblies containing soldered and/or brazed joints is given in Table 4.1. Those constraints that are directly linked to the product itself include the cost tolerance of the product to the joining process, the scale and throughput of production that will have to be satisfied, and the statutory regulations that apply. The operating environment must then be considered. Here, the stress condition and the chemical environment tend to be the critical parameters. Finally, the materials used in the components, when taken individually, will impose a limit on the maximum joining temperature that can be used, on the grounds of thermal degradation, and may restrict the atmosphere in which the joining process can be carried out. When the overall assembly is considered, any mismatch in thermal expansivity of the abutting components can force compromises with regard to the choice of materials and processes. All of these considerations will influence the design of the assembly to a greater or lesser extent.

It is worth noting that finite element analysis (FEA) and other computational methods are being increasingly used to model stress distributions in soldered and brazed assemblies. In FEA, a component or assembly is modeled in a geometrical manner in terms of a mesh of smaller units or elements, with the dimensions of each element scaling to a set of properties of interest. For example, in a two-dimensional model, the x-axis of the elements may represent a geometrical dimension of the item and the y-axis some other physical property, such as thermal expansion. Constraints are then applied to the mesh, with matrix algebra used to obtain a

Principles of Soldering and Brazing

Table 4.1 Materials systems approach to joining process development

Phase of development	Decision making	Boundary conditions
Definition of the product	Nature of the product and functional requirements	Scale of production Cost constraints Size and weight limits Statutory requirements
	Service conditions	Thermal environment Stress environment Chemical environment
Materials and process selection	Parent materials	Joining temperature Joining atmosphere Mismatch stress
	Filler alloys	Joining temperature Joining atmosphere Joint geometry Filler geometry and form
Process assessment: identification of critical materials problems	Determination of process viability	Metallurgical constraints
Identification and achievement of solutions	Mismatch → (Mechanical solutions): Interlayers, Multilayers, Graded structures, Compliant structures, Diffusion soldering/brazing Alloying (Wetting, Erosion, Phases) → (Metallurgical solutions): Surface processing, Wettable coatings, Barrier coatings, Active filler alloys, Fluxes, Diffusion soldering/brazing	
Prototyping and production	Process specified	Tolerance of process established

comprehensive solution. The large number of calculations that are needed requires a computer for this task, and there are now a number of commercial FEA software packages available to suit particular applications. An FEA prediction of the deformation that occurs in a ceramic-metal bond on cooling from the solidus temperature of the filler is given in Fig. 4.1. Despite the complexity of such modeling techniques, simplifying assumptions have to be made. When they are applied to soldering and brazing, it is normally assumed that the joints are uniform in their composition and physical properties and are infinitely thin, which can limit their usefulness in predicting responses in complex systems.

Having taken account of the more obvious constraints in the design of an assembly, a selection of filler alloys can be made, each of which will impose its own set of limiting conditions. Among the most important of these are the minimum practicable joining temperature (i.e., liquidus temperature of the filler alloy, with the addition of a margin to allow for possible temperature gradients across the joint), the geometries that the joint can assume

The Role of Materials in Defining Process Constraints

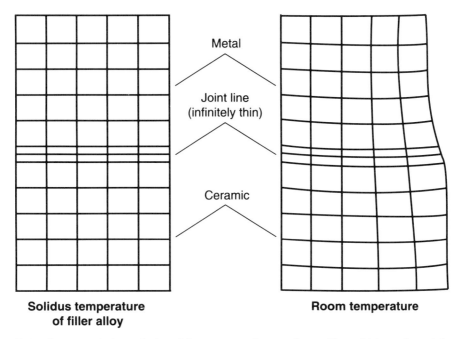

Fig. 4.1 Finite element analysis prediction of the geometry of a ceramic-metal brazed joint, at its periphery, at the solidus temperature of the filler alloy and on cooling to room temperature

and into which the filler can be fabricated, and the permissible joining atmosphere.

From the "shortlist" of filler alloys, it is usually possible to select at least one whose window of usable conditions is compatible with the other steps of manufacture. Up to this point, the selection procedure is largely a paper exercise, but the viability of the proposed joining solution needs to be established by practical trials. This is because a multiplicity of other features also enter the equation, such as wetting, erosion of the parent materials, and intermetallic phase formation within the joint. While any of these phenomena can radically affect the integrity of the product in service, much of this type of information is unavailable from the published literature and cannot be correctly surmised.

If problems are identified at this stage of the joining process development, it may be possible to obtain remedies by a number of avenues. Thus, in certain situations, catastrophic mismatch stresses may be overcome by modifying the stress distribution in the vicinity of the joint. Problems associated with the formation of deleterious intermetallic phases in the joint, lack of wetting, and, at the other extreme, excessive erosion of one or more of the parent materials can usually be circumvented by interposing a layer of a different metal between the filler and the parent material, as a coating on the joint surfaces, and thereby alter the metallurgical constitution of the joints. This changes the constituents that will alloy in the course of the joining operation and also their relative proportions. Fluxes and active filler alloys can be used to improve wetting and reduce void levels. Some of the possible remedies that may be brought to bear on these and other problems are detailed in the following sections.

If, on the other hand, no joining solution proves technically tractable, or if the solutions are not economically justifiable, then changes earlier in the decision-making chain are required. In extreme situations this may require drastic revision, perhaps even to the extent that other means of assembly have to be considered or that the functional requirements of the product must be relaxed.

4.2 *Metallurgical Constraints and Solutions*

In principle, most materials can be joined using filler alloys. However, when two different parent materials must be joined together, the available choice of fillers that are compatible with both is narrowed somewhat, especially when the constraints on the filler, processing conditions, and properties of the joints mentioned above are imposed. The problem is often made more acute by the fact that the joining processes tend to be left to a later stage of product design, which further reduces the available options. The joint should always be considered an integral part of the overall design.

Metallurgical incompatibility of materials and processing conditions will manifest itself through poor wetting, excessive erosion of the parent materials, and/or the formation of undesirable phases. Means for eliminating or suppressing these deleterious characteristics are described below.

4.2.1 Wetting of Metals

Restrictions applied to the choice of joining temperature and atmosphere, including the use of fluxes, can result in poor wetting of the component surfaces by the molten filler if the permissible process conditions are inadequate to ensure that the joint surfaces are sufficiently clean. Reactive filler metals (described in section 5.4) can often help to overcome this problem, but it is not always possible to use fillers containing active wetting agents, in which case alternative remedies are then necessary.

Poor wetting is a particularly serious problem when the parent materials are refractory metals. Their reactivity with oxygen, and in some cases with other elements in the atmosphere, and the stability of the reaction products on the surfaces of these materials cause poor wetting. Nonmetallic phases present at the surface of materials, such as graphite inclusions in cast iron and nonmetallic components in metal matrix composites, can also inhibit wetting.

These problems are not restricted to the parent metals, but can also encompass the filler metal. A prime example is provided by aluminum-bearing brazes when these need to be used without an appropriate fluxing agent. Foils and other preforms of these alloys tend to produce poor wetting and spreading over the joint surfaces. However, wetting can be improved considerably by making small additions of elements that lower the surface tension of the molten filler, as typically indicated for aluminum in Fig. 4.2, and that also help to destabilize the native surface oxide layer [Schultze and Schoer 1973]. The concentration of the individual additions is restricted to a maximum of about 0.3% in order to avoid perceptibly altering

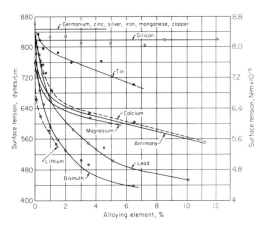

Fig. 4.2 Reduction in the surface tension of molten aluminum produced by various alloying additions. Source: After Korol'kov [1956]

the bulk metallurgical characteristics of the filler alloy.

If the joining atmosphere, which includes fluxes when these are used, cannot adequately clean and protect the component from oxidation during the process cycle, a favored solution is to apply a coating of a more noble metal to the precleaned joint surfaces. This coating may be sacrificial in that it is subsequently dissolved by the filler, which then wets the clean surfaces of the parent material. At the same time, its composition and thickness should be such that the dissolved coating does not give rise to brittle phases by reaction with the filler and parent materials.

Copper, silver, and gold are the principal elements used as wettable metallizations because of their nobility, metallurgical compatibility with most filler alloys, and ease of deposition. Tin is also widely used in conjunction with soldering processes because tin oxide is readily displaced by an advancing wave of the molten solder, which itself often contains tin. The benefits of tin are that it is relatively inexpensive and that it presents minimal risk of unfavorably altering the constitution of the solder.

Wettable metallizations are usually applied by either wet plating or vapor deposition techniques. Tin coatings are often applied by dipping methods. The choice of deposition method is determined by various factors, such as the size of the component, its geometry, the required scale of production, the capital and running costs of the deposition equipment, and the thickness of the metallization. Wet plating, chemical vapor deposition, and sputtering processes tend to be the methods favored for applying metallizations, as explained in appendix 4.1.

The metallization needs to be sufficiently thick for it to protect the underlying material from corrosion and to maintain wettability of the component over a reasonable storage period. The uniformity and density of sputter-deposited coatings of gold afford such protection at a thickness of typically 0.3 μm (12 μin.), whereas electroplated or dip coatings of this metal, as prepared conventionally, usually need to be more than an order of magnitude thicker to provide the same shelf life [Humpston and Jacobson 1990]. The alloying behavior of the filler with the coating metal may establish a maximum limit to the thickness of the metallization beyond which joint embrittlement ensues, as explained in sections 3.2.2 and 7.3.2.3. A relatively thick gold electroplating applied to surfaces to confer protection during storage before joining must be wicked off with solder immediately prior to the joining operation. The excess gold can be reclaimed from the discarded solder.

Metallizations that promote wetting can also be used to advantage to confine the molten filler to a specific area on the surfaces of components. By selectively applying the coating to prescribed areas, the filler spread can be restricted accordingly.

If the wettable coating is not adherent as applied, it can often be secured on the surface of a metal component by a subsequent heat treatment in a nonoxidizing atmosphere. This process can promote interdiffusion between the metal and the coating, resulting in a graded interface akin to that obtained using a carburizing (carbon-enriched) or sheradizing (zinc-enriched) surface treatment.

4.2.2 Wetting of Nonmetals

Nonmetals—namely ceramics, glasses, and plastics—are not wetted by most filler metals, even when their surfaces are scrupulously clean. This is because they are chemically very stable, with their atoms strongly bound to one another. Therefore, these materials will not react with and be wetted by the filler unless the latter contains an active element that can attach itself to the anionic species of the nonmetallic material, which in a compound or complex is normally either oxygen, carbon, nitrogen, or a halide element. Titanium is often used as an active constituent of brazes, as discussed in section 5.4. Active metal joining is only effective if sufficiently high temperatures, typically above 800 °C (1470 °F), can be used for the joining operation so that the active ingredient is able to react with the nonmetal. It is observed that the active element concentrates close to the interface with the nonmetallic base material.

Where this approach is incompatible with the joining process, a very similar result can be achieved via a different route. This involves coating the joint surfaces with a metal that will bond strongly to the underlying nonmetal and that is at once wetted by the filler while not entirely dissolving in the process. Any erosion through to the original component surface would result in dewetting, as exemplified by the dissolution of chromium metallizations by bismuth-containing solders observed at temperatures above 261 °C (502 °F). Bismuth and chromium react to form a low-melting-point eutectic alloy at this temperature [Humpston and Jacobson 1990].

These coatings can be applied to the nonmetallic components by methods similar to those used on metals—namely, physical vapor deposition, chemical vapor deposition, and wet plating. Also widely used are fired-on glass frits loaded with metal powder or flake, often referred to in the literature as thick-film metallization techniques. All of the methods mentioned here are described briefly in appendix 4.1.

The sputter-deposition route tends to be favored wherever high-quality thin coatings are required. It is versatile and permits a wide range of metal and alloy coatings to be applied. When metallizing a nonmetallic material, it is usual to deposit more than one layer onto the joint surface. It is essential that the layer in direct contact with the nonmetal be an active metal—usually nickel, chromium, or titanium, for the reasons explained above. The choice is largely dictated by the solubility of this metal in the filler alloy, which must be low but finite. These active elements can bond strongly to the nonmetallic materials, provided that the surfaces are clean. According to the simple picture, it might be expected that the adhesion of a metal X to an oxide is given by the thermodynamic Gibbs free energy of formation of the oxide of metal X: the more the value of the free energy is reduced when the metal reacts to form an oxide, the more active is metal X in forming an oxide or in bonding to the surface of a different oxide. That this model is an oversimplification of reality is clear from the observation that chromium forms an adherent bond to silica, even though the Gibbs free energy of formation of silica is larger in magnitude than that of chrome oxide. Evidently, the Gibbs free energy is only one factor. It is certainly necessary to also take into account the mixing energy arising from the dissolution of the metal liberated in the exchange reaction in the more active metal. This example merely highlights the complexity of the subject of metal to nonmetal bonding and accounts for the considerable literature that it has produced [Peteves 1989]. Nevertheless, the end result is virtually identical to using a filler containing an active element. A sputtering process usually incorporates a capability for cleaning surfaces prior to deposition, by being operated in a reverse-bias mode, so that the atoms of the inert carrier gas in the deposition chamber are made to bombard the surface and erode films of contaminant.

The high reactivity of titanium and the other bonding metals toward oxygen and nitrogen in air means that they rapidly lose their wettability even after a brief exposure to the atmosphere. To overcome this problem, a more noble metallization that offers good wettability to filler metals must be deposited over the active metal immediately and without opening the deposition chamber to air. A metallization system that has been found to be effective for many soldering applications is one that comprises a 0.1 μm (4 μin.) thick layer of an active metal and a 0.3 μm (12 μin.) overlay of a wettable metallization, such as gold, both applied by sputtering. At these thicknesses, the shelf life for joining is usually adequate (typically more than 6 months), while the reactive layer provides a barrier between the filler metal and the underlying nonmetal, which ensures that dewetting does not occur [Humpston and Jacobson 1990].

An alternative method of coating a nonmetallic material with an adherent metal coating is to use a two-stage joining process in which the first step is to allow an activated braze to wet and spread over the component surfaces. The braze-"tinned" surface is then usually mechanically dressed to present a flat metal surface for the actual joining step, which is performed at a temperature below the solidus point of the activated braze and may require an aggressive flux. This method is therefore a hybrid of the previous two. Owing to the number of elements that are likely to be present in the joint, the resulting alloy constitution can be somewhat complex unless the process is designed such that the reactive braze and the joining filler alloy have several constituents in common.

A somewhat different approach is to fire on a relatively thick (typically 1 to 10 μm, or 40 to 400 μin.) metal coating. This type of metallization process tends to be used only on glass and ceramic materials because of the high temperatures involved [Bever 1986]. Further details of the range of commercial processes that are available are given in the appendix to this chapter. Reference is often made in the technical literature to the active hydride process. It involves applying a metal hydride, usually of titanium or zirconium, in the form of a paste to the component surface. On heating, the hydride thermally decomposes, liberating hydrogen to leave a metal film on the component surface [Pearshall 1949]. This method has lost favor in recent years to the alternatives, as the quality of the resulting metallization tends to be variable and is reflected in the mechanical properties of the joints [Mizuhara and Mally 1985]. The unpredictable nature of the metallization quality is due to its high sensitivity to variations in the atmosphere during the firing stage. Oxygen and water vapor contents, in particular, affect the extent to which the highly reactive metal oxidizes and hence the ease of wetting by the molten filler.

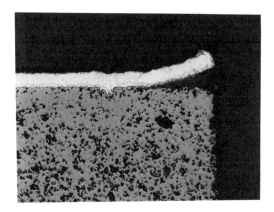

Fig. 4.3 A porous ceramic material metallized with a thick gold electroplate. The residual stress in the metallization has resulted in a peel failure through the near-surface layer of the ceramic.

An important point is that metallizations are often put down in a stressed condition. This is an inherent feature of most deposition processes and can be a source of critical weakness. An example of the manifestation of excessive stress in a metallization layer is shown in Fig. 4.3. Here, a silver layer 20 μm (800 μin.) thick has cracked away from the ceramic material through such stress. The stress concentration at the interface between the component and the metallization can be controlled by limiting the thickness of the metallization layers and by using conventional stress-relief heat treatments. However, when brazing metallized components, thicker layers are generally necessary. This is because the high process temperature means that a significant volume fraction of the metallization will react with the substrate and also be dissolved by the molten braze. Consequently, whereas a barrier layer 0.1 µm (4 µin.) thick is often adequate for a soldering process, a thickness of 1 µm (40 µin.) is more appropriate when using a braze [Hammond, David, and Santella 1988].

Molten filler alloys do not always readily wet metallic materials that have nonmetallic phases present at the surface. Two well-known examples are the graphitic phases in cast irons and lead globules in free-machining steels. The new generation of metal-matrix composites, which rely on nonmetallic reinforcement phases, present similar problems. As an alternative to the application of barrier metallizations or the use of activated filler metals it is often possible to use a chemical treatment to remove the offending minority phase from the surface of the component. A pretreatment of this type is often used when gray cast iron is to be brazed: components are "cleaned" by immersing them for several minutes in a salt bath at 400 °C (750 °F) that contains a mixture of sodium and potassium nitrides. The chemical treatment completely removes any exposed graphitic phase from the surface and leaves an iron-rich surface that is readily wetted by many filler alloys [Totty 1979].

4.2.3 Erosion

When a molten filler wets the surface of a parent material, alloying occurs, leading to a degree of dissolution of the parent material that commences at the joint interface. Dissolution of the parent materials occurs because the materials system encompassing the joint is not in thermodynamic equilibrium. This provides the driving force for wetting and spreading. The maximum solubility of the parent materials in the molten filler can be predicted by reference to the appropriate phase diagram, as explained for binary alloys in section 3.2.1. In summary, extensive erosion is likely where the liquidus surface on the phase diagram between the filler metal composition and that of the parent material has a shallow slope and where the alloying depresses the melting point of the filler in the joint. If the phase diagram exhibits either of these features, then it is only possible to limit erosion by lowering the process temperature, shortening the heating cycle, and/or restricting the volume of molten filler metal. Such changes must obviously not compromise the integrity of the joints by, for example, reducing the effective fluidity of the molten filler alloy, which in turn will impede wetting and spreading.

Erosion can be reduced somewhat if intermetallic phases form along the joint interface so as to attenuate the rate at which solid material is transported from the components into the molten filler. This approach is only successful if the intermetallic phases formed are ductile, as exemplified by the case of tin-base solders used in conjunction with silver metallizations described in Chapter 3.2.1; otherwise, joint embrittlement results. In other cases it is possible to protect the components against erosion by interposing a metallization, which will act as a barrier. An example is the application of a layer of nickel, typically 2 to 5 µm (80 to 200 µin.) thick, on components to prevent reaction with lead-tin solder.

The extent to which dissolution occurs in a given heating cycle will also depend on the kinetics of reaction. The rate of dissolution of common engineering materials and metallizations in eutectic tin-lead solder is given in Fig. 3.18. These data show that decreasing both the process cycle time and temperature can be used to reduce erosion, although not always to acceptable levels.

The propensity for dissolution of the parent metal by a filler alloy can be reduced by preloading the filler with this metal so that it is saturated on becoming molten and will not dissolve any further quantity. This approach is used for soldering to gallium arsenide devices using the Au-20Sn solder modified by the addition of 3.4% Ga. The addition reduces the solubility of gallium in the solder by half, with consequential benefits to the process yield [Humpston and Jacobson 1989].

Minor changes to the composition of the filler alloy are capable of influencing the rate at which dissolution occurs. This is exemplified by the effect of small additions of silver, typically 2%, on the rate of dissolution of silver by tin-lead solders, as can be seen from Fig. 2.31. The extent of the reduction in dissolution rate cannot be predicted from the silver-lead-tin phase diagram and must be determined empirically. Changes such as these often do not significantly alter other properties of the filler, or those of the resulting joints, and can therefore be readily implemented.

Erosion is usually most detrimental to joint quality when it occurs erratically over the joint surfaces. As the total volume of parent material that will ultimately be dissolved is fixed at the joining temperature, for a given volume of filler, the average depth of erosion is least when the wetted area is a maximum. Uniform wetting can be achieved by paying close attention to the cleanliness of the joint and filler surfaces and also to the environment in which the joining operation is carried out; in particular, thermal gradients should be minimized.

4.2.4 Phase Formation

Alloying between a molten filler and parent materials more often than not results in the formation of intermetallic compounds. This is more true of solders than brazes, because the major constituents of most solders are elements with low crystallographic symmetry and hence do not readily form solid solutions with engineering materials, which tend to be based on simple body-centered cubic (bcc), face-centered cubic (fcc), or hexagonal close-packed (hcp) crystal structures. The distribution, morphology, and proportion of these phases depend on several factors, as explained in section 3.2.

Because intermetallic compounds generally possess a higher elastic modulus than many of the filler alloys themselves, well-dispersed intermetallic phases are often beneficial to the stress-bearing capability of the filler metal. This effect is shown for gold additions to the silver-tin eutectic solder in Fig. 3.26.

By and large, agglomerations of intermetallic compounds are deleterious to the mechanical properties of joints. This is particularly true where the compound has a low elastic modulus and forms as a continuous interfacial layer between the component surfaces and the filler alloy. It is sometimes possible to restrict the coarsening of these phases by carrying out the joining operation under conditions that are unfavorable for their growth—namely, restricting the heating cycle duration and peak temperature. Alternatively, minor additions can be made to the filler alloy that will break up agglomerations of intermetallic phases present in the joint. Barrier metallizations applied to joint surfaces can be used to prevent alloying with the parent materials and the consequential formation of undesirable phases.

Some filler alloys are themselves hard because they contain intermetallic phases, but yet form high-strength joints with certain parent materials provided that the intermetallic phases are destabilized or finely dispersed within the joint. Many nickel-bearing brazes fall into this category. Further details are given in section 2.2.1.10.

4.3 Mechanical Constraints and Solutions

In an assembly composed of heterogeneous materials, there is usually a mismatch between the thermal expansion coefficients of the abutting components. This manifests itself as stress on cooling from the solidus temperature of the filler metal and is a maximum at the lowest temperature that the assembly experiences. Materials with a relatively low elastic modulus can accommodate strain and will tend to deform under the influence of this stress, while brittle materials, notably glasses and ceramics, will have a tendency to fracture, particularly if the stress distribution places the component in tension. Even if a heterogeneous assembly survives the joining operation, the stresses arising from the thermal expansion mismatch can cause it to fail by fatigue during subsequent thermal cycling in the field.

The stress in the region of the joint between two isotropic materials designated 1 and 2 with differing thermal expansivities that develops on cooling from the freezing temperature of the filler metal can be approximated by the following equation [Timoshenko 1925]:

$$\text{Stress} = \left(\frac{E_1 \cdot E_2}{E_1 + E_2}\right)(X_1 - X_2)(T_f - T_s)$$

where E is the component modulus of elasticity of 1 and 2, X is the coefficient of thermal expansion of 1 and 2, T_f is the freezing point (solidus temperature) of the filler alloy, and T_s is the temperature of the assembly corresponding to the stress. In the derivation of this equation, it is assumed that the materials are only deformed within their elastic limits and that the joint is infinitely thin. Despite these simplifying assumptions and the inaccuracies that they introduce, this expression is useful in providing an indication of whether the stress due to thermal expansion mismatch is close to or exceeds the failure stress of either of the abutting materials, that is, whether failure of the assembly is likely to occur. Some worked examples are described in Haug, Schaefer, and Schamm [1989].

In an assembly with a planar joint between two elastic but different materials 1 and 2, the magnitude of the bow distortion in one dimension can be estimated from the physical properties of the materials using a simplified model [Timoshenko 1925]. With reference to Fig. 4.4,

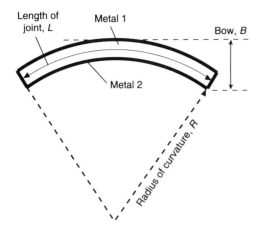

Fig. 4.4 Bow distortion of a bimetallic strip

$B = (L^2/8)R$

$$R = \frac{(A_1 + A_2)[3(1 + M)^2 + (1 + MN)(M^2 + 1/MN)]}{6(X_1 - X_2)(T_f - T_s)(1 + M)^2}$$

(Eq. 4.1)

where $M = A_1/A_2$ and $N = E_1/E_2$, B is bow distortion, L is the length of the joint, R is the radius of curvature, A_1 and A_2 are the thicknesses of components 1 and 2, and T_s is the temperature of the assembly corresponding to the bow distortion. Equation 4.1 assumes that the joint in the heterogeneous assembly is infinitely thin and totally inelastic.

From Eq 4.1, it can be seen that it is possible to effect some reduction in expansion mismatch stress, i.e. minimize R, by applying one or more of the following measures:

- Decrease the solidus temperature of the filler alloy, T_f.
- Increase the minimum service temperature of the assembly, T_s.
- Reduce the dimensions of the joint area, i.e. L.
- Change one or both materials to minimize the mismatch in thermal expansivity, $|E_1 - E_2|$.

Occasionally, it may be possible to implement one or more of these changes, but they are likely to conflict with other processing constraints if not with the intended functional requirements of the assembly. Alternative solutions must therefore be sought.

In practice, it is usually possible to obtain a small reduction in the distortion of a bowed heterogenous assembly by heat treating it at a temperature below the solidus temperature of the filler alloy to enable stress relaxation and creep to occur. However, there is a limit to the reduction in distortion that can be obtained by this means, typically 10% or less. This stems from the fact that stress-reducing mechanisms are diffusion related and become more effective as the temperature is raised toward the melting point of the joint, while mismatch stress increases as the temperature of the assembly is reduced below that at which stress relief is effective. Thus, these two tendencies act in opposition, and the optimum condition for reducing the distortion of bonded components is a compromise between the two.

Some further improvement in the residual stress level can be obtained by using a more compliant filler, especially when the joint is reasonably wide (>25 μm, or 1000 μin.). Solders, almost without exception, have lower elastic moduli than brazes at a given temperature, which are related to their lower melting points, but there are only minor differences between individual alloys within each of these categories. Therefore, the only possibility for obtaining a joint with improved compliance, short of redesigning the assembly, is to replace the braze with a solder. The benefit of such a change is largely offset by the poorer fatigue and creep resistance of solders, which means that soldered joints are less robust to manipulation than brazed joints.

Wide joint gaps (>500 μm, or 20,000 μin.) can sometimes be used to minimize the effects of expansion mismatch between two components. The filler alloy must have high viscosity in order to fill such wide joints. This is achieved by either using a filler metal with a wide melting range and performing the joining process at below the liquidus temperature, so that the alloy is not fully molten, or by mixing in metal powder. Spacers are required to control the joint gap. Wide joints can also be achieved by inserting porous shims, as described in section 2.2.1.10. One particular merit of wide joints to ceramic components is that they obviate the need to closely machine the mating surfaces of the components, which tends to be costly and can weaken the material by creating subsurface cracks. However, because the joint is wide, the mechanical properties of the joint are essentially those of the bulk filler.

A variety of mechanical schemes are available to assist in overcoming the problem of thermal ex-

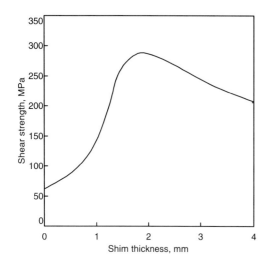

Fig. 4.5 Influence of shim thickness on the shear strength of an alumina-to-steel joint made using an Ag-Cu-Ti braze and a nickel interlayer. The joint length is 15 mm (0.6 in.). Source: After Miyazawa *et al.* [1989]

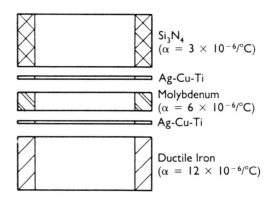

Fig. 4.6 Use of a plate of intermediate thermal expansivity to reduce the stress due to thermal expansion mismatch in a brazed assembly

pansion mismatch. Several approaches that have proved successful are described below.

4.3.1 Interlayers

One route toward reducing the mismatch stress concentration that develops in soldered and, in particular, in brazed joints involves a redesign of the joint to accommodate one or more interlayers. Two basic configurations are described in the literature.

In the first approach, a compliant interlayer is inserted that will yield when the joint is placed under stress and thereby reduce the forces acting on the components. Optimum stress reduction is normally achieved when the thickness-to-length ratio of the interlayer is in a certain range, which is determined by the combination of materials used and the dimensions of the joint [Miyazawa *et al.* 1989; Xian and Si 1992]. This is illustrated in Fig. 4.5 for a joint between alumina and steel employing a silver-copper-titanium braze and a nickel interlayer. For such a joint with a length of 15 mm (0.6 in.), the interlayer should be between about 1.5 and 3 mm (0.06 and 0.12 in.) thick. If the interlayer is much thinner than the prescribed minimum thickness, it is unable to absorb a significant proportion of the applied stress, whereas if it falls outside the upper limit, the interlayer will not yield to any extent.

Interlayers are used to relieve mismatch stresses more frequently in brazed joints than in soldered assemblies for the following reasons. First, the modulus of compliant metals that are most effective in accommodating stress is too close to that of many solders to provide much relief, because the solder will tend to yield in preference to the interlayer. Second, solders tend to form hard, interfacial phases with most engineering metals and alloys, which will confer a high modulus to the interlayer and actually exacerbate the situation. By contrast, common brazing alloys are more likely to form solid solutions, which are more ductile. High-purity copper is generally used for interlayers in brazed joints because it combines a low elastic modulus with good wetting characteristics and is also inexpensive to fabricate to the desired geometry.

An alternative approach is to redistribute the stresses across a much wider zone so that they are within tolerable levels everywhere in the assembly. One method of achieving this graduated redistribution of stress is to insert into the joint one or more thick shims or plates that have thermal expansion coefficients which are intermediate between those of the abutting components. The plates must be sufficiently thick (generally not less than 5 mm, or 0.2 in.) so that they are not significantly distorted by the imposed stresses. An assembly containing a single plate with an intermediate thermal expansivity is shown in Fig. 4.6.

This approach is particularly suitable where there is a need to join metals to ceramics and other ceramic-like nonmetals. If the intermediate plate is selected to have a thermal expansion coefficient that is close to that of the nonmetal, then it is possible to transfer the major proportion of the stress

The Role of Materials in Defining Process Constraints

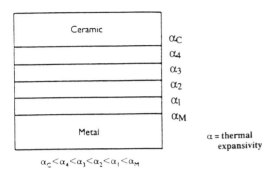

Fig. 4.7 A graded series of plates (1 to 4) designed to reduce the mismatch stress between ceramics and metals to an acceptable level. α, thermal expansivity

Fig. 4.8 Monolithic plates of graded composition, varying from essentially pure copper to about 95% tungsten. The thermal expansivities of the two surfaces differ by about $16 \times 10^{-6}/°C$ ($29 \times 10^{-6}/°F$).

to the more robust metallic part of the assembly. Where the two components have greatly different thermal expansivities, it may be necessary to use a graduated series of plates to reduce the mismatch stresses in each joint to an acceptable level. A schematic illustration of an assembly of this type is shown in Fig. 4.7. Filler metals with low and controlled expansivities are discussed in section 4.4.3.2.

A monolithic plate of graded composition and thermal expansivity can be used in place of a series of discrete homogeneous plates. Typically, these may be prepared from powder compacts. Copper/tungsten components of this type are made by infiltrating a tungsten powder compact with molten copper. By adjusting the relative proportions of copper and tungsten, the properties of the component can vary from those of essentially pure copper to about 95% tungsten. This enables one side of the component to be made tungsten-rich, with a low expansion coefficient, and the other side copper-rich, with a much higher expansion coefficient. Because there are no abrupt interfaces in such a component, it can survive thermal cycling over wide ranges of temperature almost indefinitely, without suffering distortion through creep or fatigue fracture. Graded copper-tungsten plates are shown in Fig. 4.8.

If one or both of the components are highly brittle and vulnerable to fracture under a tensile or shear stress, it is often the practice to provide reinforcement by attaching it to a mechanically robust metal plate of similar expansivity. This subassembly can then be joined via the reinforcing plate to further components of different expansivities using either of the approaches detailed above. Such a configuration involving a reinforced subassembly is used for critical wall components of fusion reactors [Jacquot *et al.* 1989].

The following disadvantages are associated with the use of graduated joint structures based on the use of intermediate plates to accommodate mismatch stresses:

- An increase in the thickness and often in the weight of the assembly, which may be significant. This modification will also introduce additional materials and fabrication costs.
- At least two soldered or brazed joints are used in place of a single joint. Because further materials are introduced into the assembly, alternative filler alloys and joining processes may need to be developed and qualified.
- The thermal and electrical conductance between the joined components is likely to be degraded. This is a consequence of the increase in the overall thickness of and number of interfaces in the assembly. Furthermore, materials with low expansion coefficients tend to be poor conductors.
- The method is difficult to apply to joints that do not have simple planar geometries.

4.3.2 Compliant Structures

Equation 4.1 given above for calculating the bow distortion of a bimetallic assembly implies that the mismatch stress is a sensitive function of joint dimension or, more precisely, joint area. Al-

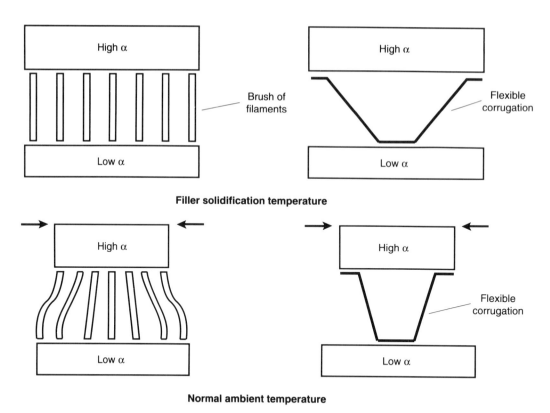

Fig. 4.9 Examples of compliant structures for mitigating mismatch expansivity (α) of the abutting components

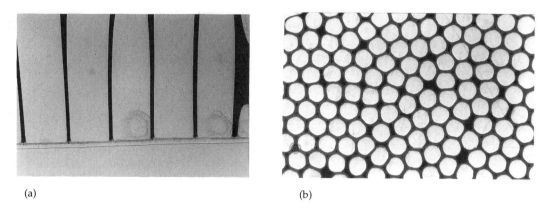

Fig. 4.10 Longitudinal (a) and transverse (b) sections through a compliant structure that is capable of accommodating a thermal expansivity difference between joined components. (a) 450×. (b) 99×

though the overall size of the assembly is likely to be fixed by the functional requirements of the product, it may be possible to replace one of the monolithic components with a filamentary, brush-like structure. Then the dimensions of each individual bimetallic joint can be made as small as necessary, thereby effectively eliminating the mismatch stress from this source, while the high aspect ratio of the filaments confers a degree of lateral compliance which can accommodate the mismatch strain. Examples of these highly compliant structures are illustrated in Fig. 4.9 and 4.10, and others are described in the scientific and technical literature [Huchisuka 1986].

The so-called flip-chip bonding process used in semiconductor assembly also results in compli-

The Role of Materials in Defining Process Constraints

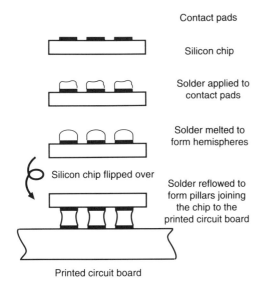

Fig. 4.11 Schematic illustration of the flip-chip joining process

ant contacts between the components, although in this case these are provided by the actual solder joint [Yung and Turlik 1991]. Flip-chip bonding involves electroplating or vapor depositing solder bumps on the contact pads of an electronic component. These bumps are heated to reform the solder as hemispherical balls, which are typically 0.1 mm (0.004 in.) high. In a subsequent stage, the components are joined to circuit boards by reflowing the solder bumps. The entire sequence is shown schematically in Fig. 4.11. The circuit boards are prepared in such a manner that solder wetting is confined to the areas of the contact pads on the boards, in order to maximize the height of the solder pillars that constitute the joints. The Pb-5Sn solder is widely used for flip-chip bonding because it has a high compliance, although its fatigue resistance is relatively low for a metal.

The use of compliant structures of a filamentary form obviously incurs a cost penalty, due to the greater complexity of manufacture. The conductance between the components via the filamentary member will also be impaired. Even with filaments having a hexagonal cross section to produce a close-packed structure, it is difficult to obtain a compliant structure that will work effectively with a packing density of greater than about 85% [Glascock and Webster 1983]. Furthermore, the ability to simultaneously make large numbers

of small-area joints is by no means a trivial exercise, but demands stringent control of tolerances and highly specified joining processes.

4.4 Constraints Imposed by the Components and Solutions

A large-area soldered or brazed joint may be defined as one where the total joint area exceeds about 20 mm^2 (0.3 in.^{2}) and the length in any direction is greater than about 5 mm (0.2 in.). This definition is based on the following practical criteria:

- The significant distortion of assemblies that have joints between materials of thermal expansivities differing by as little as 5×10^{-6}/°C (9×10^{-6}/°F). Distortion is related to the solidus temperature of the filler alloy, and this problem is therefore most acute for brazed joints as noted earlier.
- The incorporation of significant void levels (above 10%) in joints. This problem tends to be more pronounced in soldered joints and is associated with the lower joining temperatures that are used for these.

Distortion of assemblies can arise from a number of causes, some of which were discussed in section 1.4.2. Apart from mismatch in thermal expansivity, the most common sources of warping or bowing are uneven heating, which leads to temperature gradients in the components, and residual stress from earlier stages of fabrication, which is relieved in the heating cycle. Distortion from this cause can be avoided by performing the joining operations under carefully controlled conditions. For example, heating should only be carried out in furnaces that provide highly uniform temperature zones and after appropriate stress-relief routines have been performed.

4.4.1 Strength as a Function of Joint Area

The strength per unit area of a joint between components of the same material tends to reduce in proportion to the area above some lower threshold, often around 20 mm^2 (0.3 in.^{2}) for a soldered joint. This effect can be explained by the increas-

Principles of Soldering and Brazing

Fig. 4.12 Radiograph of a silicon chip (10mm × 10mm) soldered into a metallized ceramic package. Voids in the joint gap are evident as the light areas.

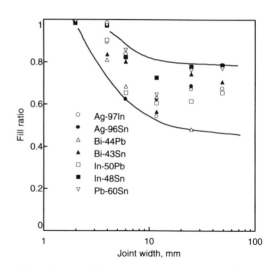

Fig. 4.14 Joint fill ratio as a function of joint width for seven common binary solder alloys. Note that essentially void-free joints are obtained provided that one dimension of the joint area is less than about 2 mm. Fill ratio = (plan area of joint)/(plan area of joint filled with solder)

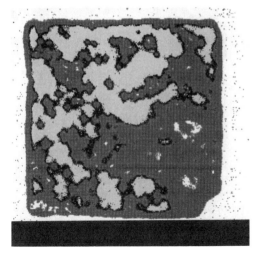

Fig. 4.13 Scanning acoustic microscope image of the soldered joint shown in Fig. 4.12. Voids in the joint gap correspond to the light areas.

ing tendency for voids to accumulate in the joint as its area is increased. The effect is less pronounced in brazed joints than in soldered joints for the reasons described below. The voids have two causes:

- Gas trapped or generated within the joint
- Solidification shrinkage of the molten filler metal

4.4.1.1 Trapped Gas

The largest source of voids in soldered and brazed joints is trapped gas, even for joints made in high vacuum using both components and filler alloys that have been given a vacuum bakeout prior to the joining cycle. Void levels in soldered joints greater than about 100 mm^2 (1.5 in.^{2}) in area and which are no narrower than 5 mm (0.2 in.) can typically reach 50% of the joint volume, which is consistent with the entrapment of air or an evolved gas. Figure 4.12 shows a radiograph of a joint between a silicon chip and a metallized ceramic package, made in high vacuum using a solder preform, that graphically demonstrates the problem. Figure 4.13 is a scanning acoustic image of the same component further enlarged and illustrates the correspondence that can be obtained between the two analytical techniques.

Air becomes trapped when the components are assembled, the mating surfaces forming an effective seal. As the temperature of the assembly is raised to the peak process temperature, the trapped air is augmented by gas evolved from the joint surfaces. The total gas volume will increase as the temperature is raised and the ambient pressure is reduced, in accordance with the gas law.

If the path length between a gas bubble and the joint periphery is small, the gas pressure can normally exceed the hydrostatic force exerted by the

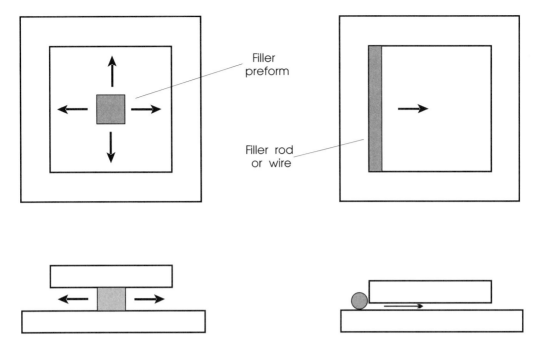

Fig. 4.15 Two configurations showing flow by a molten filler designed to sweep trapped gas out of a joint

molten filler metal, allowing the gas to escape to the surrounding atmosphere. However, the limit to the path length for this to occur is on the order of 1 mm (0.04 in.) for soldered joints (see Fig. 4.14), but it is significantly longer for brazed joints because the process temperature, and hence the pressure developed by trapped gas or vapor, is higher.

An effective method of removing trapped air is to design the joint in such a manner that the molten metal is made to flow from the center of the joint out toward the periphery or through the joint from one edge, as tends to occur when feeding the filler into the joint from a rod or a wire preform. Suitable arrangements for achieving this type of flow are illustrated schematically in Fig. 4.15. The advancing front of molten metal is then able to displace the gas and air ahead of it as it flows into the joint gap. Many of the soldering methods used in the microelectronics industry provide this type of sweeping action. Care should be taken when using this approach to ensure that the volume of the filler used is sufficient so that alloying with the components does not stifle flow of the molten alloy in its path through the joint.

Some reduction in the volume of trapped gas can be achieved by reducing the number of surfaces in the joint. Solders in particular can be applied as electroplated or "tinned" coatings to the components, thereby eliminating one free surface, but considerable care must be taken to ensure that the coated layers do not themselves contain significant volumes of gases or other volatile constituents. For some large components, it may be possible to incorporate vents through the components to provide a passage for trapped gas to escape from within the joint.

Evolved gas can originate from several sources, in particular the two discussed below.

Organic residues and adsorbed water vapor on the surfaces to be joined. These species volatilize as the temperature of the components is raised. Residues can be minimized by carefully precleaning the surfaces. A bakeout in vacuum immediately prior to joining is usually effective in removing water vapor and organic residues, provided that the temperature used exceeds about 150 °C (300 °F).

Volatile materials within the bulk of the components. This is most pronounced when the components are porous or polymeric materials and when the components (including any metallizations present), the filler metal, or flux contains constituents that volatilize during the heating cycle. A vacuum bakeout immediately prior to the joining cycle can help prevent subsequent outgassing from porous materials.

Gas evolution from polymeric materials is usually caused by thermal decomposition. The only practical remedy in this instance is to use

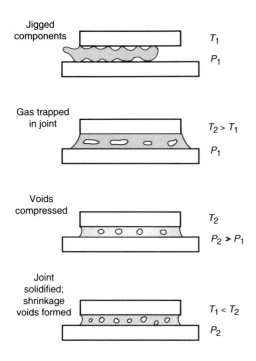

Fig. 4.16 Pressure variation method for reducing void levels due to trapped gas

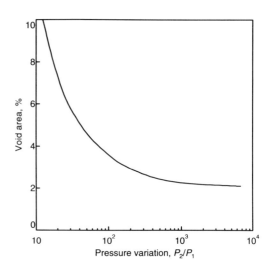

Fig. 4.17 Experimentally derived relationship between pressure variation and void level obtained in large-area soldered joints using the pressure variation method. Source: After Mizuishi, Tokuda, and Fujita [1988]

either a lower-temperature process or to change the polymeric material to one that has superior thermal stability. Some materials used for printed circuit boards are less stable than others, often in relation to their cost. Similar considerations apply to metallic components and filler metals, where these contain volatile elements such as zinc, magnesium, cadmium, and, to a lesser extent, manganese, and also to fluxes and pastes, which generally contain volatiles.

Regardless of the source of the trapped gas, the measures described for reducing void levels are directed toward excluding and expelling gas from the joint. An alternative approach is to compress the trapped gas so that it occupies a smaller volume fraction of the joint. The "pressure variation" method was developed specifically to reduce void levels arising from trapped gas in adhesively bonded joints [Bascom and Bitner 1975]. It can be used to make large-area joints using solder preforms that have void levels consistently below 5% [Mizuishi, Tokuda, and Fujita 1988]. Most of the residual porosity is caused by solidification shrinkage, a phenomenon that is considered below. The principle of the pressure variation method is to use external gas pressure to compress the air trapped in the joint. The procedure, as applied to vacuum joining, is as follows (see Fig. 4.16):

- The components to be joined are located in a jig and placed in the bonding enclosure, which is then pumped to a reduced pressure, P_1.
- The temperature of the assembly is raised to melt the filler metal, while keeping pressure P_1 constant.
- The pressure in the enclosure is raised from P_1 to a value of P_2, which is several orders of magnitude higher.
- The assembly is allowed to cool so that the filler solidifies while the pressure is maintained at P_2.

The voids corresponding to trapped gas are reduced in volume roughly according to the gas law (with suitable corrections applied to take account of the nonideality of the particular gas used). In the simplest case, corresponding to ideality, if the initial void volume at pressure P_1 is V_1, then, at constant temperature, the volume at pressure P_2, $V_2 = V_1(P_1/P_2)$. Hence, the greater is P_2 in relation to P_1, the more effective is the method. This condition is also found to apply qualitatively to practical situations.

The experimentally derived relationship between pressure variation (P_2/P_1) and the volume of voids, V_2, obtained using this process is shown in Fig. 4.17. The nonlinearity of the relationship at

Table 4.2 Solidification shrinkage of selected elements common to widely used solders and brazes

Element	Volume contraction on solidification, % of solid
Zinc	6.9
Aluminum	6.6
Gold	5.2
Silver	5.0
Copper	4.8
Cadmium	4.7
Lead	3.6
Tin	2.6
Indium	2.5
Antimony	–0.9
Bismuth	–3.3

large P_2/P_1 ratios is due to both the departure from ideality of the gas and the contribution to the void volume from solidification shrinkage. The pressure variation method for minimizing voids is obviously not suitable for situations where vapor is continually being evolved from volatile constituents.

4.4.1.2 Solidification Shrinkage

In any soldered or brazed joint, a fraction of the joint volume will comprise residual voids that do not derive from trapped air or gas. These residual voids are extremely difficult if not impossible to remove because they are intrinsic to the filler metal, being caused by the shrinkage when it solidifies. Table 4.2 lists values for the shrinkage volume contraction of elements common to many filler metals. The reservoir of filler metal represented by fillets and edge spillage is seldom able to feed large-area joints and compensate for the contraction because the outer extremities of joints usually solidify first through radiative heat losses to the surroundings.

The magnitude of solidification shrinkage, as given in Table 4.2, accounts for the fact that it is difficult to make joints of large area that contain less than about 3 to 5% voids. Shrinkage voids do not usually occur in small or narrow joints (<2 mm, or 0.08 in., in one of the joint area dimensions). This is because relatively large thermal gradients usually develop along the joint when the assembly is cooled from the joining temperature; this causes the filler to directionally solidify from one end to the other, thereby preventing voids from forming within the joint.

It is possible to achieve the same effect in large-area and wide joints by imposing a temperature gradient on the assembly, from either center-to-edge or edge-to-edge such that the periphery of the joint is always the last portion to solidify. However, this is not always easy to achieve, especially when large numbers of components are involved. Furthermore, the imposition of a temperature gradient on a large assembly may produce stress gradients and thereby dimensional distortion of the components, which becomes fixed when the filler solidifies.

Where the parent materials and filler metal are of similar composition, maintaining the assembly at elevated temperature but below the solidus temperature of the filler can result in a gradual reduction in void levels arising from solidification shrinkage by vacancy diffusion. This is the mechanism by which dry joint interfaces are removed in diffusion bonding. The application of pressure has proved effective in removing voids when joining nickel-containing turbine components using a nickel-boron braze [Duvall, Owczarski, and Paulonis 1974]. The magnitude of the stress that is effective depends on the particular joining process and materials associated with the joint. However, void levels can actually increase for certain combinations of materials if there is nonsymmetrical diffusion of different elements across an interface (the Kirkendall effect). Voids arising from this source can be suppressed by applying a compressive force. Although interdiffusion between tin and copper gives rise to high levels of Kirkendall porosity at temperatures above 400 °C (750 °F), a process has been developed successfully for joining copper components using a thin layer of tin solder at 500 to 800 °C (930 to 1470 °F) which can see subsequent service at temperatures of up to 900 °C (1650 °F). Provided that a pressure of at least 5 MPa (700 psi) is applied, voiding is eliminated. This diffusion soldering process is referred to in section 4.4.2 below.

Bismuth and, to a lesser extent, antimony are exceptional among metals in having a volume expansion rather than a volume contraction on freezing, as shown in Table 4.2. Therefore, by combining bismuth and/or antimony with other elements, it is possible to produce filler metals with essentially zero volume change on solidification. The Bi-43Sn solder (melting point = 139 °C, or 282 °F) is an example of one such alloy that finds application in the electronics industry, where joints of guaranteed hermeticity are required [Dogra 1985]. Additions of bismuth and antimony are not usually made to brazes, as they can cause intergranular embrittlement in many base metals at typical brazing temperatures.

A general approach that has been found to be effective in producing well-filled and hermetic joints is one in which strong metallurgical reactions occur across the joint during the heating cycle while a compressive force is applied. The void-free joints obtained using the diffusion soldering and diffusion brazing processes described below are associated with such reactions.

4.4.2 Diffusion Soldering and Brazing

In soldering and brazing, wetting of the component surfaces is not always easy to achieve and, when it does occur, the resulting interalloying between the filler and components can cause excessive erosion of the parent materials and embrittlement of joints due to the formation of phases with inferior mechanical properties. These problems notwithstanding, solders and brazes have the singular merit of being able to fill joints of irregular dimensions and produce well-rounded fillets at the edges of the joint.

Diffusion bonding sidesteps the need for wetting and spreading of fillers. Because this type of joining process can be carried out at relatively low temperatures compared with those used for brazing, thermal expansion mismatch stresses, which can present a major problem in the brazing of heterogeneous assemblies, are considerably diminished. Once formed, diffusion-bonded joints are stable to much higher temperatures so that the service temperature of the assembly can actually exceed the peak temperature of the joining process without risk of the joint remelting. Although the formation of undesirable intermetallic phases can also occur in diffusion bonding, because the filler is dispensed with there are fewer constituents involved, and it is thus easier to select a safe combination of materials.

However, diffusion bonding tends to be limited as a production process because it is not tolerant to joints of variable width and, moreover, its reliability is highly sensitive to surface cleanliness. High loads (typically 10 to 100 MPa, or 1.5 to 15 ksi) have to be applied during the bonding cycle to ensure good metal-metal contact across the joint interface. The duration of the heating cycle is typically hours compared with minutes for soldering and brazing, because solid-state diffusion is that much slower than liquid-solid reactions. These factors, and the absence of any significant fillets to minimize stress concentrations at the edges of joints, considerably limit the applications of diffusion bonding.

Hybrid joining processes have been developed that combine the good joint filling and tolerance to surface preparation of conventional soldering and brazing with the greater flexibility with regard to service temperature and reduced thermal expansion mismatch stress that is obtainable from diffusion bonding. Such processes use a molten filler metal to initially fill the joint gap, but, during the process cycle, the filler alloys with the component materials to form solid phases and raise the remelt temperature of the joint. Other advantages of this type of process are:

- The necessary applied pressures are much less than those required for diffusion bonding and are typically in the range of 0.5 to 5 MPa (70 to 700 psi).
- Exceptionally good joint filling can be achieved over large areas.
- The joints are very thin, usually less than 10 µm (400 µin.), which benefits the mechanical properties, particularly when the filler alloy is a solder.
- Crisp joint edges with small rounded fillets but little gross spillage can be obtained. The need for finishing operations is thereby considerably reduced.

The first two of these features derive from the generation of liquid in the joint during the bonding process, and the second pair are a consequence of the volume of liquid being made much lower than in conventional filler metal joining. Two further benefits stem from the fact that the process involves isothermal solidification:

- The joining process can be integrated with superplastic forming into a single operation similar to fabrication based on diffusion bonding. Following solidification at the joining temperature, superplastic forming can be carried out *in situ* and, if necessary, at a different temperature.
- The strength of joints made with brazes containing metalloids that depress the melting point such as boron and phosphorus, can be considerably boosted because isothermal solidification can prevent the formation of embrittling metalloid compounds, as explained below with regard to nickel-boron.

Table 4.3 Metal combinations used for diffusion brazing/soldering

Substrate	Filler metal	Process temperature °C	°F	Remelt temperature °C	°F	Ref
Aluminum alloys	Cu	550	1020	550	1020	Niemann and Wille [1978]
	Zn-1Cu	525	975	525	975	Ricks et al. [1989]
Cobalt alloys	Ni-4B	1175	2145	1475	2685	Duvall, Owczarski, and Paulonis [1974]
Copper alloys	Au	1000	1830	1050	1920	Teng [1989]
Nickel alloys	Ni-4B	1175	2145	1450	2640	Duvall, Owczarski, and Paulonis [1974]
	Ni-12P	1100	2010	1450	2640	Ikawa, Nakao, and Isai [1979]
Silver alloys	Ag-30Cu	825	1515	950	1740	Tuah-Poku, Dollar, and Massalski [1988]
Steel (including stainless)	Fe-12Cr-4B	1050	1920	1400	2550	Nakahashi et al. [1985]
Titanium alloys	Cu-50Ni	975	1785	1700	3090	Norris [1986]
	Ag-15Cu-15Zn	700	1290	700	1290	Elahi and Fenn [1981]
Copper metallizations	Sn	500	930	>900	>1650	Hieber et al. [1989]
Silver metallizations	Sn	250	480	>480	>895	Jacobson and Humpston [1992]
	In	175	345	>660	>1220	Jacobson and Humpston [1992]

Several processes of this type have been developed and are known as diffusion brazing/soldering or transient liquid phase bonding. A list of representative diffusion brazing and soldering systems taken from the literature is given in Table 4.3.

One of the first diffusion brazing processes to be exploited was developed for joining nickel-base superalloys using the Ni-4B braze (melting point = 1090 °C, or 2084 °F) [Duvall, Owczarski, and Paulonis 1974]. The sequence of steps for making a diffusion-brazed joint is illustrated schematically in Fig. 4.18. The joint configuration comprises the two component parts and a filler preform inserted between them (stage 1). The latter is much thinner (<25 μm, or 1000 μin.) than a typical braze or solder preform. The components are clamped together with a compressive stress of a few megapascals, and the assembly is then heated to above the liquidus temperature of the filler metal, which melts, wets the joint surfaces, and fills the gap (stage 2).

If the assembly were then cooled, the joint would resemble a thin, but conventional, brazed joint in its properties and metallurgical characteristics. However, by maintaining the assembly at the process temperature, the composition of the filler metal will change with time because the boron in the filler diffuses away from the joint and into the nickel matrix of the superalloy, eventually leaving a solid layer of essentially pure nickel between the components (stage 3). In other words, solidification is achieved isothermally, with the advantage that the precipitation of the embrittling second phase is prevented. The formation of

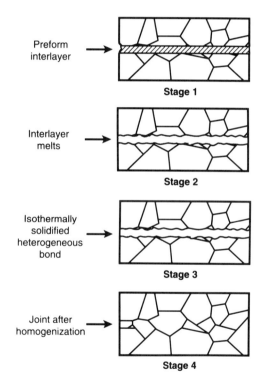

Fig. 4.18 Schematic illustration of the steps involved in making a diffusion-brazed joint

nickel-boron intermetallic phases in conventionally brazed joints using the Ni-4B filler alloy reduces their fracture toughness. Nickel has a melting point that is about 360 °C (630 °F) higher than the original braze, as shown by the phase diagram

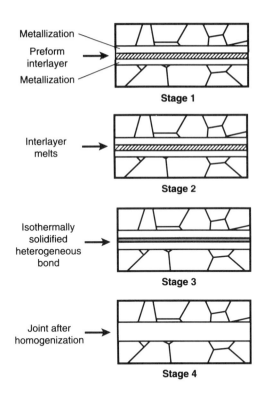

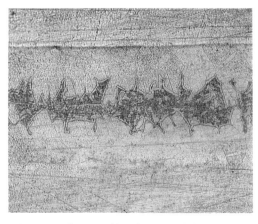

Fig. 4.20 Diffusion-soldered joint made between two engineering copper alloy substrates using a tin interlayer. 530×

Fig. 4.19 Schematic illustration of the steps involved in making a diffusion-soldered joint

in Fig. 2.15, raising the remelt temperature of the joint by that amount. By continuing the heat treatment for an extended period (typically more than 100 h), diffusion will continue until the joint is homogenized in both composition and structure, making it largely indistinguishable from the flanking components (stage 4).

It is not always necessary to add a filler to the joint for diffusion brazing. If components of two different metals or alloys are to be joined together, then a liquid phase can form at a temperature well below the melting point of either of the component materials, provided that there is a eutectic reaction between them. An isothermal diffusion brazing process for joining Zircaloy 2 [Zr-1.5Sn-0.25(Fe,Ni,Cr)] to stainless steel at approximately 950 °C (1740 °F) has been described in the literature. This takes advantage of a eutectic reaction that occurs between iron and zirconium at 934 °C (1713 °F) [Owczarski 1962].

The kinetics of the reactions of the more common diffusion brazing systems have been extensively studied by experiment and mathematical modeling and numerous examples can be found in the scientific and technical literature [Isaac, Dollar, and Massalski 1988; Nakagawa, Lee, and North 1991; Ramirez and Liu 1992].

Diffusion solders differ from diffusion brazes by virtue of the fact that the products of the interdiffusion process tend to be intermetallic compounds rather than solid solutions. This is because the kinetics of the reactions, which are essentially solid state, are much slower, owing to the lower process temperatures, and primary solid solutions are therefore not achieved over practical timescales. For the same reason, the depth of reaction with the parent materials is considerably reduced.

Most diffusion soldering processes are based on copper, silver, or gold. These metals are not prohibitive in cost for this application, because the depth of interaction with the filler metal is shallow and thus they can be applied as thin metallizations (<10 μm, or 400 μin.) to the component surfaces, without the danger of the component materials entering the reaction. The use of copper, silver, and gold as surface coatings confers the particular advantage that, being relatively noble, they are readily wetted even in slightly oxidizing atmospheres. This makes diffusion soldering processes relatively tolerant to the condition of the atmosphere in which the joining operation is carried out. The sequence of steps involved in making a diffusion-soldered joint is shown schematically in Fig. 4.19.

An alloy system that has furnished an effective and reliable diffusion soldering process is copper-tin [Hieber *et al.* 1989]. The alloy phase diagram is given in Fig. 3.15. The Cu_6Sn_5 and Cu_3Sn intermetallic phases that form by reaction between molten tin and copper are widely regarded as unacceptable by electronic engineers because they are hard and brittle. However, robust joints that are

The Role of Materials in Defining Process Constraints

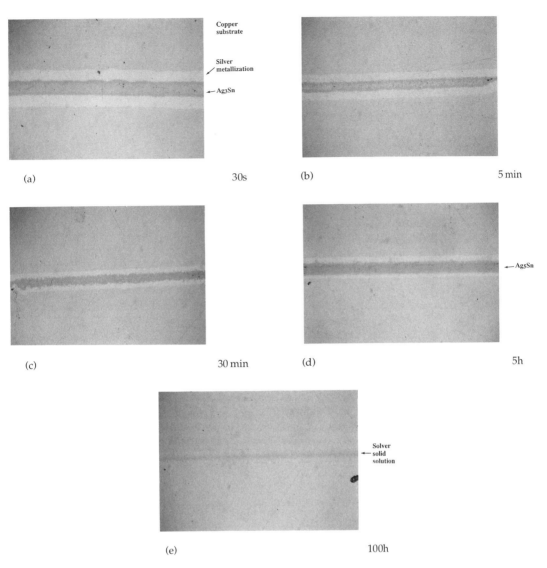

Fig. 4.21 Series of micrographs showing the progressive change in joint microstructure that occurs on making a diffusion-soldered joint using the silver-tin system. All 24×

stable to 900 °C (1650 °F) or above can be produced if the assembly is subjected to prolonged heating at elevated temperature (500 to 800 °C, or 930 to 1470 °F) under an adequate compressive load (above 5 MPa, or 700 psi). A typical joint prepared in this way is shown in Fig. 4.20.

The authors have found the silver-tin system to be one of the most versatile for diffusion soldering of components for use in the electronics industry, because well-filled joints can be produced using compressive loads of as little as 0.5 MPa (70 psi) [Jacobson and Humpston 1992]. Silver-tin solder of eutectic composition reacts with silver to form the Ag_3Sn phase as a continuous interfacial layer, which is both tough and strongly adherent to the other phases in this alloy system. The rate of reaction between silver and molten tin has been characterized and is represented graphically in Fig. 3.7. The controllable nature of the alloying reaction in the conventional soldering system is indicated by the general profile of the erosion curves, which show that the reaction is self-limiting in character within the context of realistic processing cycle times and temperatures. If a thin layer of tin, typically 5 μm (200 μin.) thick, is sandwiched between two components, each covered with a 10 μm (400 μin.) thick layer of silver, and heated to 250 °C (480 °F), the tin will melt and react with the

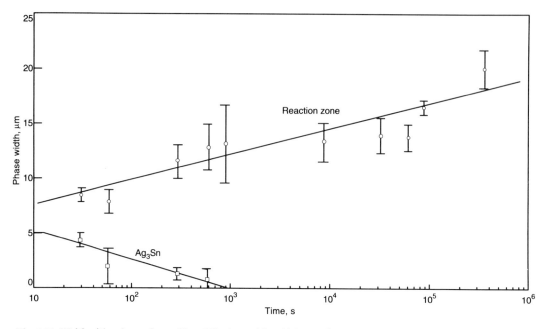

Fig. 4.22 Width of the phases formed in a diffusion-soldered joint made using a silver-tin couple as a function of the process cycle time at 250 °C (480 °F)

silver to form Ag_3Sn. On continued heating, the tin is progressively converted to this compound until no liquid tin remains. By keeping the tin layer thin, it completely reacts to form solid phases at the joining temperature in less than 1 min.

The remelt temperature of a silver-tin diffusion-soldered joint is determined by the phases that are present. Immediately after the liquid tin has been consumed by reaction, the remelt temperature is that of the Ag_3Sn compound, which is 480 °C (895 °F). Longer heating times promote continued diffusion of tin from the Ag_3Sn reaction zone into the silver. Consequently, the width of this zone decreases as it is replaced first by $\zeta(Ag_5Sn)$ and, ultimately, by a solid solution of tin in silver, as anticipated from the silver-tin phase diagram that is given in Fig. 2.26. This progression is illustrated by the series of micrographs shown in Fig. 4.21. The change in the width of phases in the joint with time is indicated in Fig. 4.22.

As the reaction with the silver proceeds, the remelt temperature rises progressively toward the melting point of silver (962 °C, or 1764 °F). Mechanical property measurements have shown that the shear strength of diffusion-soldered joints containing the Ag_3Sn phase are close to the 25 MPa (4 ksi) value for conventional soldered joints made with the Ag-96.5Sn eutectic solder to silver-coated components. As the joint microstructure converts to a silver solid solution, the mechanical properties shift in tandem toward those of pure silver, with the shear strength increasing toward 75 MPa (11 ksi). This can be significant from an applications point of view, because the strength of silver is approximately three times that of the Ag-96.5Sn eutectic alloy, which itself is superior in this respect to the common Pb-60Sn solder by a factor of two to three [Harada and Satoh 1990].

Another attractive feature of the silver-tin alloy system for diffusion soldering is that there is negligible volume contraction as the reaction proceeds, which is a fortuitous consequence of the closely similar specific volumes of the various phases. Therefore, the tendency to form voids or cracks as a result of volume change is minimal. The same is not true of the copper-tin system, where volume changes during diffusion soldering are sufficient to generate voids, unless high compressive stresses, of at least an order of magnitude higher than those needed with the silver-tin system, are applied during the process cycle.

4.4.3 Joints to Strong Materials

New materials with enhanced strengths are continually coming onto the market. These are either composite materials, such as metal-matrix composites (MMCs), or precipitation-strengthened alloys. There is a desire to exploit these materials in a range of applications, but their widespread adoption is contingent on being able to utilize the favorable strength levels in joined assemblies. In general, the strength of joints, even when welded, is inferior to that of the materials in monolithic form. Moreover, the heating cycle used in the joining process can itself degrade the properties of these materials. For example, aluminum/SiC MMCs are susceptible to degradation when heated above about 500 °C (930 °F) due to reaction between the constituents, which results in the formation of a brittle interfacial layer of Al_3C_4 [Iseki, Kameda, and Maruyama 1984]. Usually this problem is more severe in welding, owing to the higher temperatures involved, and thus most efforts have focused on attempts to devise joining methods based on brazing and diffusion bonding.

It is possible to produce brazed joints to these new materials that are metallurgically sound, that is, which have a low incidence of voids and are free from embrittling intermetallic phases. Such joints can be sufficiently strong for fracture to occur through the parent materials. However, the fracture stresses in the joined assemblies are usually lower than those involving conventional engineering alloys! This apparent paradox can be explained by the combination of relatively high elastic modulus coupled with the low ductility and fracture toughness of the stiff parent materials. The stresses arising from the bonding operation cannot find relief through plastic deformation of the materials, owing to their high elastic modulus, and remain localized, their concentration depending on the joint geometry and mode of stressing, as described below. The assemblies are therefore vulnerable to brittle fracture at low apparent stress levels.

The joint and its surroundings are inevitably the weak link in an assembly of high-strength materials. In order to maximize the mechanical resilience of the assembly, it is necessary to devise means of distributing the stresses within it and to prevent the development of stress concentrations in the vicinity of the joint. At the same time, the strength of the joint itself, which usually means the filler alloy, must be optimized. These aspects will be considered in turn.

4.4.3.1 *Joint Design to Minimize Concentration of Stresses*

Analysis of the stress distribution in joints is relatively straightforward, provided that the joints have simple geometry. Some common examples are considered below.

Axially loaded butt joints and the effect of voids and cracks. An axially loaded butt joint in a cylindrical assembly is shown schematically in Fig. 4.23(a). The tensile stress σ_z produces an axial strain ε_z and lateral contractions $-\varepsilon_x$ and $-\varepsilon_y$, with accompanying shear stresses σ_x and σ_y. It is assumed that the material is randomly polycrystalline and therefore isotropic, in which case $\varepsilon_x = \varepsilon_y$ and $\sigma_x = \sigma_y$.

On the central axis of the assembly, the shear stresses σ_x and σ_y, which are physically manifested as necking when homogeneous materials are loaded in tension, act equally in the radial and circumferential directions. Figure 4.23(b) represents a solid element of the joint lying on the central axis of symmetry. Provided that the joint is thin, the filler is constrained by the higher Young's modulus of the components from deforming and the shear and tensile strengths of the joint will increase with reducing joint gap, up to a maximum. This effect, which is illustrated for both brazes and solders in Fig. 4.24 and 4.25, respectively, can be explained as follows. A very thin layer of filler in a joint to components that are considerably stronger than the filler itself will deform in a manner very different from that when it is in bulk form. When the stress applied to a joint with a small joint gap reaches the yield point of the filler, the latter cannot deform plastically, owing to the adjacent layers of the component materials which are still stressed within their elastic range. Consequently, the filler is subjected to a triaxial (hydrostatic) tension and does not fail until the applied stress has reached the brittle fracture strength of the filler. Thus, for example, silver, which does not alloy significantly with iron and whose ultimate tensile strength as measured in a standard tensile test is 150 MPa (25 ksi), will sustain a stress of up to 680 MPa (100 ksi) when in the form of a filler in a joint to high-strength steel. As the width of the joint is increased beyond the optimum value, the effect described diminishes and the strength declines towards that of the bulk filler. Narrower joints are degraded by incomplete filling.

It has so far been assumed that the strain ε_z parallel to the applied load σ_z is uniform throughout the assembly. In practice this is not the case, because the mechanical properties and, in particular, the Poisson's ratio, ν, where

$$\nu = \frac{\varepsilon_x}{\varepsilon_z} = \frac{\varepsilon_y}{\varepsilon_z}$$

Principles of Soldering and Brazing

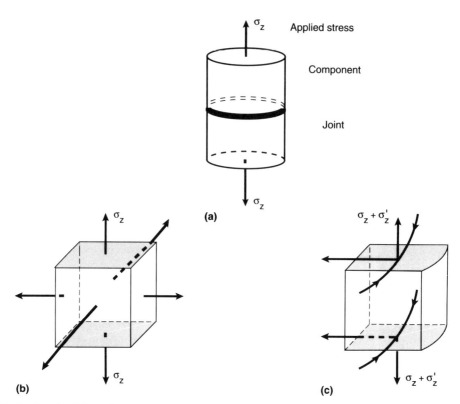

Fig. 4.23 Butt joint loaded axially in tension. (a) A stress concentration exists at the periphery of the joint due to a difference in the Poisson's ratios of the filler and of the components. (b) Stress distribution in the axial center of the joint. Deformation of joint surfaces is constrained by the components, and thus the filler is subject to triaxial tension. (c) Stress distribution at the periphery of the joint. The difference in Poisson's ratio between the components and the filler generates a shear stress at the joint interface and an additional normal tensile stress, σ_z'.

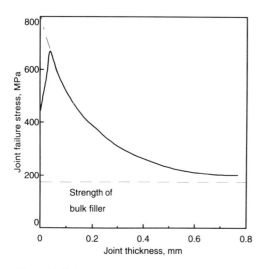

Fig. 4.24 Relationship between the fracture stress and joint thickness of butt joints in medium-carbon steel testpieces made with a silver-base braze. Very narrow joint gaps tend to be inadequately filled, thereby causing the measured joint strengths to decline. Source: After Sloboda [1961]

of the filler alloy and the components differ. This means that the tendency for lateral deformation in response to the normal strain, ε_z, will be unequal in the two materials and give rise to an additional stress at the filler/component interface in the vicinity of the joint periphery. The magnitude of the resulting stress is dependent on the radius of the joint and its thickness and is least for thin joints in small-diameter components. The perpendicular component of the additional stress at the outer circumference of the joint, σ_z', is aligned with the applied tensile stress, σ_z, and is depicted schematically in Fig. 4.23(c), where the curved surface of the element lies on the cylindrical face of the assembly.

Hence, the effective tensile stress acting at the periphery of the joint is higher than in the center by the ratio K, where:

$$K = \frac{\sigma_z + \sigma_z'}{\sigma_z}$$

The Role of Materials in Defining Process Constraints

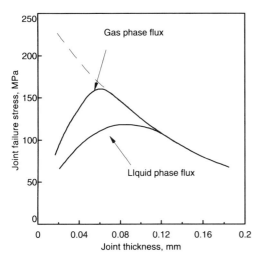

Fig. 4.25 Shear strength of soldered joints to brass testpieces as a function of joint thickness. Narrow joints are progressively more difficult to fill, thus decreasing the measured shear strength of thin joints. A gaseous flux is able to penetrate narrower joint gaps than a liquid flux; consequently, thinner joints can be made before the joint-filling problem appears. Source: After Manko [1992]

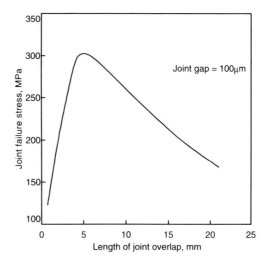

Fig. 4.27 Effect of joint length on shear stress for a simple lap joint between mild steel components brazed using Ag-Cu-Cd-Zn. Source: After Sloboda [1961]

Consider an elliptical hole, that is, a crack. The maximum stress at the end of a crack is given by (see Fig. 4.26):

$$\sigma_z' \triangleq 2\sigma_z\left(\frac{a}{b}\right) \text{ and } K \triangleq 2\left(\frac{a}{b}\right) \gg 1$$

where a is half the crack length and b is half the crack width and $a \gg b$. Thus, at the end of a very narrow crack, the stress concentration can be extremely high, so that a low applied stress can easily exceed the tensile stress of the filler in that region, enabling the crack to propagate and cause the joint to fail. Note that for a circular hole, $a = b$ and hence $K = 3$. Therefore, voids caused by solidification shrinkage or trapped gas are not nearly as detrimental to the mechanical properties of the assembly as cracks [Dieter 1976, p 66-70].

Longitudinally loaded lap joints and the effect of fillets. In a single lap joint loaded in tension, stress concentrations arise from two sources: namely, the differential straining of the components and the filler and the eccentricity of the loading path.

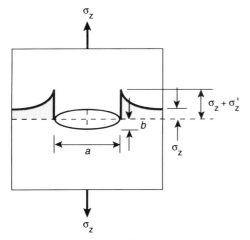

Fig. 4.26 Stress concentrations caused by an elliptical hole in a component. For a crack, $a \gg b$, giving a high stress concentration at its ends. For a circular void, $a = b$, so that $\sigma_z' = 2$. Thus, voids are not as critical as cracks to the mechanical properties of joints.

In simple butt joints that are well filled, the stress concentration, K, is usually small and less than about 1.2.

In contrast with tensile strength, shear strength decreases with increasing joint area (length), as shown by the data in Fig. 4.27. This apparent anomaly can be explained by the fact that the shear stress is highest toward the ends of the joint, so that if the length of the joint is increased beyond a certain value, the filler in the central portion of the joint will carry little or no stress, with the applied

Principles of Soldering and Brazing

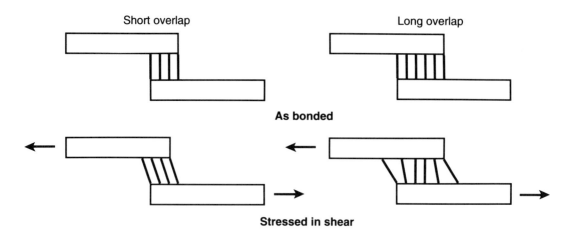

Fig. 4.28 Schematic illustration of the stress distribution in the filler metal of lap joints of long and short overlap. When stressed in shear, the central portion of a long lap joint carries little or no load.

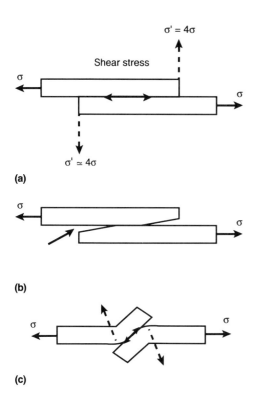

Fig. 4.29 Failure in a simple lap joint loaded in tension. (a) Stress concentrations. (b) Initiation of failure. Edge-opening crack (free arrow) formed and propagated by the high normal stress concentration. (c) Progression of joint to fracture. Plastic bending of the joint region results in the majority of the failure being due to peel-type debonding. Source: After Dunford and Partridge 1990

stress concentrating at both ends (see Fig. 4.28). This explains why simply increasing the length of the overlap does not improve the strength of this type of joint beyond a certain level.

Thus, the stress concentration in the filler from this source is proportional to the length of overlap and inversely proportional to the thickness of the joint and is therefore least in thin joints of short overlap. In a lap joint of simple geometry, the stress concentration, K, is usually relatively small (< 2).

Far more relevant to the strength of single lap joints are the tensile or "peeling" stresses that act normal to the ends of the joint and originate from the eccentricity of the loading of the assembly. The elastic analysis is relatively complex, but the result obtained is that longitudinal loading of a single lap joint effectively applies a perpendicular tensile stress of approximately four times that amount to the ends of the overlaps [Harris and Adams 1984]. These perpendicular tensile forces initiate failure of the joint by peel. With the continued application of stress, the sample rotates in an attempt to correct for the axial misalignment and the fracture continues to propagate due to peel-type debonding. The stress concentrations in a simple lap joint and their influence on its resulting failure mode are illustrated in Fig. 4.29. The influence of overlap length on the failure mode of simple lap joints made with an adhesive has been verified experimentally and is indicated in Fig. 4.30.

Fillets at the edges of a joint, and particularly at the ends of a single lap joint, act to reduce the

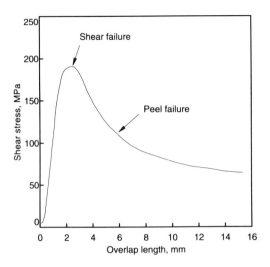

Fig. 4.30 Effect of overlap length of a shear joint on adhesive joint strength when the assembly is stressed in tension. As the overlap length increases, the forces in the joint change from shear to peel as the specimen fails under the applied stress. Source: After Dunford and Partridge [1990]

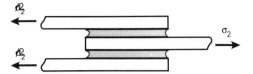

Fig. 4.31 Double lap joint

stress concentration in that region. They accomplish this by coupling some of the applied stress into the ends of the laps, thereby reducing the differential straining between the components and the filler in the joint and also shifting the position of the maximum perpendicular tensile stress (originating from the eccentricity of loading) to outside the joint. The magnitude of these effects depends on the radius of the fillets (R), the step height (H), and the elastic properties of the filler. To be effective, the radius of the fillets must exceed the step height (i.e., $R > H$), and hence it is desirable for soldered and brazed joints to have large and well-rounded fillets at their peripheries.

In a double lap joint of the form shown in Fig. 4.31, some bending moment still exists, but the symmetry of the configuration results in the failure load of the joint being increased by a factor of four. This improvement has been verified by experiment [Kinloch 1982].

Other joint configurations. Stress concentrations can be reduced by using joint configurations that distribute the load away from the joint. The approach should encompass the entire assembly and may require special overall design. The scarf and step joints are good examples of configurations that can make for "strong" joints, particularly as they pose no weight or thickness penalties.

Scarfing results in the differential strains and thus the stress concentrations at the ends of the joint being considerably reduced, while the step joint relies on the step sizes being small to achieve the same effect. Both configurations are symmetrical, thus balancing axial stresses over the joint.

Strap joints, both single and double, are widely used by industry because little or no machining of the components is required, although the thickness and weight of the assembly are increased and its aero/fluid dynamic performance is often impaired. Strap joints are often considered to be strong because, when loaded in tension, it is the parent material that fails. However, the fracture stress is substantially lower than in a monolithic body due to the stresses being concentrated in the components close to the edges of the straps. The origin of this stress concentration is indicated in Fig. 4.32.

For both strap and lap joints, the stress concentration in the components can be reduced by tapering the ends of the overlapping material. This modification is equivalent to having very large fillets in that region. A point to be aware of is that tapering of the strap and lap ends only substantially boosts strength when the tapering is taken right to the edge of the strap or lap, or at least to a thickness where fillets can complete the graduation [Thamm 1976].

An example of a joint configuration that has been developed to achieve high-strength brazed joints in long-fiber-reinforced aluminum-boron composite materials is shown in Fig. 4.33. This comprises a butt joint that is reinforced with a profiled double strap. The ends of the straps are tapered in a manner so as to graduate the stress levels, although the direction of the tapers are not what might be expected intuitively. By using this joint configuration, joined assemblies have been produced that have strengths reaching up to 90% that of long-fiber-reinforced parent materials, which can exceed 1500 MPa (220 ksi) [Breinan and Kreider 1969]. Clearly, the angle of taper should be low, but this is not specified in the paper cited.

In the preceding discussion, the assemblies were considered to be loaded solely in uniaxial tension. The location of stress concentration and

Principles of Soldering and Brazing

its magnitude will change as the stressing mode is altered, and thus the optimum style of joint varies depending on the stress environment in which the component is required to operate.

4.4.3.2 Reinforced Filler Alloys to Enhance Joint Strength

One of the limiting parameters of joint strength is that of the filler itself. Filler alloys can be strengthened by metallurgical mechanisms involving elements placed in solid solution, microscopic second-phase particles (of either intermetallic precipitates or a dispersed refractory phase and usually less than 1 µm in size), and refinement of the grains of the filler. These mechanisms have all been described previously and examples of their application given in Chapter 2.

Another approach is to load the filler with a uniform distribution of coarse particles or fibers (typically 100 µm to 1 mm, or 4000 µin. to 0.04

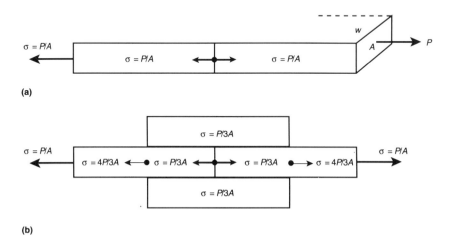

Fig. 4.32 Stress concentrations in butt and strap joint configurations. (a) Simple butt joint. (b) Butt joint with a double strap. The reinforcing straps transfer the stresses from the butt joint to the edges of the straps. This arises from the reduced load in the reinforcing strap(s) constraining Poisson contraction parallel to the edges of the straps. Source: After Breinan and Kreider [1969]

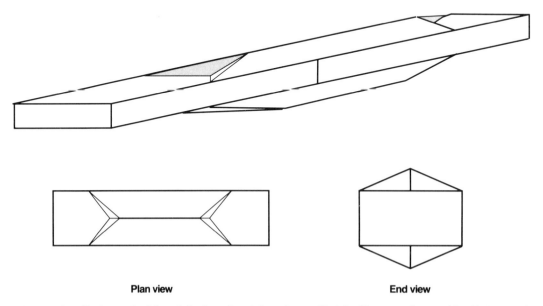

Fig. 4.33 An effective method for reinforcing a butt joint using profiled double straps. Source: After Breinan and Kreider [1969]

in., in size) of a nonmetallic material. Attempts have also been made to use particles of refractory metals, but these tend to agglomerate when the filler is molten, owing to their higher density [Ho and Chung, 1990]. A more successful avenue has been to incorporate into the filler chopped carbon fibers electroplated with nickel or copper. This has been achieved with a braze (Ag-28Cu) and also with a solder (Pb-60Sn), and the results show a threefold enhancement of the shear and tensile strengths of the joints with respect to the unmodified fillers and, more particularly, a significant reduction in the thermal expansivity of the filler [Ho and Chung 1990; Cao and Chung 1992]. The fraction of fibers was up to 55% of the volume of the composite filler. At 42 vol%, the thermal expansivity of the solder composite declined to zero over the temperature range of 25 to 100 °C (77 to 212 °F). These new composite filler materials thus offer benefits where there is a requirement to join low-expansivity materials, such as alumina (thermal expansivity = 6.4×10^{-6}/C) to kovar (thermal expansivity = 5.3×10^{-6}/°C). The high thermal expansivity of a conventional filler alloy (typically greater than 18×10^{-6}/°C) introduces a shear stress at the component/filler interface, which is overcome by using a carbon-fiber-loaded filler. The reduction in the thermal expansivity brought about by the fiber addition accounts for a threefold enhancement in fatigue life on thermal cycling that is observed in the bonded assemblies. Clearly, there is need for further research in this area.

Appendix 4.1: A Brief Survey of the Main Metallization Techniques

Four main techniques are used for applying metal coatings to metallic and nonmetallic materials and are discussed below:

Physical Vapor Deposition

Physical vapor deposition (PVD) embraces all methods where the coating material is physically converted into a vapor and then made to condense onto the surface of the substrate without undergoing any fundamental chemical change in the process. The various methods are distinguished by the means used to generate and deposit the vapor of the coating material.

Vacuum evaporation covers those methods where the coating material is thermally vaporized. This is commonly accomplished by resistance heating or by electron beam bombardment.

In sputtering, by contrast, the surface of a solid target is vaporized by bombarding it with inert gas ions, accelerated by a potential of 500 to 5000 V. A glow discharge in a low-pressure atmosphere of the inert gas—either self-sustained, as in cathodic sputtering, or supported thermionically, as in triode sputtering—is normally set up for this purpose. The rate of sputtering may be increased by magnetically intensifying the glow discharge, as in magnetron sputtering. Reverse bias sputtering or fast atom bombardment is normally available as a built-in facility for cleaning of the substrate surfaces immediately prior to the sputtering operation. This can considerably enhance the adhesion of coatings.

Where the deposition process takes advantage of the ionized fraction of the condensing vapor, the process is described as "ion-aided." Ion plating and ionized-cluster beam deposition are two techniques based on this principle. Instead of generating a discharge around a target, energetic ion beams may be aimed directly at the surface of the substrate when either a surface coating will be obtained, as in ion beam deposition, or embedded within the surface, as in ion implantation, at higher incident energies. Owing to the greater sophistication of these ion-aided processes and the higher costs involved, they are not widely used for metallizing components as part of a joining process.

Chemical Vapor Deposition

The deposition of a coating by means of a chemical reaction occurring from a gaseous phase on or immediately adjacent to the surface of a substrate is known as chemical vapor deposition (CVD). The substrate is usually heated to generate the reaction.

Chemical vapor deposition may be classified according to the type of chemical reaction involved. In a decomposition reaction, a gaseous compound AB may be decomposed into a solid condensate A and a gaseous product B when placed in contact with a colder substrate. If, for example, the compound AB instead dissociates into a solid phase A and a gas phase AB_2, then the CVD process is referred to as one involving a disproportion reaction. Oxidation and reduction of halides constitute the two other types of reactions that are widely employed.

Wet Plating

Wet plating of metallic layers encompasses processes where the coatings are deposited on a substrate through immersion of the substrate in a liquid, usually aqueous, containing the appropriate metallic ions. The deposition often functions by ionic discharge, with the metal deposited onto an electronegatively charged conductive substrate (cathode). The plating process can introduce organic compounds into the metal coatings, although these can often be minimized by judicious choice of bath formulation.

Thin coatings can be grown autocatalytically (i.e., without an applied electric field) through a reduction of the metal ions in the plating bath by the immersed substrate; this process is known as chemical displacement. Another autocatalytic method, commonly referred to as electroless plating, involves the deposition of metal from a plating bath containing the metal ions together with a reductant. This process differs from chemical displacement in that no significant reaction occurs within the volume of the liquid and the depositing metal catalyzes further deposition, so that thicker films can be grown. It is usually necessary to activate nonmetallic substrates by chemical treatment for them to generate the catalytic reaction. Nonmetallic elements, principally phosphorus and boron, tend to be incorporated into the metallic coating from the reductant.

Thick Film

Thick-film formulations usually comprise a slurry, containing the metals or metal compounds in an organic carrier, which is intended to be applied by painting or screen printing onto the desired areas. Subsequent firing drives off the organic fraction and stabilizes the metallization by producing a diffused interface with the nonmetal. It is usual practice to apply and fire each thick-film metallization separately, although processes have been developed whereby at least two thick-film layers are fired together. Common thick-film metallizations are discussed below.

Systems based on reactive metals (zirconium, tungsten, titanium, manganese, molybdenum). These formulations are fired at about 1600 °C (2900 °F) in a reducing atmosphere. Because of the relatively refractory nature of the resulting metal surface, either a strongly reducing environment is required to effect wetting or a wettable metallization should be applied, as described above. Alternatively, the wettable surface layer may be applied over the reactive metal layer and the two layers fired together. An example is a tungsten-loaded frit overcoated with a nickel paste which are cofired at 1300 °C (2370 °F). Even at this temperature, the interdiffusion between the two metals is slight, so that a discrete layer of nickel forms on the surface after the heating cycle [Kon-ya *et al.* 1990].

Systems based on noble metals (copper, silver, gold, palladium, platinum). These materials are fired between 850 and 950 °C (1560 and 1740 °F). The silver, gold, and platinum metallizations can be fired in air, whereas a reducing atmosphere is generally required for the less noble metals.

Metal-loaded glass frits are fired on the surfaces of components above 400 °C (750 °F) to form a glaze that is strongly adherent to the nonmetal. The concentration of the glass at the interface with the component means that the outer layer of the coating is sufficiently metallic in character for it to be readily electroplated or directly soldered or brazed.

Metallizations designed for firing are supplied as complex proprietary formulations and are available in different physical forms. It is advisable to consult the supplier on their conditions of use.

As might be expected, there are advantages and disadvantages associated with the different

Table 4A.1 Techniques for applying metallizations: characteristic features

Process	Metallic materials capable of deposition	Suitable substrates	Throwing power	Film thickness achievable	Film thickness control	Throughput of process
Vacuum evaporation	Elemental metals and some alloys	Most nonvolatile materials	Line-of-sight process	nm-μm	Good	Low, batch
Sputtering	Wide range of elemental metals and alloys	Most nonvolatile materials	Moderate (function of target size, gas pressure, and target-substrate distance)	nm-μm	Excellent	Low, batch
Chemical vapor deposition	Elemental metals	Materials that can withstand the high temperatures required	Good	μm-mm	Good, but need to stringently control several process variables simultaneously	High, many items at a time; batch or continuous
Electroplating	Elemental metals and some binary and ternary alloys	Electrical conductors	Moderate	μm-mm	Generally less precise than for vapor deposition	Very high, can be continuous
Electroless plating	Elemental metals and a few binary alloys	Wide range of materials	Good	μm	Generally poor	High, can be continuous
Thick film	Wide range of elemental metals and alloys	Materials that can withstand the firing temperatures	Physical access to surfaces is required	μm-mm	Moderate	High, batch or conveyor belt

Table 4A.2 Metallization techniques: relative merits

Process	Advantages	Disadvantages
Vacuum evaporation	Relatively simple equipment required for resistance heating evaporation, which is suitable for coatings of most elemental metals.	Not suitable for alloys that have constituents with greatly differing vapor pressures. Meticulous substrate cleaning prior to deposition is required.
Sputtering	Possible to coat a wide range of compositions. Dense coatings and good adhesion obtainable.	Requires sophisticated equipment. Low throughput. Heating of substrates and low deposition rates in conventional diode or triode sputtering.
Chemical vapor deposition	High-quality coatings are obtainable. Output is generally high.	Equipment is sophisticated and is usually specific to particular coatings.
Electroplating	High throughput. Large areas can be coated with uniform thickness. Limited only by the size of the plating bath. Relatively easy to control.	Chemical handling, vapor, and effluent problems. Film impurities and imperfections can also present problems. Can only apply coatings to electrically conductive materials. Thorough cleaning and chemical activation of substrates are required prior to plating.
Electroless plating	Large areas can be coated with uniform thickness. Good throwing power. Only very basic equipment is required.	As above, except that nonconducting materials can be plated. Range of available coatings is restricted.
Thick film	Requires simple equipment. Lends itself to high volume production using screen printing and firing in belt furnaces.	Relies on manufacturers' proprietary formulations. Relatively high process temperatures are used. Only thick films can be applied by this technique.

Table 4A.3 Metallization techniques: important process parameters

Process	Rate of deposition, μm/min	Pressure in deposition chamber, mPa	Substrate temperature during coating process
Vacuum evaporation	0.001-5	0.01-10	Substate is often heated to 200 °C (390 °F) to promote adhesion
Sputtering	0.005-1	100-10,000	Mostly below 100 °C (212 °F)
Chemical vapor deposition	5-100	10,000-100,000	200-2000 °C (390-3630 °F), but usually 400-800 °C (750-1470 °F)
Electroplating	0.1-100	Ambient	10-100 °C (50-212 °F)
Electroless plating	0.1-1	Ambient	10-100 °C (50-212 °F)
Thick film	1000-10,000 (does not include firing times)	Ambient	400-1800 °C (750-3270 °F)

Table 4A.4 Metallization techniques: coating quality

Process	Coating thickness uniformity	Coating continuity	Coating purity	Coating adhesion to substrate
Vacuum evaporation	Variable; determined by source-substrate geometry	Moderate to low porosity	Purity limited by source materials and deposition atmosphere	Fair
Sputtering	Higher uniformity possible than for vacuum evaporation	Low porosity	Purity limited by source materials and deposition atmosphere	Generally excellent
Chemical vapor deposition	Good uniformity possible; depends on design of the deposition chamber	Dense and essentially pore free	Purity is that of the starting materials or even better	Variable; dependent on materials and processing conditions
Electroplating	Good uniformity on flats, nonuniform at edges	Susceptible to porosity and blistering	May incorporate salts and gaseous inclusions	Variable; often excellent
Electroless plating	Fair uniformity	Susceptible to porosity and blistering	May incorporate salts and gaseous inclusions	Variable; often excellent
Thick film	Variable	Dense coatings are achievable	Often contain glass and possibly organic residues	Variable; dependent on materials and processing conditions

metallization techniques discussed above. Vapor deposition is generally superior to wet plating in offering better control of impurities and reduced porosity in thin coatings. Wet coating, by comparison, tends to be faster and cheaper and can provide thicker coatings. The technique that will normally be chosen will be the one best suited to the particular application on the grounds of its fitness for purpose and cost in terms of plant and processing cost. A brief comparison of the characteristics of the principal methods used for applying metallizations together with those of the coatings that they are capable of producing is presented in Tables 4A.1 to 4A.4. It must be pointed out that the entries in the tables represent the general situation; particular cases may be out of the ranges indicated.

Further information on these metallization techniques can be found in the literature, which is extensive. A well-presented overview of wet and vacuum coating techniques has been listed by Missel and Platakis [1978]. Other useful sources of information are given in the selected bibliography appended to the Preface.

References

Bascom, W.D. and Bitner, J.L., 1975. Void Reduction in Large Area Bonding of IC Components, *Solid State Technol.*, Vol 9, p 37-39

Bever, M.B., Ed., 1986. *Encyclopedia of Materials Science and Engineering*, Pergamon Press, p 2463-2475

Breinan, E.M. and Kreider, K.G., 1969. Braze Bonding and Joining of Aluminum Boron Composites, *Met. Eng. Quart.*, Vol 9 (No. 11), p 192-202

Cao, J. and Chung, D.D.L., 1992. Carbon Fiber Silver-Copper Brazing Filler Composites for Brazing Ceramics, *Weld. J.*, Vol 71 (No. 1), p 21s-24s

Dieter, G.E., 1976. *Mechanical Metallurgy*, 2nd ed., McGraw-Hill

Dogra, K.S., 1985. A Bismuth Tin Alloy for Hermetic Seals, *Brazing Soldering*, Vol 9 (No. 3), p 28-30

Dunford, D.V. and Partridge, P.G., 1990. Strength and Fracture Behaviour of Diffusion-Bonded Joints in Al-Li (8090) Alloy. Part 1: Shear Strength, *J. Mater. Sci.,* Vol 25, p 4957-4964; Part 2: Fracture Behaviour, 1991, Vol 26, p 2625-2629

Duvall, D.S., Owczarski, W.A., and Paulonis, D.F., 1974. TLP Bonding: A New Method for Joining Heat Resistant Alloys, *Weld. J.*, Vol 53 (No. 4), p 203-214

Elahi, M. and Fenn, R., 1981. The Joining of a Titanium Alloy Using a Copper/Silver Intermediate Layer, *Proc. Conf. Joining of Metals: Practice and Performance*, Coventry, U.K., April, p 137-144

Glascock, H.H. and Webster, H.F., 1983. Structured Copper: A Pliable High Conductance Material for Bonding Silicon Power Devices, *IEEE Trans. Components Hybrids Manuf. Technol.*, Vol 6 (No. 4), p 460-466

Hammond, J.P., David, S.A., and Santella, M.L., 1988. Brazing Ceramic Oxides to Metals at Low Temperatures, *Weld. J.*, Vol 67 (No. 10), p 227s-232s

Harada, M. and Satoh, R., 1990. Mechanical Characteristics of 96.5Sn/3.5Ag Solder in Microbonding, *IEEE Trans. Components Hybrids Manuf. Technol.*, Vol 13 (No. 4), p 736-742

Harris, J.A. and Adams, R.D., 1984. Strength Prediction of Bonded Single Lap Joints by Non-linear Finite Element Methods, *Int. J. Adhesion Adhesives,* Vol 4 (No. 2), p 65-78

Haug, T., Schaefer, W., and Schamm, R., 1989. Joining Electrochemical High Temperature Components, *Proc. Conf. Joining Ceramics, Glass and Metals*, Bad Nauheim, Germany, p 171-178

Hieber, H. *et al.*, 1989. Heat-Resistant Contacts With Use of Liquid Phase Transition, *Proc. Conf. 7th European Hybrid Microelectronic Conf.*, Hamburg, May, Session 1.4, p 7-13

Ho, C.T. and Chung, D.D.L., 1990. Carbon Fiber Reinforced Tin-Lead Alloy as a Low Thermal Expansion Solder Preform, *J. Mater. Res.*, Vol 5 (No. 6), p 1266-1270

Huchisuka, T., 1986. Bonding of Sintered Alloys, *Met. Technol.*, Vol 56 (No. 5), p 21-27

Humpston, G. and Jacobson, D.M., 1989. Gold in Gallium Arsenide Die-Attach Technology, *Gold Bull.*, Vol 22 (No. 3), p 79-91

Humpston, G. and Jacobson, D.M., 1990. Solder Spread: A Criterion for Evaluation of Soldering, *Gold Bull.*, Vol 23 (No. 3), p 83-95

Ikawa, H., Nakao, Y., and Isai, T., 1979. Theoretical Considerations on the Metallurgical Processes in TLP Bonding of Nickel-Based Superalloys, *Trans. Jpn. Weld. Soc.*, Vol 10 (No. 1), p 24-29

Isaac, T., Dollar, M., and Massalski, T.B., 1988. A Study of the Transient Liquid Phase Bonding Process Applied to a Ag/Cu/Ag Sandwich Joint, *Metall. Trans. A*, Vol 19A (No. 3), p 675-686

Iseki, T., Kameda, T., and Maruyama, T., 1984. Interfacial Reactions Between SiC and Aluminum During Joining of MMCs, *J. Mater. Sci.*, Vol 19, p 1692-1698

Jacobson, D.M. and Humpston, G., 1992. Diffusion Soldering, *Soldering Surface Mount Technol.*, Vol 10 (No. 2), p 27-32

Jacquot, P. *et al.*, 1989. Vacuum Brazing of Graphite-Metals, *Proc. Conf. Joining Ceramics, Glass and Metals*, Bad Nauheim, Germany, p 45-54

Kinloch, A.J., 1982. The Science of Adhesion, Part 2, Mechanics and Mechanisms of Failure, *J. Mater. Sci.*, Vol 17 (No. 3), p 617-651

Kon-ya, S. *et al.*, 1990. New Metallizing Process of Alumina Ceramics for Hermetic Sealing, *Proc. 3rd Electronic Materials and Processing Congress*, 20-23 Aug, San Francisco, CA, p 19-24

Korol'kov, A.M., 1956. Effect of Added Elements on the Surface Tension of Aluminum at 700 to 740 °C, *Otdelenie Teknicheskik Nauk*, Vol 2, p 35-42

Missel, L. and Platakis, N.S., 1978. Wet and Vacuum Coating Processes, *Met. Finish.*, Vol 5, p 93-97; Vol 6, p 65-69; Vol 7, p 50-54; Vol 8, p 56-61; Vol 9, p 65-68

Miyazawa, Y. *et al.*, 1989. Effect of Precompression on the Strength of Ceramic/Steel Joints, *Proc. Conf. Materials Research Society International Meeting on Advanced Materials*, Tokyo, Vol 8, *Metal-Ceramic Joints*, p 131-137

Mizuhara, H. and Mally, K., 1985. Ceramic-to-Metal Joining with Active Braze Filler Metal, *Weld. J.*, Vol 64 (No. 10), p 27-32

Mizuishi, K., Tokuda, M., and Fujita, Y., 1988. Fluxless and Virtually Voidless Soldering for Semiconductor Chips, *IEEE Trans. Components Hybrids Manuf. Technol.*, Vol 11 (No. 4), p 447-451

Nakagawa, H., Lee, C.H., and North, T.H., 1991. Modeling of Base Metal Dissolution Behavior During Transient Liquid-Phase Brazing, *Metall. Trans. A*, Vol 22A (No. 2), p 543-555

Nakahashi, M. *et al.*, 1985. Transient Liquid Phase Bonding for Heat Resistant Steels, *J. Jpn. Inst. Met.*, Vol 49 (No. 4), p 285-290

Niemann, J.T. and Wille, G.W., 1978. Fluxless Brazing of Aluminum Castings, *Weld. J.*, Vol 57 (No. 10), p 285S-291S

Norris, B., 1986. Liquid Interface Diffusion (LID) Bonding of Titanium Structures, *Proc. Conf. Designing With Titanium*, Bristol, UK, July, p 83-86

Ornellas, D.L. and Catalano, E., 1974. Diffusion Bonding of Gold to Gold, *Rev. Sci. Instrument.*, Vol 45 (No. 7), p 955

Owczarski, W.A., 1962. Eutectic Brazing of Zircaloy 2 to Type 304 Stainless Steel, *Weld. J.*, Vol 41 (No. 2), p 78s-83s

Pearshall, C.S., 1949. New Brazing Methods for Joining Non-metallic Materials to Metals, *Mater. Meth.*, Vol 30 (No. 6), p 61-62

Peteves, S.D., 1989. *Designing Interfaces for Technological Applications: Ceramic-Ceramic, Ceramic-Metal Joining*, Elsevier

Ramirez, J.E. and Liu, S., 1992. Diffusion Brazing in the Nickel-Boron System, *Weld. J.*, Vol 71 (No. 10), p 365s-375s

Ricks, R.A. *et al.*, 1989. Transient Liquid Phase Bonding of Aluminum-Lithium Base Alloy AA8090 Using Roll-Clad Zn Based Interlayers, *Proc. Conf. Aluminum Lithium Alloys*, Williamsburg, VA, March, p 441-449

Schultze, W. and Schoer, H., 1973. Fluxless Brazing of Aluminum Using Protective Gas, *Weld. J.*, Vol 52 (No. 9), p 644-651

Sloboda, M.H., 1961. Design and Strength of Brazed Joints, *Weld. Met. Fabr.*, Vol 7, p 291-296

Teng, T.G., 1989. Precision Fabrication of MMW TWT Circuit With Solid-Liquid Transition Braze, *Proc. Conf. Tomorrow's Materials Today*, Reno, NV, May, p 790-797

Thamm, F., 1976. Stress Distribution in Lap Joints with Partially Thinned Adherends, *J. Adhesion*, Vol 7, p 301-309

Timoshenko, S., 1925. Analysis of Bi-metal Thermostats, *J. Opt. Soc. Am. Rev. Sci. Instrum.*, Vol 11 (No. 9), p 233-255

Totty, D.R., 1979. Brazing Cast Iron—No Longer a Problem?, Paper 7, *Proc. 3rd Int. Brazing and Soldering Conf.*, London

Tuah-Poku, I., Dollar, M., and Massalski, T.B., 1988. A Study of the Transient Liquid Phase Bonding Process Applied to a Ag/Cu/Ag Sandwich Joint, *Metall. Trans. A*, Vol 19A (No. 3), p 675-686

Xian, A. and Si, Z., 1992. Interlayer Design for Joining Pressureless Sintered Sialon Ceramic and 40Cr Steel Brazing With Ag57Cu38Ti5 Filler Metal, *J. Mater. Sci.*, Vol 27 (No. 3), p 1560-1566

Yung, E.K. and Turlik, I., 1991. Electroplated Solder Joints for Flip-Chip Applications, *IEEE Trans. Components Hybrids Manuf. Technol.*, Vol 14 (No. 3), p 549-559

Chapter 5

The Joining Environment

5.1 Introduction

When considering the metallurgical aspects of soldering and brazing in Chapters 2, 3, and 4, it was assumed that the components and the filler are perfectly clean and remain so throughout the process cycle, enabling the constituents to freely interact and causing the filler metal to spread over the component surfaces. However, this situation represents the ideal case. In reality, oxides and other nonmetallic compounds are usually present on surfaces that have been exposed to ambient atmospheres, and these will interfere with or inhibit wetting and alloying. Any oxygen or moisture present in the joining environment will further exacerbate this effect, particularly as the kinetics of oxidation reactions are highly temperature dependent. Thus, the nature and quality of joints depend not only on alloying reactions but also on the processing environment—in particular, on whether the surroundings are oxidizing, reducing, or neutral. The term "surroundings" refers to both the gas atmosphere itself and any chemicals, such as fluxes, that are in the vicinity of the workpiece. These aspects are considered in sections 5.2 and 5.3.

Materials used in joining, whether filler alloys, fluxes, or atmospheres, are becoming increasingly subjected to restrictions on the grounds of health, safety, and pollution concerns. These regulations can limit the choice of materials and processes that are deemed acceptable for industrial use. This issue is addressed in section 5.5.

Most nonmetallic materials, even when their surfaces are clean, are not wetted by conventional filler metals. Where wetting does occur, the contact angle between the molten filler and the parent material is often high and thus the filler does not spread over the component surfaces. This situation cannot be remedied with the help of chemical fluxes, which are unable to change the physical properties of the intrinsic materials that govern the wetting characteristics, as explained in Chapter 1.

Wetting and spreading on nonmetals can be induced by incorporating within the filler alloy highly active elements, such as titanium, which react chemically with the base materials to form interfacial compounds that the filler can wet. Although the manner in which reactive fillers promote wetting is normally different from that of chemical fluxes, they can also be used to promote wetting of oxidized metal surfaces and thereby provide a fluxing action. Owing to the fact that the active constituent of the filler metal can reduce the oxides of less refractory metals, it will remove this surface film, enabling wetting to proceed in a conventional manner through alloying. Reactive fillers are described in section 5.4, and a process that utilizes reactive alloys as self-fluxing filler metals is described in Chapter 7.

5.2 Joining Atmospheres

Many types of assemblies demand furnace brazing under a protective atmosphere, including

Principles of Soldering and Brazing

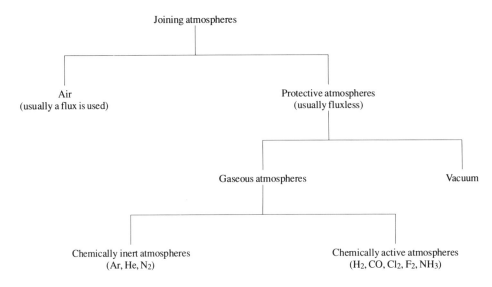

Fig. 5.1 Joining atmospheres

assemblies intended for service in a vacuum environment, which must be free from volatile contaminants, and parent metal components that are disfigured by oxide scale. The categories of joining atmospheres that are available and their interrelationships are shown in Fig. 5.1. Generally, fluxes are needed only when carrying out the joining operation in air or other oxidizing environments.

Two distinct types of gaseous atmosphere are used for joining:

- Chemically inert gas atmospheres (e.g., argon, nitrogen, helium, vacuum). These function by excluding oxygen and other gaseous elements that might react with the components to form surface films and inhibit flowing of and wetting by the filler metal.
- Chemically active atmospheres, both gases and fluxes, which are designed to react with surface films present on the components and/or the filler metal during the joining cycle and remove them in the process. These atmospheres may either decompose surface films (as does hydrogen when acting on certain oxide or sulfide layers, for example) or react with the films to produce compounds that can be displaced by the molten filler metal. An example of the latter is magnesium vapor that is introduced during the furnace brazing of aluminum. The vapor reacts with the alumina surface coating to form a complex aluminum-magnesium oxide spinel that is readily broken up by molten filler metals.

Controlled gas atmospheres require a confining vessel, which invariably means a furnace of some type. Furnace joining also offers other advantages:

- The process may be easily automated for either batch or continuous production, because the heating conditions can be accurately controlled and reproduced without the need for much operator skill.
- Furnace joining allows uniform heating of components of almost any geometry and is suitable for parts that are likely to distort if heated locally.
- The atmospheric protection afforded leads to economies with regard to the use of flux and finishing operations, such as cleaning and the removal of flux residues.

Against this must be considered the following potential disadvantages:

- Capital costs of the equipment, including the associated gas atmosphere handling or vacuum system, may be significant in relation to processing costs.
- The entire assembly is heated during the process cycle, which can result in a loss of me-

chanical properties, even to components removed from the joint area.
- The range of permissible parent materials and filler metals tends to be restricted to elements and chemicals of low volatility to avoid contamination of the furnace. For a similar reason, most fluxes are undesirable.

Certain metals are embrittled on heating in the presence of standard gas atmospheres (oxygen, nitrogen, hydrogen, and carbon-containing gases) and must therefore be joined in a vacuum furnace. These are principally the refractory metals beryllium, molybdenum, niobium (columbium), tantalum, titanium, vanadium, and zirconium. On the other hand, tungsten, which is also a refractory metal, can be brazed in air under cover of mild fluxes. Thus, the requirements of each metal or alloy must be individually assessed.

5.2.1 Atmospheres and Reduction of Oxide Films

A principal process requirement for successful soldering and brazing is to ensure that surfaces are free from oxides and other films that can inhibit wetting by the molten filler metal and the formation of strong metallic bonds. The ability to remove a layer of oxide from a given metal depends on the ease of either physically detaching the film from the underlying metal or of chemically separating the oxygen ions from the metallic ions present in the oxide, that is, the strength of the relevant chemical bonds. Chemical reduction of metal oxide by atmospheres will be considered first.

A measure of the strength of the metal-oxygen chemical bond is given by the change in the Gibbs free energy when that metal reacts to form the oxide, as detailed in Appendix 5.1 at the end of this chapter. Table 5.1 shows the Gibbs free energy of formation of oxides for a selection of metals at room temperature. This formation energy is sometimes referred to reciprocally as the dissociation potential of the oxide. The least stable metal oxides are those of the noble metals—gold, silver, and members of the platinum group. These metals are therefore the most readily soldered or brazed to, whereas the refractory metals and aluminum, beryllium, and magnesium, which have particularly stable oxides, are the most difficult to join. It is precisely because gold will not form a stable oxide in air that it is widely used as a surface coating for fluxless joining processes (see section 4.2).

Table 5.1 Gibbs free energy of formation of selected metal oxides at room temperature (25 °C, or 77 °F). The more negative the value, the more stable the oxide.

Element	Oxide	Gibbs free energy of formation, kJ/mol O_2
Gold	Au_2O_3	+111
Silver	Ag_2O	–22
Lead	PdO	–38
Copper	CuO	–180
	Cu_2O	–333
Nickel	NiO	–432
Iron	FeO	–488
	Fe_2O_3	–493
	Fe_3O_4	–506
Tin	SnO_2	–518
Indium	In_2O_3	–622
Chromium	Cr_2O_3	–697
Titanium	TiO_2	–912
Aluminum	Al_2O_3	–1048
Beryllium	BeO	–1163
Magnesium	MgO	–1138

Other factors must be considered in connection with oxide reduction. In particular, many metals form different oxides of varying stability—for example, cuprous oxide (Cu_2O) and cupric oxide (CuO). Furthermore, oxides formed on alloy surfaces are not generally pure metal oxides but rather compounds or other forms of mixed oxides. Because iron and chromium can have isomorphous oxides, Fe_2O_3 and Cr_2O_3, a solid solution oxide, $(Fe,Cr)_2O_3$, is formed on chrome steels over a certain range of temperatures. This mixed oxide is more difficult to reduce than Fe_2O_3, but is easier than Cr_2O_3. Many alloys are covered by oxides of nonuniform composition and structure, adding further complexity to the subject. In its present state of development, chemical thermodynamics is not able to predict accurately conditions under which dissociation of oxides will occur, but can only provide a semiquantitative indication, particularly when the kinetics of reaction are taken into account. For this reason, the thermodynamic principles for analyzing oxide reduction will be considered only for pure metals.

5.2.2 Thermodynamic Aspects of Oxide Reduction

All chemical reactions are reversible, including oxidation reactions. In general, the oxidation of any metal can be described by an equation of the form:

$$nM + \left(\frac{m}{2}\right)O_2 \leftrightarrow M_nO_m \quad \text{(Eq 5.1)}$$

The reaction will proceed spontaneously in either direction—namely, oxidation of the metal, or, conversely, reduction of the oxide, if it is energetically favorable to do so. In determining the energy balance in a chemical reaction, one enters the realm of thermodynamics. From the name, one might understand thermodynamics to be concerned specifically with the flow of heat, but it has come to denote the study of the energy of material systems and their surroundings. A condensed treatment of relevant thermodynamic functions and their relationships, which has by necessity required a degree of oversimplification, is given in Appendix 5.1 at the end of this chapter. A more rigorous treatment is given in standard textbooks on thermodynamics, such as those listed in the bibliography given in the Preface.

In the appendix, it is shown that the thermodynamic relationship among (1) the driving force, which is represented by the Gibbs free energy change (ΔG) per molecule of oxygen produced in the oxidation reaction, (2) temperature, and (3) the dissociation pressure of the oxide in question, $P_{O_2}^M$, is:

$$\Delta G_{(2/m) M_nO_m} = RT \ln P_{O_2}^M \quad \text{(See Eq 5A.12)}$$

where R is the universal gas constant and T is temperature in degrees Kelvin. As pointed out in the appendix, derivation of this relationship required some simplifying assumptions, including insolubility of oxygen in the metal, but it is substantially valid in most situations involving pure metals.

The free energy change (ΔG) for oxidation reactions involving a series of metals can be charted on a diagram as a function of temperature, as shown in Fig. 5.2. This representation is known as an Ellingham diagram. It may be noted that the slopes of the curves on the Ellingham diagram, for solid metals at atmospheric pressure, are largely identical. This is because the slope is a measure of the entropy increment $\Delta S(T)$ for the designated reaction, as defined in the appendix, according to the relationship:

$$\left[\frac{\Delta G(T)}{T}\right]_P = -\Delta S(T)$$

The free energy curves are essentially straight lines below and above the melting point. The discontinuous changes in slope that occur at the melting point are due to changes in ΔS between the solid and liquid states. Note that the slopes of the free energy curves characterizing oxidation reactions where the product is a vapor (e.g., $2C + O_2 = 2CO$) are radically different from those where the product is a solid, as can be seen from Fig. 5.3. The difference reflects the entropy changes being different in the two cases.

The Ellingham diagram can be used to determine whether, in principle, an atmosphere is capable of reducing surface oxides, although it does not provide any indication of the kinetics of the reactions. The use of the Ellingham diagram in soldering and brazing practice is described below.

At any given temperature, the smaller the equilibrium partial pressure of oxygen in the metal oxide, the stronger the bond between the oxide and the parent metal, that is, the greater the stability of the oxide. The partial pressure of a gas in an atmosphere is defined in section 1.4.2.5. Thus, the tendency for the oxide to decompose will be greater the lower the oxygen content of the atmosphere and the higher the temperature. Expressed mathematically, at a given temperature: If $P_{O_2}^M < P_{O_2}^A$, where $P_{O_2}^M$ is the dissociation pressure of the oxide and $P_{O_2}^A$ is the partial pressure of oxygen in the atmosphere, the oxide is stable. However, if $P_{O_2}^M > P_{O_2}^A$, the oxide will spontaneously reduce to the metal, which will then be exposed to the filler alloy.

As shown on the Ellingham diagram (Fig. 5.3), the dissociation pressure $P_{O_2}^M$ increases with rising temperature for all metals. Thus, for an atmosphere containing a given partial pressure of oxygen, there exists a critical temperature, T_c, at which the boundary condition will apply, whereby:

$$P_{O_2}^M = P_{O_2}^A$$

and the oxide will commence to dissociate spontaneously, according to thermodynamic theory.

This condition can, in principle, be achieved by varying either the oxygen partial pressure or the temperature, or both, as will be explained in the following section. However, there are practical limits to this, notably when the temperature required is so high that the material melts or some physical property degrades irreversibly or if the

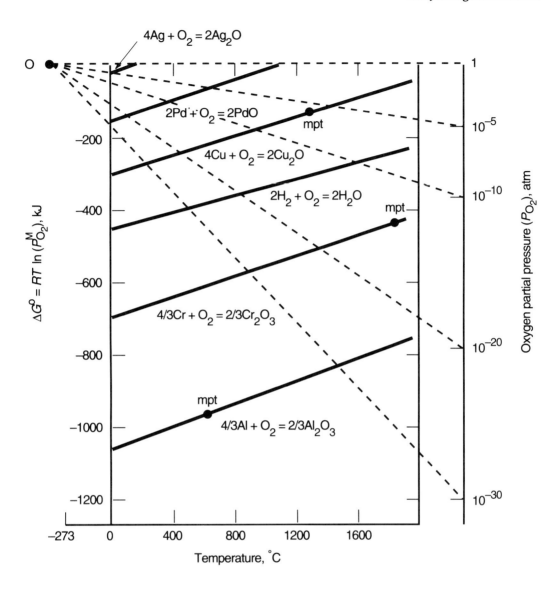

Fig. 5.2 Simplified Ellingham diagram showing the free energy change for oxidation of several metals. Oxide stability is reduced by elevated temperature and decreased oxygen partial pressure. Each dashed line corresponds to the Gibbs free energy change as a function of temperature, relating to a particular oxygen partial pressure.

partial pressure of oxygen needs to be so low that it is practically unattainable. The farther down the diagram a particular metal-oxygen reaction curve lies, the more stable is the oxide and the more difficult it is to reduce (i.e., higher temperatures and atmospheres of lower oxygen content are required to effect reduction).

When applying the Ellingham diagram to joining, and particularly to brazing, the following points should be borne in mind:

- The oxides present on the surfaces are of pure composition, corresponding to those represented on the diagram.

- The diagram can be used only to establish whether the reduction of the oxide in question is thermodynamically possible. It does not provide any information about the rates of oxide removal. Even when conditions are favorable for reduction of the oxide, the rate might not be sufficiently fast to make a process

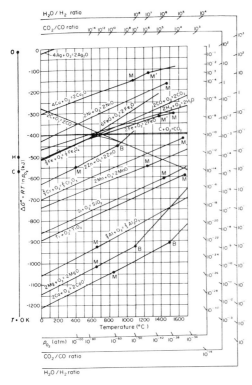

Fig. 5.3 Ellingham diagram for selected oxides. M, melting point of metal; B, boiling point of metal; M', melting point of oxide

cess temperature is required to drive the reaction at a reasonable rate, and the oxide is seldom pure.

For many other metals, heating alone in air is not adequate to reduce the oxide, because the components are degraded or even melt before the critical temperature, T_c, is reached. Moreover, the rate of oxidation roughly doubles with each 25 °C (45 °F) rise in temperature. Thus, stable oxides become progressively thicker and tenacious, and consequently more difficult to remove, over the time interval that the component is being heated to the critical temperature. Excessive oxidation can damage component surfaces, particularly if the film spalls off locally, because the rate of oxidation will be nonuniform over the surface, producing an unsightly finish. For these reasons, it is usual practice to heat the components in a suitable protective atmosphere or vacuum, which will both protect the surfaces from further oxidation and reduce the partial oxygen pressure and hence the critical temperature.

The conditions of temperature and oxygen partial pressure required to spontaneously reduce a metal oxide can be deduced from the Ellingham diagram, a simplified version of which is given in Fig. 5.2. Reduction will occur when the free energy curve for metal-oxide formation lies above the oxygen partial pressure curve at the temperature of interest, that is, when the oxygen pressure in the atmosphere is less than that which will cause the metal under consideration to oxidize. Thus, the critical temperature for the reduction of PdO decreases from 1020 °C (1868 °F) in pure oxygen at atmospheric pressure to 380 °C (715 °F) if the oxygen partial pressure is decreased to 10^{-10} atm (10^{-5} mPa). It can be seen from the more detailed Ellingham diagram given in Fig. 5.3 that oxide reduction in vacuum is practicable for copper, palladium, and silver. For metals having oxidizing reaction curves that are located below the 10^{-10} atm (10^{-2} mPa) oxygen partial pressure curve, such as chromium and aluminum, it will be energetically favorable for the metal to oxidize by reaction with the residual oxygen and any water vapor present in the furnace atmosphere in most industrial plants.

As mentioned above, care must be taken to select an atmosphere that is inert toward all of the metals in the assembly being joined. Vacuum can degrade certain materials, notably brass, even at soldering temperatures, due to the loss of zinc through volatilization—a consequence of the high vapor pressure of this element. Likewise, lead-containing solders are unstable in high vacuum at temperatures much above 300 °C (570 °F) and are not recommended for use under these conditions.

based on atmosphere control alone economically viable.

- The diagram indicates only the conditions under which the oxide is spontaneously reduced to the metal. In practice, it is often found that oxide removal will occur at higher partial oxygen pressures than the limiting value required to satisfy thermodynamic criteria, as pointed out in section 5.2.3.2.

5.2.3 Practical Application of the Ellingham Diagram

5.2.3.1 Joining in Inert Atmospheres and Vacuum

As seen in the simplified Ellingham diagram shown in Fig. 5.2, silver oxide (Ag_2O) and palladium oxide (PdO) can be reduced to metal below their melting points by heating in air to 180 and 920 °C (356 and 1688 °F), respectively. In practice, silver oxide is not considered to dissociate until it is heated to about 190 °C (375 °F); the ex-

The Joining Environment

Table 5.2 Boiling/sublimation temperature of selected elements at a pressure of 10^{-10} atm (10^{-2} mPa), in rounded values

Element	Boiling/sublimation temperature	
	°C	°F
Cadmium	100	212
Zinc	150	302
Magnesium	210	410
Antimony	300	572
Bismuth	350	662
Lead	375	707
Indium	525	977
Manganese	550	1022
Silver	630	1166
Aluminum	725	1337
Tin	730	1346
Copper	780	1436
Chromium	800	1472
Gold	880	1616
Palladium	905	1661
Iron	950	1742
Cobalt	1020	1868
Nickel	1025	1877
Titanium	1130	2066
Molybdenum	1680	3056
Tungsten	2230	4046

Table 5.2 lists the boiling/sublimation temperatures of selected elements at 10^{-10} atm (10^{-2} mPa). For metals to be joined under reduced pressure, the process temperature must be considerably less than the boiling/sublimation temperature (by a factor of less than ½ in K/K), if volatilization is not to be significant.

The oxygen partial pressure in a vacuum furnace can be reduced substantially below the gas pressure in the vacuum by repeatedly pumping out and backfilling the chamber with a dry, oxygen-free gas (see section 1.4.2.5). Care must be taken to ensure that the inlet system is completely leak-tight. Otherwise, some oxygen will be bled into the furnace, and this will impair or even nullify the benefit of the inert atmosphere. A periodic flushing of the chamber with the gas will also serve to prevent any buildup of oxygen released in the dissociation of oxides during the heating cycle. The minimum partial oxygen pressure that can be achieved using high-quality industrial equipment is on the order of 10^{-10} atm (10^{-2} mPa). Note that it is convenient to use an atmosphere as the unit of pressure in thermodynamic calculations, and this convention is also applied to Ellingham diagrams.

The effectiveness of using the process temperature and oxygen partial pressure to control oxide reduction, or at least prevent oxidation, is limited by the presence of adsorbed water vapor on the walls of the vacuum chamber and on other free surfaces. The desorption of water vapor effectively increases the oxygen partial pressure in the chamber, as will be shown below, and this can have a deleterious effect on the oxide removal process. Therefore, it is good practice to heat the walls of the chamber to promote desorption, while simultaneously removing the vapor from the chamber by alternately pumping out and/or flushing with dry, inert gas before commencing the heating cycle.

Owing to increasingly stringent environmental legislation, joining in inert atmospheres is gaining in popularity. This is particularly true in the electronics industry, where the trend is toward nitrogen atmosphere furnaces for both wave and reflow soldering processes. In commercial systems, the nitrogen ambient atmosphere typically contains less than 10 ppm of other species. The running costs associated with the large volumes of nitrogen that are required to achieve this quality of atmosphere are offset by the ability to dispense with postjoining treatments.

5.2.3.2 Joining in Reducing Atmospheres

If the partial oxygen pressure surrounding the workpiece cannot be sufficiently lowered by introducing a vacuum or inert gas environment, then a reducing atmosphere might be able to remove the oxide. The three most widely used reducing gases are hydrogen, carbon monoxide, and "cracked" or dissociated ammonia.

For the case of a hydrogen atmosphere, the boundary condition for oxide reduction is governed by two reactions:

$$n\text{M} + \left(\frac{m}{2}\right)\text{O}_2 \leftrightarrow \text{M}_n\text{O}_m$$

determined by $\Delta G_{\text{M}_n\text{O}_m}$, and:

$$2\text{H}_2 + \text{O}_2 \leftrightarrow 2\text{H}_2\text{O}$$

determined by $\Delta G_{2\text{H}_2\text{O}}$. Subtracting these equations gives:

$$n\text{M} + m\text{H}_2\text{O} \leftrightarrow \text{M}_n\text{O}_m + m\text{H}_2$$

determined by $\Delta G_{\text{M}_n\text{O}_m + m\text{H}_2}$, and

$$\Delta G_{\text{M}_n\text{O}_m + m\text{H}_2} = \Delta G_{\text{M}_n\text{O}_m} - \left(\frac{m}{2}\right)\Delta G_{2\text{H}_2\text{O}}$$

Now, as shown in Appendix 5.1:

$$\Delta G_{M_nO_m + mH_2} = mRT \ln\left(\frac{P^A_{H_2}}{P^A_{H_2O}}\right)$$

and

$$\Delta G_{M_nO_m} = \frac{m}{2} RT \ln P^M_{O_2}$$

$$= \frac{m}{2} RT \ln P^A_{O_2}$$

under the boundary condition for oxidation. Therefore:

$$RT \ln P^A_{O_2} = 2RT \ln \frac{P^A_{H_2}}{P^A_{H_2O}} + \Delta G_{2H_2O} \quad (Eq. 5.2)$$

The term $RT \ln P^A_{O_2}$ is the oxygen potential, ΔG_{O_2}, which provides a measure of the "oxidizing strength" of the atmosphere for converting a metal to its oxide via the reaction given in Eq 5.1. According to Eq 5.2, the oxygen potential of the atmosphere can be reduced by increasing the hydrogen content and/or by lowering the fraction of water vapor present, other factors being equal.

If instead of hydrogen, a reducing atmosphere of carbon monoxide is used, the reduction equation is then:

$$2CO + O_2 \leftrightarrow 2CO_2$$

which, by similar reasoning, leads to the equality:

$$RT \ln P^A_{O_2} = 2RT \ln\left(\frac{P^A_{CO}}{P^A_{CO_2}}\right) + \Delta G_{2CO_2}$$

In this case, the oxygen potential can be reduced by increasing the carbon monoxide fraction of the atmosphere and/or reducing the carbon dioxide content.

Rather than performing the analysis algebraically to determine the temperature and concentration of gases in the atmosphere to effect reduction, it is possible to obtain the solution more simply by a graphical procedure using the Ellingham diagram. For convenience to the user, the Ellingham diagram is provided with a series of side-scales, most usually the partial oxygen pressure and frequently also the corresponding ratios of H_2/H_2O and CO/CO_2. On the left-hand side is shown an axis for $T = -273$ °C (0 K), with points marked at values of the free energies, ΔG, at this temperature for the hydrogen/oxygen (point H), the carbon monoxide/oxygen (point C), and other reactions. Each of these points is associated with one of the side-scales shown in Fig. 5.3.

As an example of the use of the Ellingham diagram, we shall consider the conditions for reduction of chromium oxide. The free energy of formation of Cr_2O_3 as a function of temperature is represented by curve AB in Fig. 5.4. Values of the free energy of formation of water vapor from the reaction of hydrogen with oxygen are represented by a family of curves diverging from the point H, each curve corresponding to a different ratio of the partial pressures $P^A_{H_2}/P^A_{H_2O}$ and the molar ratio H_2/H_2O in the atmosphere. When curve AB crosses PQ, belonging to the family of water vapor curves and corresponding to a water vapor/hydrogen partial pressure ratio of 10^{-5}, the chromium/oxygen and hydrogen/oxygen reactions are in equilibrium, because their respective free energies are the same. This means that the oxygen potentials for the two reactions are identical. When curve AB lies above PQ at a particular temperature, chromium oxide will be spontaneously reduced by the hydrogen to form chromium and water vapor, because the latter is more stable than Cr_2O_3. The reverse is true when curve AB lies below PQ.

Therefore, at any particular temperature, it is possible to use this graphical representation to identify the H_2/H_2O partial pressure ratio that is required to reduce the surface oxide on a metal. For example, a partial pressure ratio of H_2/H_2O greater than or equal to 10^{-5} should reduce Cr_2O_3 to chromium at 800 °C (1470 °F) and above. The calculation of the conditions needed to reduce oxides using carbon monoxide is similar, except that point C on the 0 K axis and the CO/CO_2 partial pressure ratio scale are now employed (Fig. 5.3). A graphic demonstration of the improvement in wetting by improving the quality of the joining atmosphere is provided by Fig. 5.5, which shows the area over which a pellet of molten copper spreads at 1120 °C (2050 °F) as a function of the H_2O/H_2 partial pressure ratio.

In joining technology, the fraction of water vapor in a hydrogen atmosphere is generally expressed in terms of the dew point, a parameter that can be directly measured. By definition, the dew point is the temperature at which water vapor in a given enclosure is saturated. The dew point may

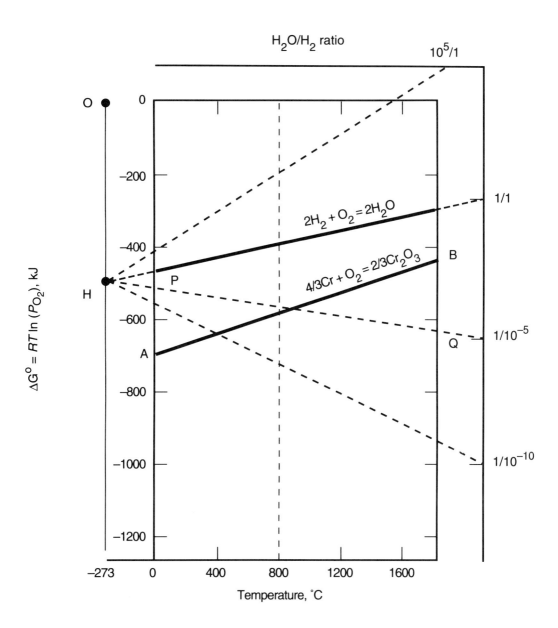

Fig. 5.4 Simplified Ellingham diagram illustrating the graphical method for determining the temperature and H_2O/H_2 ratio that will spontaneously reduce a metal oxide to metal. The set of dashed lines corresponds to the Gibbs free energy change as a function of temperature for the reaction of hydrogen with oxygen to produce water vapor for different H_2O/H_2 ratios.

be measured simply by observing the onset of condensation of moisture on a polished metal surface as the temperature is lowered. The relationship between dew point and the logarithm of the partial pressure of water vapor is shown in Fig. 5.6. The dryness that can be obtained with conventional drying agents at room temperature is marked on this curve. It is clear that both silica gel and phosphorus pentoxide (P_2O_5) are considerably less effective than chilling the hydrogen to the liquid nitrogen boiling point (–196 °C, or –321 °F) at removing water from a gas supply that is piped into a furnace.

Commercial supplies of hydrogen, nitrogen, and other gases will inevitably contain some oxygen and water vapor. Both species must be moni-

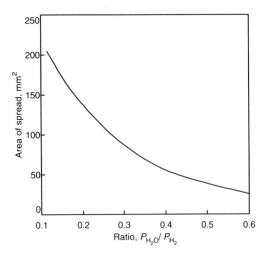

Fig. 5.5 Area of spread by molten copper on mild steel, at 1120 °C (2050 °F), in controlled H_2O/H_2 ratio atmospheres. Source: Bannos [1984]

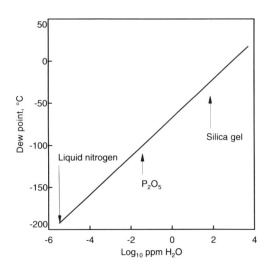

Fig. 5.6 Relationship between dew point and fraction of water vapor in an atmosphere. Drying agents are considerably less effective than low temperatures at reducing the moisture level. Note: vol% = ppm × 10^{-4}

tored and removed as necessary to meet the conditions for oxide reduction. Highly specified joining processes often use bottled gases because of their certified purity. Large-scale industrial processes, on the other hand, tend to rely on cheaper gas supplies. Widely used in these situations are liquid nitrogen that is allowed to vaporize on site; cracked ammonia, consisting of hydrogen and nitrogen in a 3:1 atomic ratio; and burnt natural gas, which contains a mixture of hydrogen, nitrogen, carbon monoxide, carbon dioxide, and some unreacted methane. Nitrogen boiled off from a cryogenic tank containing the liquefied gas possesses lower levels of oxygen and water vapor than all but the purest grades of bottled nitrogen. It is also relatively inexpensive, being comparable in price per liter to bottled mineral water.

It is evident from the Ellingham diagram that the stability of metal oxides decreases as the temperature is increased and the oxygen partial pressure is reduced. Consequently, reducing atmospheres are not usually effective toward the oxides of common industrial metals below about 750 °C (1380 °F) and therefore cannot be usefully exploited in soldering processes.

Normal chemical reduction is not the only mechanism that can remove surface oxides. It is often possible to perform flux-free joining at oxygen pressures that are significantly higher than the dissociation pressure of the oxide concerned. For example, the dissociation pressure for titanium di-

oxide at 1000 °C (1830 °F) is 10^{-30} atm, whereas fluxless brazing and even diffusion bonding of titanium is possible at this temperature under moderate vacuum conditions (10^{-9} atm = 0.1 mPa). This is because titanium has a capacity for absorbing large volumes of gases such as hydrogen, nitrogen, and oxygen on heating above about 800 °C (1470 °F), and this, in effect, means that its surface remains free of an oxide coating during a typical joining operation [Stubbington 1988]. This characteristic should not be relied on to accommodate a poor-quality atmosphere, because the oxidation is detrimental to the mechanical properties of titanium and its alloys.

Other gases such as chlorine and fluorine are more effective than hydrogen and carbon monoxide at removing surface oxides of particular metals, as is clearly indicated on the relevant Ellingham diagrams [Wicks and Block 1963]. The power of chlorides and fluorides as cleaning agents accounts for the widespread use of chlorofluorocarbons (CFCs) by industry and for the fact that they are proving to be difficult to replace by more environmentally friendly chemicals. Such gases partly operate by converting the oxide to a halide, which is volatile at the joining temperature and thus vaporizes. These halide atmospheres also chemically attack the underlying metal and physically undermine the oxide, as occurs in the fluxing of aluminum. The relative effectiveness of these two mechanisms depends on the metals involved and the specific process environment. Accordingly, the halide gases are considered further in the next section, which deals specifically with the "fluxless" brazing of aluminum and its alloys.

5.2.3.3 "Fluxless" Brazing of Aluminum

Fluxless soldering and brazing processes are widely used, and several examples are referred to elsewhere in this book, particularly in Chapter 7. Owing to the highly refractory character of its native oxide, aluminum is harder to braze without fluxes than are most other metals; however, special processes have been developed successfully for aluminum. Because aluminum and its alloys are widely used in engineering, methods of brazing these materials merit special attention. Fluxless brazing of aluminum is described below, and processes involving fluxes are detailed in section 5.3.3.

The native oxide coating of aluminum oxide (alumina) and hydroxides on aluminum cannot be removed by reducing the oxygen partial pressure in the joining atmosphere alone. A favored approach for brazing aluminum without using a flux involves adding an agent to the furnace that is capable of reducing the oxide layer on the surface and so promotes wetting by the braze [Sugiyama 1989, p 705-706; Singleton 1970, p 845-846; Terrill et al. 1971, p 835-837]. Magnesium is one element that satisfies this requirement. Apart from being able to reduce alumina to aluminum metal, magnesium also preferentially scavenges oxygen and acts as a getter for water vapor in the brazing environment. Other effective activating agents include many of the rare earth metals, such as yttrium, beryllium, and scandium, and the alkali metals, including calcium, strontium, and lithium. Mischmetal, which is a mixture of rare earth elements in a ratio that occurs naturally in ores, can be used as a cheaper alternative to a pure rare earth metal.

Magnesium is favored over the other activating metals, because it is less expensive and has the highest vapor pressure. This means that a relatively high proportion of the magnesium vaporizes into the furnace atmosphere and is thereby active in removing the oxide layer. Metal activators with low vapor pressures must be alloyed directly to the components being joined in order for them to be similarly effective. However, this is not always feasible.

There is a fundamental problem with volatile activators, including magnesium; they are not compatible with high-vacuum systems. The vapor fouls the pumping system and spreads throughout the vacuum chamber, coating all surfaces. Furthermore, the vapor reduces the efficiency of radiant heat transfer to the assemblies being brazed and thereby increases the heating energy that is required.

The fact that the activators both clean and protect the surfaces of the aluminum alloy components means that they are by definition true fluxes, in contrast to the so-called aluminum fluxes (see section 5.3.3). For this reason, the description of this type of process as "fluxless" is a misnomer.

Instead of using activators that behave like true fluxes, it is possible to carry out the brazing of aluminum under a protective gas atmosphere by adding small amounts of certain metals (<0.1%) to the brazing alloy that destabilize the oxide and thereby promote wetting. It has been claimed that effective elements include arsenic, barium, bismuth, cobalt, iron, nickel, silver, and strontium in parts per thousand by weight to the Al-13Si eutectic braze [Schultze and Schoer, 1973]. The minor additions do not significantly affect the corrosion properties of the braze and can even slightly enhance the mechanical strength of joints made with them. At the same time, all surfaces have to be cleaned according to a procedure that replaces the native oxide with a thin layer of a stable oxihydride that resists rehydration, but which is readily displaced by the molten braze [Staniek and Wefers 1991]. The furnace atmosphere, which can be either an inert gas (such as nitrogen or argon) or a high vacuum, must have a combined oxygen and water vapor content of less than 5 ppm. Despite these stringent requirements, this method is used industrially to braze intricate assemblies, such as radiators for automotive applications on a commercial scale. It is capable of producing consistently strong joints with excellent fillets [VAW 1982].

Successful cleaning procedures are highly specific to both the substrate alloys and the brazing alloys. Effective cleaning agents for the Al-13Si braze include a dilute sodium hydroxide solution heated to 65 °C (150 °F) in which the material must be immersed for a closely defined period, followed by a "smut" removal step involving immersion in dilute nitric acid or a sulfuric-chromic acid mixture, again used under stringent conditions. Proprietary additives are often introduced to widen the tolerance of the cleaning conditions [Aluminum Association 1990].

5.3 Chemical Fluxes

Successful soldering and brazing are largely dependent on the ability of the filler to wet and spread on component surfaces. A major barrier to wetting is presented by stable nonmetallic films and coatings on the surfaces, particularly oxides and carbonaceous residues. Fluxes are chemical agents that are used to remove these layers and thereby promote wetting by the molten filler. In order to be effective in exposing a bare metal surface to the filler, a flux must be capable of fulfilling the following functions:

- Remove oxides and other films that exist on surfaces to be joined by either chemical or physical means
- Protect the cleaned joint from oxidation during the joining cycle
- Be displaced by the molten filler as the latter spreads through the joint

Ideally, the flux should leave no residues or should produce residues that are easily removed by, for example, being soluble in water. It should also be compatible with the filler and substrate materials. For example, ammonia-containing solder fluxes are not suitable for brass components, because intergranular corrosion can result through chemical reaction.

Chemical fluxes always function while in a gaseous or liquid form, although they are frequently solid at room temperature. If the flux is liquid at the joining temperature, it has to wet the joint surfaces in order to be effective. A flux that is liquid can beneficially suppress the volatilization of high-vapor-pressure constituents of filler metals and thereby improve the quality of joints.

Fluxes can be introduced to the joint in a number of ways, which will be discussed here. A flux can be applied in the form of a powder, paste, or liquid immediately prior to the heating cycle. The joint is then heated to the required bonding temperature, by which point solid fluxes have become molten, and then the filler is introduced.

A flux can also be placed within or adjacent to the joint together with the filler metal as a preform and the assembly heated to the bonding temperature. As a properly chosen flux will melt at a temperature below the melting point of the filler, the molten flux is able to spread over the joint surfaces and clean them before the filler metal melts and displaces the flux.

Another method involves introducing the flux together with the filler into a joint already held at the bonding temperature, in the form of flux-cored solder wire or flux-coated brazing rod. Although this technique is widely practiced because it is fast and convenient, it is not recommended because the heated component surfaces are unprotected until the filler is applied. More aggressive fluxes are then required, which in turn tend to accentuate corrosion and cleaning problems.

Alternatively fluxes can be applied together with the filler, prior to the heating cycle, in the form of pastes and creams, which are normally proprietary formulations. They comprise mixtures of the filler metal, which is present as a powder of a prescribed grain size range, together with a flux and an organic binder that is selected to produce the desired viscosity and to burn off without leaving contaminating residues. These pastes and creams are often used in automated soldering and brazing processes because they can be screen printed or dispensed using syringes. Because of the large surface area of the powdered filler metal in contact with the flux, corrosion is inevitable; therefore, these products have a finite shelf life.

The mechanisms of flux action are almost as diverse as the flux formulations that are commercially available. In many cases the mechanisms have not been fully elucidated, which is in no short measure due to the commercial secrecy surrounding flux compositions. Several different fluxing mechanisms cover the majority of soldering and brazing operations that are encountered. Even these are sufficiently complex not to be understood in detail at the present time. However, fluxing mechanisms can be classified according to whether they remove the nonmetallic surface coating by physical or chemical means.

A flux can chemically remove a surface coating by:

- Dissolving the coating
- Reacting with the coating to form a product that is unstable at the bonding temperature
- Reducing the oxide to metal in an exchange reaction, such as occurs when reducing gases are effective

A surface coating can also be physically removed. This usually occurs through:

- Erosion of the underlying metal. In this mechanism, the flux does not react with the surface coating itself, but is able to percolate through it and react with the metal, thereby causing detachment of the coating.
- Wetting of the coating in a manner that causes it to spall off. This mode of fluxing applies to joining processes where components are subjected to the thermal shock that cracks the coating due to the relatively brittle nature of oxide layers. Immersion in molten salts and fluidized-bed furnaces are examples of this type of process.

In addition, physical fluxing action can be achieved by applying mechanical agitation, without the need for a material fluxing agent, as discussed in section 5.3.5.

Many fluxes function by a combination of mechanisms, and thus fluxing action is best illustrated with reference to specific soldering and brazing processes. Environmental considerations are impinging heavily on the use of certain fluxes, particularly those incorporating organic materials, because these have tended to rely on CFCs to remove their residues. Hence, there is currently a move toward fluxes whose residues are soluble in water or cleaning agents that do not contain environmentally harmful substances [Lea 1991; Ellis 1991].

5.3.1 Soldering Fluxes

The overwhelming majority of soldered joints made are used to interconnect electronic components. The principal material that is joined in these applications is copper, due to its high electrical conductivity; the solder is almost invariably a tin-lead alloy. Because electronic components tend to be manufactured and stored under reasonably cool, clean, and dry conditions, they are likely to have only a thin layer of copper oxides as a barrier to wetting. This, coupled with the fact that soldering is usually performed rapidly and generally below 300 °C (570 °F), means that the fluxes required tend not to be highly aggressive chemicals. Major considerations pertaining to fluxes intended for electrical/electronic applications, other than their cleaning ability, are the nature of the residues and the ease of their removal. These factors are of concern owing to the need to avoid subsequent deterioration of the joints and ultimately failure of the circuitry through corrosion [Turbini et al. 1991].

Soldering fluxes suitable for most applications contain at least four basic ingredients, each of which has an identified role [Klein Wassink 1989, p 205-262; Manko 1992, p 21-54; Minges 1989, p 643-650]:

- Acids or halides to provide the cleaning action
- An ingredient that is liquid at the soldering temperature that seals and protects the cleaned surfaces against reoxidation
- A surfactant that promotes wetting of the joint surfaces by the active constituents
- A rheological additive to match application requirements

In practice, commercial fluxes often contain more than these four ingredients in order to meet the requirements of the soldering process. Fluxes have been formulated that are satisfactory for most pure metals and alloys, including stainless steel. For the latter material, a solution of zinc chloride in hydrochloric acid constitutes the chemically active ingredient [Mei and Morris 1992]. Noted exceptions are beryllium, chromium, magnesium, titanium, and some aluminum alloys, which are presently classified as "unsolderable" in air, unless coated with a different metal (see section 4.2.1). When selecting a flux for a particular application, it is usually good practice to follow the manufacturer's guidelines, because the effectiveness of a particular formulation tends to be highly sensitive to the combination of metals and process conditions with which it is used.

The acid ingredient of a solder flux can be either inorganic (hydrochloric acid is commonly used) or organic (e.g., carboxylic acids). A common carboxylic acid used in fluxes is abietic acid (R_3COOH, where R is an organic radical), which is a major constituent of the rosin fluxes. In both cases, the acid reacts with the surface cupric oxide and converts it to compounds that are readily removed, either chemically and/or physically, from the joint surfaces. The basic form of the chemical reactions between these fluxes and copper oxide can be described by the following equations:

$$CuO + 2HCl = CuCl_2 + H_2O$$
$$CuO + 2R_3COOH = Cu(R_3COO)_2 + H_2O$$

Copper chloride is soluble in water, while copper abiet [$Cu(R_3COO)_2$] is miscible with rosin. Hence, the compounds that now "contain" the copper oxide dissolve in the excess liquid flux to leave a clean metal surface when the flux is displaced by the molten filler metal.

Like most active flux constituents, these acids become progressively more active with increasing temperature until a point is reached when their effectiveness ceases, because they either thermally decompose or boil off. The corrosive properties of acids at room temperature can present handling problems and application difficulties, especially when the flux must either be introduced directly to the components or mixed with solder to form pastes some time before the joining operation. However, there are salts that liberate acid only when heated, thus allowing these types of problems to be avoided. One of these is zinc chloride, a halide that produces hydrochloric acid by reaction with moisture at elevated temperatures, according to the reaction:

$$ZnCl_2 + H_2O = Zn(OH)Cl + HCl$$

By using a mixture of similar salts, the activation temperature of the flux and its corrosivity can be adjusted over a wide range. In all cases the residues are highly corrosive. Other halide fluxes operate in a very similar manner to zinc chloride. For example, the amine hydrohalides, such as hydrazine hydrochloride (R_2NH_2Cl), thermally decompose with the liberation of hydrochloric acid:

$$R_2NH_2Cl = R_2NH + HCl$$

The flux constituent that protects the clean metal surface from reoxidation also serves as the carrier for the other ingredients. It need only be effective for a few seconds in many soldering processes. Alcohols, oils, esters, glycol, and even water are capable of fulfilling this function at the relatively low temperatures used for most soldering operations (<250 °C, or 480 °F). As the flux carrier constitutes by far the largest volume fraction of the formulation, it has a strong influence on the aggregate properties, even though it is not intended to be chemically active. The flux carrier is usually a mixture of chemicals formulated partly to prevent the flux from boiling violently when the workpiece reaches a specific temperature.

Surfactants are added to lower the surface tension of the liquid flux. The effect of minute additions of detergents, soaps, and soluble oils to water is well known, and these are often added to water-base fluxes to ensure wetting when either oil or grease is likely to be present on the component surfaces. The surfactant needs to be inert toward the other constituents of the flux and also to the clean metal surface. For this reason, ion-free organic complexes, similar to those widely used in the electroplating industry, tend to be favored.

The rheological agent is provided to impart the correct degree of "body" to the flux to suit the application method, whether syringe dispensing, screen printing, or stenciling.

A common commercial designation of fluxes used for soldering is as follows:

- R: rosin
- RMA: rosin mildly activated
- RA: rosin activated
- OA: organic acid
- IA: inorganic acid
- WS: water soluble
- SA: synthetically activated

This classification is somewhat loose and can be misleading rather than helpful. It does not strictly indicate the chemical characteristics of the flux, particularly its aggressiveness to specific oxides. However, because this classification is widely used, it warrants consideration here. An alternative classification of soldering flux types, devised by the International Organization for Standardization, is given in Table 5.3. Although this classification is scientifically more precise, it is less descriptive; therefore, the traditional terms tend to be preferred. The different categories can be grouped as shown below.

R, RMA, RA. Fluxes with these designations contain rosin as the principal active chemical ingredient, the difference between them being a progressive increase in the level of chemical activity from R through to RA.

OA stands for organic acid, which provides the cleaning action in this type of flux. The organic acid used will tend to be one of those described above. These fluxes are generally more aggressive than RA fluxes.

IA fluxes are based on inorganic acids, usually hydrochloric acid, and are among the most aggressive of the available fluxes.

SA, WS. Synthetically activated fluxes are formulated to have residues that are soluble in CFC solvents. With concern growing about atmospheric pollution, and CFCs in particular, the SA fluxes are likely to be superseded by newer formulations that have water-soluble residues. The WS fluxes that are currently available are comparable to the RMA fluxes in terms of their activity.

Table 5.3 Classification of soldering fluxes using the method adopted by the International Organization for Standardization

Flux type	Flux basis	Flux activation	Flux form
1 Resin	1 Rosin 2 Resin	1 Nonactivated 2 Halogen activated	A Liquid
2 Organic	1 Water soluble 2 Not water soluble	3 Not halogen activated	B Solid
3 Inorganic	1 Salts	1 With ammonium chloride 2 Without ammonium chloride	
	2 Acids	1 Phosphoric acid, 2 Other acids	
	3 Alkalis	1 Ammonia and/or amines	C Paste

Thus, a resin-base paste flux with a halogen activator is classed as type 122C.

Because flux residues, like other contaminants, can easily affect the performance and service life of electronic circuits, effective cleaning procedures assume a critical importance. Cleaning agents and schedules that are used in the electronics industry are reviewed in Klein Wassink [1989, p 241-255], Manko [1992, p 293-326] and Minges [1989, p 658-669].

A new generation of "no clean" fluxes is coming into use, where virtually all the active ingredients are volatized during soldering.

5.3.2 Brazing Fluxes

The requirements for fluxes used in brazing processes are much the same as those for soldering [Baskin 1992]. The considerably higher temperatures involved in brazing account for the compositions of brazing fluxes being rather different from those of soldering fluxes. Whereas soldering fluxes can employ water and organic liquids as their carrier, these rapidly volatilize at the higher temperatures used for brazing and are therefore ineffective. Instead, glass complexes must be used above 600 °C (1110 °F). These possess low volatility and generally also have a low permeability to air and so can provide the cleaned surfaces of the joint with the necessary protection against reoxidation.

At the lower end of the brazing temperature range (i.e., 600 to 800 °C, or 1110 to 1470 °F), the glass carrier is based on borates ($-B_xO_y$), with the ratio of oxygen to boron optimized to provide a balance between viscosity and permeability to oxygen. In general, the higher the oxygen to boron ratio, the higher the viscosity, and the permeability is correspondingly reduced. The requirement for the flux to be displaced by the molten braze places a limit on the viscosity of the flux. Above 800 °C (1470 °F), borates alone are too permeable and need to be partly replaced by silicates (Si_xO_y) so that the resulting glass is a borosilicate, which has a higher viscosity and is therefore able to protect workpieces at higher temperatures. However, borosilicate residues are largely insoluble in water, necessitating more rigorous cleaning procedures; this contrasts with those of the simple borates, which can be dissolved in water and are thus more convenient to use.

In addition to providing protection against oxidation, borates have the ability to dissolve a limited fraction of oxides from the surfaces of steel and copper components. This accounts for the effectiveness of borax (sodium tetraborate) as a flux at temperatures above about 750 °C (1380 °F). Adding boric acid or boric oxide to borax lowers the melting point of the flux, because it reduces the oxygen to boron ratio.

When heated by a torch, sodium salts, including borax, produce a bright yellow glare that is unpleasant to operators. The glare is substantially reduced by partly or totally substituting the borax with the corresponding potassium salt (potassium tetraborate), albeit with a small cost penalty.

Borates on their own are insufficiently active to clean the surfaces of many metals. Therefore, fluxes intended for general use contain a proportion of halides. Glare considerations have favored potassium halides, particularly potassium fluorides, over the equivalent sodium salts. The improved fluxing action results largely from the greater oxide dissolution ability of the fluorides. Indeed, the dominant mechanism responsible for

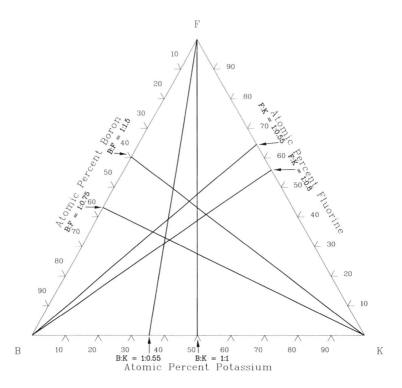

Fig. 5.7 B:K:F atom ratio of common brazing fluxes

surface cleaning by the fluoborate-type fluxes is believed to be direct dissolution of the oxides, and there is little evidence for any of the other recognized types of cleaning action referred to above. At the same time, the fluorides reduce the melting point of the flux, because they disrupt the cross-linkages in the borate atomic lattice. The fluoborate fluxes especially suit the low-melting-point silver-base quaternary alloy brazes, which melt at temperatures down to about 590 °C (1095 °F), because they can be used at comparable temperatures. The fluoride addition does, however, reduce the upper working temperature limit of these fluxes, because it increases their permeability to oxygen and reduces their thermal stability, owing to the formation of hydrogen fluoride on heating, which is volatile. Commercial fluoborate fluxes fall into the elemental composition range of B:F = 1:0.75 to 1.5; B:K = 1:0.55 to 1; F:K = 1:0.55 to 0.8 (by atom ratio). This range is represented in Fig. 5.7.

Wetting agents are not strictly necessary for brazing fluxes because, at the elevated joining temperatures used, organic residues will have decomposed, leaving carbonaceous deposits that will be either eliminated through oxidation or cleaned off the surface by the flux. Nevertheless, wetting and rheological agents are added to flux pastes to produce a smooth consistency, which aids application to the workpiece.

Commercial fluxes are proprietary formulations that contain specific ingredients tailored to application requirements and incorporate various subtleties. By way of example, many brazing operators require a flux that will coat a heated rod of the braze when dipped into a tub of the flux powder. This flux-coated rod is then applied to the workpiece, and the brazing operation is carried out using a torch in air. To satisfy this mode of application, which considerably speeds the joining procedure, fluxes have been formulated that contain close to 70% of a hydrated potassium fluoborate compound. This compound releases sufficient moisture when heated to form a sticky paste that will adhere to the metal rod. This example illustrates the finer points affecting user preference and helps to explain why it is best to consult suppliers when considering a flux for a particular requirement.

5.3.3 Fluxes for Aluminum and its Alloys

Aluminum forms a natural refractory oxide that is remarkably stable and tenacious. It is mechanically durable, with a hardness that is inferior only to that of diamond, and its high melting point (2050 °C, or 3722 °F)) reflects its high degree of physical stability. Alumina is also chemically stable to the extent that it cannot be directly reduced to the metal by aqueous reagents. On exposure to air, a layer of alumina will form almost instantaneously on the surface of aluminum and will grow to an equilibrium thickness of between 2 and 5 nm at ambient temperature. On heating to 500 to 600 °C (930 to 1110 °F), the thickness of this surface coating will increase to about 1 μm. Therefore, special fluxes have been formulated for use with aluminum alloys. These have to be particularly effective in protecting the metal from oxidation.

The aluminum fluxes divide into two categories: those that are suitable for use with solders at temperatures below 450 °C (840 °F), and those that can be used at higher temperatures with brazes. A commonality between the aluminum soldering and brazing fluxes is that they all contain halide compounds. These are highly corrosive, especially in the presence of moisture, including humid atmospheres. Therefore, all flux residues must be removed as completely as possible. The cleaning processes are very laborious and costly, and there is always a danger that some residues will survive the cleaning procedures, resulting in corrosion in the vicinity of the joint.

5.3.3.1 *Aluminum Soldering Fluxes*

The fluxes used for soldering of aluminum and its alloys are of two types: organic and chloride based [The Aluminium Association 1990, p 27-28].

Organic fluxes contain amines, fluoborates, and a heavy metal compound in an organic carrier, and they come in the form of viscous liquids or powders. A typical example of this type of flux has a composition of 83% triethanolamine, 10% fluoboric acid, and 7% cadmium fluoborate (a viscous liquid). Its operating range is 180 to 280 °C (355 to 535 °F).

The fluxing action relies on disrupting the oxide, which cracks and crazes during the heating operation due to the differential thermal expansion between the metal and oxide. This enables the flux to come into direct contact with the aluminum and deposit a film of the metal ion in the flux (in this instance, cadmium) onto the aluminum surface via an exchange reaction. Organic fluxes must not be exposed to a torch or flame; otherwise, they will char and this will impede solder flow. Aluminum alloys containing more than about 1% magnesium cannot be satisfactorily soldered using these fluxes, because magnesia is more refractory than alumina and the flux is correspondingly less effective.

Chloride-base fluxes contain zinc or tin chlorides with ammonium chloride and fluoride and are generally applied as a water-base slurry or paste to precleaned component surfaces. An example of such a flux has a formulation of 88% tin chloride, 10% ammonium chloride, and 2% sodium fluoride (powder), with a working range of 300 to 400 °C (570 to 750 °F). By substituting the tin chloride with zinc chloride, the temperature of operation can be raised to 380 to 450 °C (715 to 840 °F).

The fluxing mechanism is essentially the same as that for the organic fluxes—namely, one involving an exchange reaction whereby aluminum on the surface is replaced by zinc or tin. The effectiveness of these fluxes is reduced by the presence of silicon in the aluminum alloys, because silicon is not as amenable to the exchange reaction as is aluminum.

Because both of these fluxes operate by substituting for aluminum a metal that has reasonable oxidation resistance and that is more readily wetted by the filler metal, they are fundamentally different from conventional soldering and brazing fluxes. The latter act simply by cleaning and protecting the original surfaces of the components. Because the replacement metal is denser than aluminum, this type of fluxing process is commonly referred to as "heavy metal deposition."

The quantity of flux that needs to be applied is a function of the humidity of the ambient atmosphere. In moist atmospheres, a proportion of the flux is rendered ineffective through hydrolization by reaction with water vapor. It has been found that the quantities of flux that need to be applied can be considerably reduced by carrying out the soldering operation in a completely dry environment.

5.3.3.2 *Aluminum Brazing Fluxes*

Two principal types of fluxes are used for brazing aluminum: chloride and fluoride based [The Aluminium Association 1990, p 27-28].

Chloride formulations. The active ingredients are chlorides of the alkali earth metals. These fluxes operate by infiltrating cracks in the alumina; upon reaching the metal, they proceed to undermine the oxide layer and mechanically displace it. The flux residues left on the workpiece surfaces are highly corrosive and must be completely removed.

Fluoride formulations. Many of the well-known fluxes of this type contain a mixed sodium aluminum fluoride that, when molten, can dissolve alumina, but the residues are the source of severe corrosion if left on the components surfaces. However, by using potassium rather than sodium in the formulation, the flux can be made neutral, without compromising its ability to dissolve alumina. The proprietary Nocolok® flux comprises a eutectic between K_3AlF_6 and $KAlF_4$ which melts at 562 °C (1044 °F). The Nocolok® flux is not hygroscopic, and its residues do not corrode aluminum. Therefore, it can be applied to joint surfaces and left there. It does not spall off during thermal or other forms of stressing [Cooke, Wright, and Hirschfield 1978].

Several of the fluxes used for the brazing of aluminum and its alloys are associated with the dip method of brazing, also referred to as salt bath brazing. These fluxes are composed of chloride and fluoride mixtures of the alkaline earth elements and of aluminum. The fluxing action involves detachment of the alumina layer, resulting in the accumulation of oxide platelets in the flux and thus increasing its viscosity. This platelet formation has the undesirable feature of enhancing corrosion of the brazed components and of limiting the effectiveness of the flux, because partly adhered platelets tend to trap flux and obstruct cleaning.

There has been some disagreement about the fluxing mechanism, although the consensus is that the active constituents are the fluoride constituents, with the chlorides acting essentially as melting point depressants. Jordan and Milner [1951] claimed that the fluxing mechanism involves a cell action dependent on the presence of oxygen. More recently, Terrill *et al.* [1971] have claimed that hydrogen fluoride plays an important role. They have suggested that the fluorides in the fluxes generate hydrogen fluoride by reaction with dissolved moisture in the salt bath and with moisture picked up from the surrounding atmosphere. The hydrogen fluoride reacts with areas of aluminum exposed by crazing of the oxide surface layer, thereby liberating hydrogen, which prevents the aluminum from reoxidizing and, presumably, the gas also helps detach the remaining layer of oxide. Certainly, there is evidence for the evolution of hydrogen fluoride from the flux bath. The chlorides in the salt bath can react with the moisture to form oxychlorides. Oxides and hydrides are also formed that react with the fluxes to turn the bath alkaline and render it ineffective. These reactions progressively exhaust the flux of its activity. For this reason, it is good practice to regularly purge the bath of moisture by dipping in aluminum sheets. The aluminum reacts with the moisture to release hydrogen, as mentioned above, and the gas bubbles to the surface, where it burns with a yellow flame. The disappearance of the flame indicates that the moisture level is sufficiently low to resume brazing.

5.3.3.3 *Aluminum Gaseous Fluxes*

Gaseous fluxing of aluminum can be achieved by using halide vapors as the furnace atmosphere, under conditions that can be approximately determined from the appropriate Ellingham diagram [Terrill et al. 1971, 837-9]. This approach is based on the reduction reaction:

$$3MHa_2 + Al_2O_3 = 2AlHa_3 + 3MO$$

where Ha is the halide of metal M. Chlorides, bromides, and iodides of boron and phosphorus have been found to be particularly effective. The use of fluorides does not lead to the formation of good fillets, possibly because the aluminum fluoride is nonvolatile at typical aluminum brazing temperatures and is not completely displaced from the area of the joint by the molten braze. It has been established by experiment that the primary mechanism responsible for the fluxing action in this case is the undermining and detachment of the oxide rather than a reduction reaction, although a reduction reaction stage is probably responsible for initiating wetting [Milner 1958].

5.3.4 Self-Fluxing Filler Alloys

Certain brazing alloys have been formulated to provide self-fluxing action in the heating cycle used for bonding [Schwartz 1987, 285]. The fluxing agent is an element that has a high affinity for oxygen, such as lithium or phosphorus. Brazes of this type include copper-phosphorus, silver-copper-phosphorus, silver-copper-lithium, and copper-tin-phosphorus alloys. These brazes are used with copper-base alloys, stainless steels, and other heat-resistant alloys. The phosphorus-containing brazes produce weak joints when used to join steels due to the formation of a near-continuous interfacial layer of the brittle phase Fe_3P. However, the addition of small quantities of nickel to the phosphorus-containing brazes converts Fe_3P to Fe_2P, which is a more ductile phase.

The fluxing action occurs in the following manner (see Fig. 5.8). The liquid filler reacts with the oxide on the surface of the component to produce a molten slag. This slag, which is rich in the active element (phosphorus or lithium), then floats to the free surface of the filler and protects the joint and the filler from further oxidation. The filler wets and spreads over the clean metal surface to form a satisfactory joint. The slag itself also has a limited capability to directly dissolve surface oxides of certain metals.

Phosphorus usually represents no more than 5% of the self-fluxing alloys, and the other active elements may be present in even lower proportions. In comparison, the slag contains typically five times that amount, leaving the filler with correspondingly less of the fluxing elements, and these are predominantly concentrated in intermetallic phases such as Fe_3P.

5.3.5 Ultrasonic Fluxing

Ultrasonic soldering and brazing, which is more correctly called ultrasonic fluxing, provides a mechanical alternative to a flux for disrupting the oxide layer. Ultrasonic agitation is used to break up and displace surface films of oxide. This technique is highly effective in removing even the most tenacious oxides and is therefore applicable to virtually all metals, but is most widely used with aluminum [Jones and Thomas 1956]. The combination of ultrasonic agitation and hydrostatic pressure has been used to force molten tin-lead solder to "wet" bare alumina [Naka and Taniguchi 1992]. The bonding mechanism is most probably mechanical keying of the solder onto the rough surface of the ceramic, and the joint strengths are correspondingly low (typically 10 MPa, or 1.5 ksi).

The normal mode of operation involves applying an ultrasonically activated soldering iron to the workpiece with a metal foil or shim wedged between them [Schaffer et al. 1962]. In a variant arrangement, a solder bath is ultrasonically excited and the workpiece is dip coated with the solder. Ultrasonic fluxing is in essence a more sophisticated form of rub or scrub soldering that is used to bond silicon chips to gold-plated ceramic carriers using a gold-silicon alloy, whereby one of the components is mechanically rubbed over the molten filler metal against the mating component.

Ultrasonic fluxing is a fluxing and tinning operation for applying a metal coating to a clean sur-

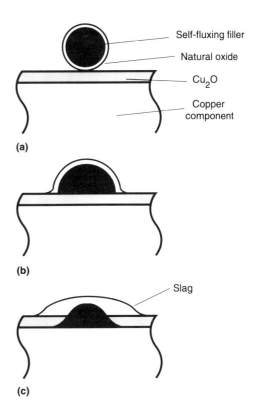

Fig. 5.8 Wetting mechanism of self-fluxing filler metals. (a) Self-fluxing filler applied to copper component. (b) Filler and its oxide melt and wet the oxide film on the component surface. (c) Oxide film on the component dissolves in the molten braze to form a slag that floats to the free surface. The filler then wets and spreads over the clean substrate surface.

face of the workpiece. Once the metal coating has been applied, it protects the workpiece surface against reoxidation, and for that reason is akin to chemical fluxing, as described above. The coated part can subsequently be soldered or brazed using a conventional process.

Ultrasonic fluxing requires special soldering tips and transducer assemblies tailored for specific applications. The need for special ultrasonic equipment and for separate tinning and joining operations makes this method relatively costly, so that it is generally limited to applications where conventional soldering and brazing, with and without fluxes, cannot be used. However, it is attractive as a combined method of fluxing and tinning, because it is residue free. With the progressive improvements being made to ultrasonic technology, this technique is entering into use for applying metallizations to refractory materials such as alumina [Naka and Okamoto 1989].

5.4 Reactive Filler Alloys

For a molten alloy to wet and spread over the joint surfaces, there must be a degree of chemical interaction between the filler metal and the parent materials. Where the parent materials are metals with clean surfaces, the interaction generally means alloying, with some associated erosion of the joint surfaces and compound formation. The wetting of nonmetallic components by filler metals is more difficult, but can be accomplished by two principal routes. One of these is to apply metallizations to the joint surfaces so as to render them essentially metallic in character. The chosen metallization must obviously not be significantly soluble in the filler alloy or dewetting can occur when the molten solder or braze contacts the nonmetal. The application of metallizations that are wettable by fillers is discussed in section 4.2. Another approach involves incorporating small quantities of highly reactive elements into the filler. Provided that at least one of the products of reaction with the base material is metallic and remains as a layer on the surface of the nonmetal, the filler alloy can wet it and form sound joints.

When selecting a reactive filler alloy, the active ingredient, its optimum concentration, and the appropriate processing conditions need to be considered. For the active element incorporated in the filler to be effective, it must be reactive with the nonmetal. One of the most commonly used active constituents is titanium which, when added to the silver-copper-base brazes, can facilitate wetting of the majority of engineering ceramics. Less reactive elements, such as chromium, and more reactive elements, such as hafnium, are also used, the choice depending on the parent material and other factors, as will be explained below.

There is often a wide temperature margin between the melting point of the activated filler and its temperature of application, particularly solders [Xian and Si 1991]. The process temperature is governed by the reactivity of the filler, which increases with temperature. If this temperature margin is substantial, the heating cycle should be kept short in order to prevent extensive erosion of the joint surfaces or the growth of thick interfacial phases. The nature of the interfacial phases, their morphology, and the characteristics of the boundaries between these phases and the adjacent ones in the joint frequently govern the mechanical resilience of the joints that form.

The three key considerations affecting the active element—namely, the choice of the element, its concentration in the filler alloy, and the minimum process temperature—can all be ascertained from chemical thermodynamics. The relevant calculations presuppose that thermodynamic equilibrium is achieved, which is not unreasonable given the relatively high temperatures involved. The principles used in such an exercise are outlined below.

Reactive filler alloys are primarily used for joining to engineering ceramics and refractory metals. Most other materials tend to degrade, if not melt, at the relatively high process temperatures required to achieve wetting by the active constituent (at least 800 °C, or 1470 °F). There are many different types of engineering ceramics, oxides, carbides and nitrides being the most common [Schwartz 1990]. Reactive fillers are also finding use with metal-matrix composite materials in which the nonmetallic reinforcement phase, often silicon carbide or carbon/graphite, can constitute more than 50 vol%. The refractory reinforcement phase inhibits wetting by conventional filler alloys. The joining of ceramics is treated extensively by Nicholas [1990] and Schwartz [1990], and only a brief review of some of the critical considerations follows below.

5.4.1 Wetting of Ceramics

The factors that govern the wetting of ceramics by metal fillers will now be considered, using as an example a combination comprising a silicon nitride ceramic and titanium-bearing filler alloys.

Nonactivated brazes do not wet silicon nitride, even in the presence of fluxes: measured contact angles are typically greater than 130°. The addition of titanium promotes wetting and spreading with low contact angles (<15°), provided that the concentration is sufficiently high. This is shown by the data given in Fig. 5.9, which relates to the wetting of three nitride ceramics by copper with different titanium contents.

Titanium can react with silicon nitride in two possible ways, to form either titanium silicides or titanium nitride (there is only one stable nitride of titanium):

$$Si_3N_4 + Ti = \text{titanium silicides} + N_2 \text{ (gas)}$$
$$Si_3N_4 + Ti = \text{titanium nitride (TiN)} + Si$$

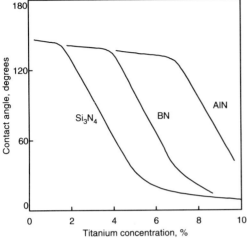

Fig. 5.9 Effect of titanium concentration on the wetting of some nitride ceramics by Cu-Ti-activated brazes, as measured by the contact angle. Source: After Nicholas [1989]

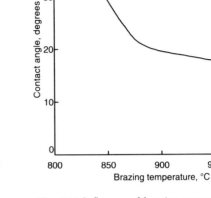

Fig. 5.10 Influence of brazing temperature on the wetting of Si_3N_4 by the Ag-27Cu-2Ti alloy, as measured by the contact angle. Source: After Nicholas and Peteves [1991]

Table 5.4 Gibbs free energy of possible reactions between silicon nitride and titanium

Chemical reaction	Gibbs free energy of reaction as a function of temperature (in K), kJ/mol Ti	Energy of reaction at 1000 K (727 °C, or 1341 °F) kJ
$\frac{2}{3}Si_3N_4 + Ti = TiSi_2 + \frac{4}{3}N_2$	$G = 350 - 0.205T$	+145
$\frac{1}{3}Si_3N_4 + Ti = TiSi + \frac{2}{3}N_2$	$G = 125 - 0.122T$	+3
$\frac{1}{5}Si_3N_4 + Ti = \frac{1}{5}Ti_5Si_3 + \frac{2}{5}N_2$	$G = 40 - 0.080T$	−40
$\frac{1}{4}Si_3N_4 + Ti = TiN + \frac{3}{4}Si$	$G = -150 + 0.011T$	−139

Which of these two possibilities is more favorable from an energy point of view may be ascertained by thermodynamic analysis. The Gibbs free energy change resulting from these reactions as a function of temperature can be calculated; the result is given in Table 5.4. These data show that the preferred reaction is the one resulting in titanium nitride and free silicon. Thermodynamic calculations suggest that this reaction should proceed at all temperatures. However, there is a minimum temperature threshold below which the reaction effectively ceases, as shown by the plot in Fig. 5.10, relating contact angle to temperature.

The empirical data represented in Fig. 5.10 were obtained in wetting experiments employing the sessile drop test, which involves placing a measure of a solid alloy onto a substrate of interest, heating it to above its melting point, and then measuring the contact angle visually. Details of this test are given in section 6.3.

Analysis of joint microstructures supports the thermodynamic prediction that a layer of the compound TiN is formed adjacent to the ceramic interface [Nicholas and Peteves 1991]. It is observed that a second reaction layer of Ti_5Si_3 is present between the TiN layer and the filler alloy, formed by reaction of titanium with silicon that is freed in the decomposition of silicon nitride.

The formation of a layer of a titanium silicide above the nitride is beneficial from the point of view of the brazing process. This is because titanium nitrides are essentially nonmetallic and are not wetted by molten filler metals. On the other hand, titanium silicides have a metallic character and are readily wetted by silver-copper and other brazing alloys [Nicholas 1989].

The metallic character of the interfacial reaction products is important in ensuring wetting by the braze. This can be seen from the decrease in wetting angle that occurs as the heating cycle is

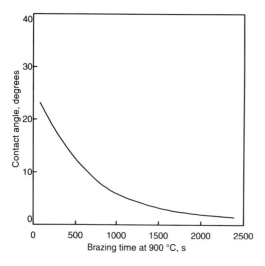

Fig. 5.11 Variation in contact angle with brazing time for Ag-27Cu-2Ti on Si_3N_4. Source: After Loehman [1988]

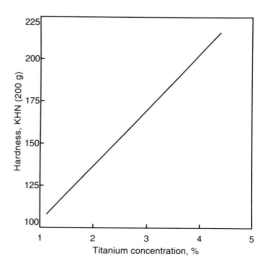

Fig. 5.12 Knoop hardness of Ag-Cu eutectic alloys containing titanium. Source: After Mizuhara and Mally [1985]

extended (Fig. 5.11), which results from the formation and growth of titanium silicides that become progressively rich in titanium as the reaction proceeds. A more dramatic illustration of this point is provided by the contact angle of silver-copper-titanium brazes on substrates of silicon nitride, boron nitride, and aluminum nitride, respectively, under identical conditions. Figure 5.9 shows that much higher concentrations of titanium are required to wet boron nitride; neither of the reaction products—namely, TiN or TiB_2—are wetted by plain silver-copper eutectic. It is only when titanium is added and raised to a concentration that is sufficient to force the composition of these reaction products off stoichiometry and make them titanium-rich that wetting occurs. An even higher concentration of this element is needed to induce wetting of aluminum nitride. In all cases, the titanium concentrates at the bridging compound/filler interface, leaving the solidified filler denuded of the reactive constituent.

5.4.2 Influence of Concentration of the Reactive Constituent

From the preceding discussion it might appear that a high concentration of the active element is desirable. However, for various reasons, many filler alloys have concentration limits. In the case of silver-copper filler metals, alloying with titanium increases the hardness of the braze to a point where it becomes unworkable, as shown by Fig. 5.12. This limitation can be overcome by either preparing the alloy in a ductile form by rapid solidification, producing a composite preform comprising a core of titanium and a cladding of silver-copper alloy, or applying titanium as a metallization to the component surfaces prior to joining. All of these preparation techniques are described sections 2.2.1.3 and 4.2.2.

With silver-copper-titanium brazes there are additional reasons for restricting the titanium concentration to about 2%. The first of these is the effect on the melting temperature of the alloy; the addition of more than 2% titanium substantially widens the melting range of the alloys, which is an undesirable characteristic for a filler alloy. The Ag-Cu 5Ti composition has a melting range of 775 to 927 °C (1425 to 1700 °F). Moreover, it can be seen from the silver-copper-titanium phase diagram shown in Fig. 5.13 that at concentrations of titanium above 2%, the molten braze separates into two distinct liquid compositions. The solidified filler alloy will then be inhomogeneous; this will result in the composition and properties of the joint varying in an unpredictable manner, if the titanium concentration is much above 2%.

At first sight, a 2% titanium concentration in a filler metal might appear to be too low for the braze to be effective with many ceramics. The data in Fig. 5.9 certainly indicate this to be the case with regard to copper-titanium alloys. In fact, it is not the percentage of the reactive element *per se*

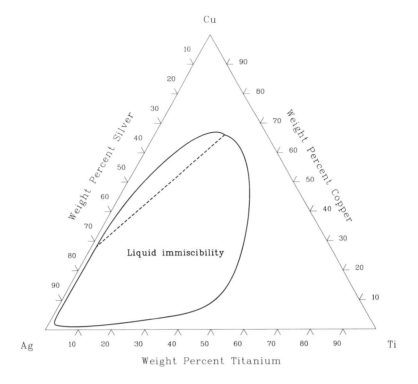

Fig. 5.13 Liquidus surface of the Ag-Cu-Ti phase diagram showing the region of liquid immiscibility. The critical tie line that links the two liquid phases of Ag-27Cu-2Ti and Ag-66Cu-22Ti is marked.

that governs the efficacy of the filler metal, but rather its chemical activity.

The activity of metal A in solution in metal B can be considered as the equivalent concentration of pure A in terms of chemical activity. Exceeding the solubility limit results in the formation of intermetallic phases that tie up a proportion of the solute A, so that this fraction is not free to react unless the phases in question can be dissociated.

The addition of silver to copper-titanium alloys substantially increases the effective concentration of the titanium, because the solubility limit of titanium in copper at 1150 °C (2100 °F) is 67%, whereas it is only 3% in silver at the same temperature. That is, the addition of silver reduces the tendency of the titanium to associate with the molten alloy, and this means that the activity of the titanium is correspondingly increased. Furthermore, as noted above, at a 2% concentration titanium causes the molten silver-copper alloy to separate into two fractions, one with a composition of Ag-27Cu-2Ti and the other being Ag-66Cu-22Ti. The 22% Ti fraction is more than adequately active to ensure wetting of even highly refractory ceramics such as alumina (Al_2O_3). Concomitantly, the activity of the alloy falls away sharply with reduction in the titanium concentration to below 2%.

Nicholas [1988] has shown that by making additions of elements such as indium and tin to silver-copper-titanium alloys it is possible to increase the activity of the titanium, even while reducing its concentration. The enhancement is due to the low solubility of titanium in these metals. For example, it is only 6.7% in tin at 1150 °C (2100 °F). This point is illustrated by Fig. 5.14.

It is also possible to boost the effectiveness of the reactive constituent by enhancing its natural tendency to concentrate at the filler/ceramic interface. This can be achieved by applying an electric field across the joint gap while the braze is molten [Minegishi, Sakurai, and Morozumi 1991]. Under the influence of the applied voltage, ions of the element in the braze with the highest induced charge will tend to migrate toward the cathodic (negatively biased) side of the joint. Because the principal constituents of brazes intended for use at elevated temperatures tend to be relatively noble elements, it will be ions of the active constituent that are driven toward the cathode. This process,

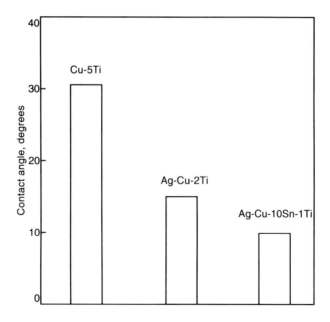

Fig. 5.14 Comparison of the influence of composition on the wetting of Si_3N_4 ceramic by titanium-activated brazes under comparable process conditions, as measured by the contact angle. Source: After Nicholas and Peteves [1991]

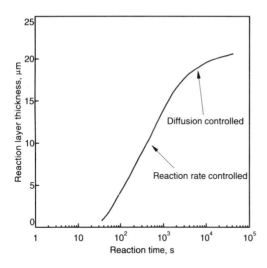

Fig. 5.15 Reaction layer thickness as a function of brazing time for Si_3N_4 wetted by Cu-5Ti at 1125 °C (2055 °F). Source: After Nakao, Nishimoto, and Saida [1989]

known as field-assisted brazing, requires the ceramic to be electrically conductive, such as zirconia and silicon nitride. It should be noted that many ceramics that are considered to be electrical insulators at room temperature have significant conductivity when heated to 1000 °C (1830 °F), as their volumetric resistivity reduces by six orders of magnitude over that temperature range [Morrell 1985].

5.4.3 Formation and Nature of the Reaction Products

Reaction between a reactive filler alloy and a nonmetal results in modification of the wetted surface with the formation of one or more interfacial compounds. As discussed in Chapter 2, when designing a joining process involving an active filler alloy, due consideration must be given to all of the reaction products. Volatile elements can generate voids in a joint through the evolution of vapor, while other products can, either individually or in combination with other species, form low-melting-point phases or produce liquid immiscibility and other undesirable features.

The thickness of an interfacial layer formed between the Cu-5Ti braze on silicon nitride as a function of the brazing cycle duration and peak temperature are given in Fig. 5.15 and 5.16, respectively. As with most chemical reactions, the thickness of the layer increases with the available thermal energy. The initially linear slope of the graph given in Fig. 5.15 indicates that the reaction between the molten braze and the ceramic is at

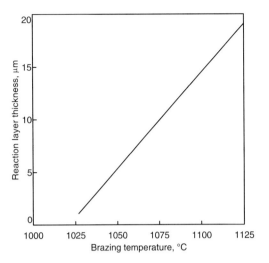
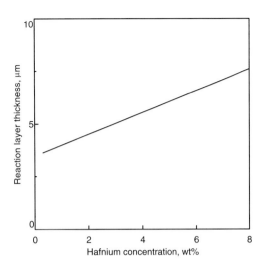

Fig. 5.16 Reaction layer thickness as a function of the brazing temperature for Si_3N_4 wetted by Cu-5Ti for 1000 s. Source: After Nakao, Nishimoto, and Saida [1989]

Fig. 5.17 Reaction layer thickness as a function of the concentration of the active metal for SiC brazed with Ag-Cu-Hf alloys. Source: After Lugscheider and Tillmann [1991]

first reaction rate controlled. When the reaction product zone has reached a certain thickness, solid-state diffusion through it then determines the rate of subsequent growth of the layer. At this crossover, the growth slows down significantly, as can be seen in Fig. 5.15. The transition between these two growth regimes often is not sharp and, overall, the thickness of the reaction layer tends to increase asymptotically with time. The concentration of the active constituent also affects the thickness of the reaction layer formed under fixed process conditions, as can be seen in Fig. 5.17.

Silver-copper alloys activated with hafnium are generally preferred to their titanium equivalents for joining to nitride ceramics, because the reaction product, HfN, is a barrier to silicon and nitrogen diffusion, whereas TiN is not [Lugscheider and Tillmann 1991]. Therefore, once a continuous layer of HfN has formed, the reaction effectively ceases, so that there is no falloff in the mechanical properties through a thickening of the reaction zone on extended heating. However, there is a penalty for this benefit; namely, more stringent furnace atmospheres are necessary due to the greater reactivity of hafnium with oxygen.

The sensitivity of active brazes to the brazing atmosphere is reflected by the data given in Fig. 5.18, which shows the strength of ceramic-metal assemblies brazed using a silver-copper-titanium alloy in different atmospheres. The highest and most consistent joint strengths have been achieved in atmospheres of high-purity nitrogen and argon.

The thickness of the reaction layer formed between titanium-containing brazes and nitride ceramics can be greatly reduced by selected alloying additions. Kuzumaki, Ariga, and Miyamoto [1990] have shown that the addition of niobium (columbium) to silver-copper-titanium alloys is effective in restricting the width of the interfacial layer. The reason for this is not entirely clear, and there are at least two possible explanations. One is that because titanium and niobium form a solid solution, the TiN reaction layer is replaced by (Ti,Nb)N, which may be a more effective barrier to silicon and nitrogen diffusion than is TiN. Alternatively, the presence of the niobium may reduce the activity of titanium in the alloy, due to the solubility of titanium in niobium [Akselsen 1992]. The influence of niobium additions on the thickness of the reaction layer can be seen in Fig. 5.19.

5.4.4 Mechanical Properties of Joints

As might be expected, the strength of a ceramic-metal assembly will depend on the nature of the interfacial reaction layer and any stresses in the assembly arising from a mismatch in thermal expansivity. The strength of copper-alumina assemblies joined using the Cu/CuO_2 eutectic process is shown in Fig. 5.20 as a function of the thickness of the reaction layer. As the layer forms and grows, the joint strength increases progressively up to a maximum value for a reaction layer 5 μm (200

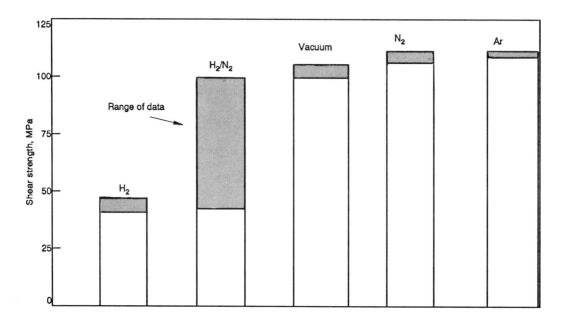

Fig. 5.18 Influence of the brazing atmosphere on the shear strength of ZrO_2/mild steel joints made with Ag-Cu-3Ti filler alloy. Source: After Weise, Malikowski, and Krappitz [1989]

μin.) thick. Further growth of the layer causes the strength to decline as cracks and voids develop within it. This characteristic is typical of metal-ceramic assemblies. Unlike metal-metal joints, thick reaction layers are often not catastrophic to joint strength of metal-ceramic and ceramic-ceramic assemblies. Although the interfacial compounds tend to have high elastic modulus and low fracture toughness, they are not greatly different from glasses and most ceramics in these respects. Therefore, it is possible to form reactively brazed joints to such materials that have mechanical properties comparable to those of monoliths of the same materials, despite the presence of interfacial reaction layers.

In joints to ceramics, the presence of the interfacial reaction layer often has a beneficial role in reducing thermal mismatch stresses. The bridging compounds frequently have coefficients of thermal expansivity that are intermediate between those of ceramic and metal components. For example, the $AlCuO_2$ layer (CTE = 11×10^{-6}/°C) formed in a copper-alumina joint is equivalent to having a thin intermediate plate that reduces the stress concentration between the abutting components (copper, CTE = 17×10^{-6}/°C; alumina, CTE = 5×10^{-6}/°C).

The effect of thermal expansion mismatch on the strength of brazed joints to silicon nitride is shown in Fig. 5.21. Failure of such assemblies generally occurs through the near-surface layer of the ceramic, with the fracture being initiated by small defects in that material. By applying fracture mechanics modeling, it is possible to calculate the minimum size of a defect inside the ceramic that will cause spontaneous failure of a ceramic-metal brazed joint, as a function of the mismatch in thermal expansivity. The calculated curve given in Fig. 5.22 shows the extreme sensitivity of critical defect size to thermal expansivity mismatch, highlighting the need to minimize stresses from this source. Methods for reducing or redistributing stress concentrations in joints have been described in section 4.3. For similar reasons, the method used to shape the ceramic component prior to joining has a profound influence on the joint strength. Any machining process tends to generate subsurface cracks and other flaws that degrade the intrinsic strength of the material. Accordingly, it is recommended that either the ceramic be resintered following machining, if this is possible, or the damaged surfaces (50 to 100 μm, or 2000 to 4000 μin., in depth) be removed by gentle lapping or chemical etching prior to brazing [Mizuhara and Mally 1985].

The Joining Environment

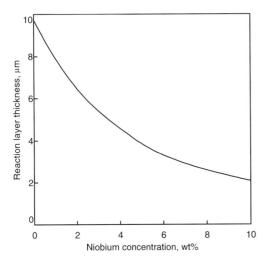

Fig. 5.19 Reduction in the thickness of the reaction layer formed by the addition of niobium to the Ag-Cu-5Ti braze wetted onto aluminum nitride. Source: After Kuzumaki, Ariga, and Miyamoto [1990]

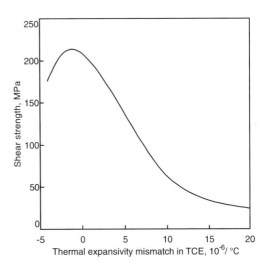

Fig. 5.21 Effect of thermal expansivity mismatch, relative to that of the ceramic, on the shear strength of Si_3N_4-metal brazed joints. Source: After Naka, Kubo, and Okamoto [1989]

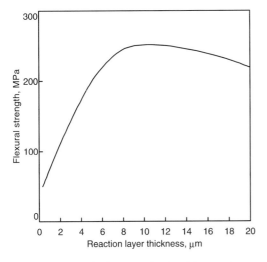

Fig. 5.20 Strength of copper-alumina joints made using the Cu/CuO_2 eutectic braze, as a function of the thickness of the $AlCuO_2$ interfacial reaction layer formed. Source: After Kim and Kim [1992]

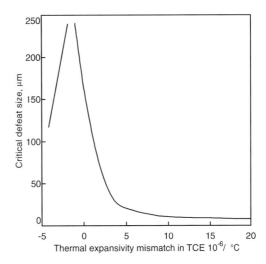

Fig. 5.22 Relationship between thermal expansivity mismatch, relative to that of the ceramic, and the critical defect size that will cause failure of the ceramic due to imposed stress. Calculated data obtained from fracture mechanics modeling. Source: After Akselsen [1992]

5.5 Health, Safety, and Environmental Aspects of Soldering and Brazing

Soldering and brazing involve the use of a large number of different materials, including metallic and nonmetallic elements for the fillers and the parent materials, and organic and inorganic chemicals used in fluxes, in controlled atmospheres, and for removing flux residues. Several of these materials are hazardous in varying degrees to operators or to the environment [Sax and Lewis Sr. 1989]. Accordingly, they must be handled, used, and disposed of according to national codes of practices or regulations governing hazardous substances. Official listings produced by national health and safety authorities classify materials according to their toxicity level and set the exposure limits for hazardous vapors and dusts.

All materials that are likely to be encountered in a joining context will have an assigned value of maximum exposure limit, usually in weight per unit volume (normally expressed in mg/m^3). The form of the material is also relevant. Powders and dusts are more hazardous than nonvolatile liquids and monolithic solids; they are ranked according to the maximum inhalable quantity in mg/m^3 and time weighted over a period of time, either short term (minutes) or a longer period of many hours. Ventilation and personal protection through the use of respiratory equipment, specialist gloves, and other items of protective clothing may need to be provided in order to comply with the relevant regulations. Preventing exposure to the hazard by appropriate measures should always be given higher priority than protective measures.

Care must be taken in both the storage of materials prior to use and in the subsequent disposal of residues, exhaust emissions, and other associated effluent, such as solutions containing rinsed fluxes. These are usually subject to statutory controls. Many organic chemicals and gases used in joining, such as solder fluxes, binders used in pastes, and halogenated gases, also pose fire risks. The flammability is rated according to flash point, which is the lowest temperature at which the substance can be spontaneously ignited when it is in a saturated condition. Regulations and codes of practices must be consulted at all times before commencing work on a new joining process.

Health and safety information sheets are now available for all hazardous substances that are supplied commercially, and in many countries it is encumbent on the suppliers of these materials to provide hazard information and precautions appropriate to their foreseeable use. For example, a United Kingdom supplier's data sheet for proprietary rosin flux in the form of a liquid specifies a threshold limit value (TLV) for the solvent of 400 ppm. For the fume of the same flux, the TLV is 0.1 mg/m^3. The solvent is highly flammable and is stated as having a flash point of 12 °C (54 °F). The same data sheet also specifies suitable precautions to be taken when handling and using the flux and the appropriate authorities and regulations that govern the disposal of waste and residues.

As an illustrative example of the range of information that is now being provided by materials suppliers, the set of data sheets for a proprietary cleaning agent for removing rosin-containing flux residues from soldered circuit boards is reproduced in Table 5.5. It is given in the form appropriate to users in the European Community. The abbreviations used for specifying the hazard data are explained in the appropriate European regulations and guidance documents. The product, Dowanol PX-16S, which is supplied by the Dow Chemical Company, has been formulated as an alternative to CFC-113, the use of which is being progressively restricted on environmental grounds.

Table 5.5 Data sheet for a defluxing agent, listing health, safety, and environmental information

Product name:
Dowanol PX-16S

Ingredients
Alkoxypropanol with additives

Physical data

Main boiling point, °C:	186
Freezing point, °C:	<–70
Solubility in water, wt%:	Completely miscible
Vapor pressure, mbar:	0.7 (at 25 °C)
Specific gravity (water = 1):	0.94
Surface tension, mN/m:	25.7 (at 25 °C)
Appearance:	Clear, colorless liquid

Designed usage
Defluxing agent to remove rosin-containing flux residues from soldered circuit boards. Either a water-free rinse (pure Dowanol PX-16S) process or a process with a solvent dip and a subsequent water rinse (semiaqueous process) can be used.

Notice: The information herein is presented in good faith, but no warranty, express or implied, is given nor is freedom from any patent owned by the Dow Chemical Company or by others to be inferred. Note: Dowanol PX-16S is a trademark of the Dow Chemical Company.

Table 5.5   Data sheet for a defluxing agent, listing health, safety, and environmental information (continued)

Reactivity data
Flash point, °C: 79 (Pensky Martens closed cup)
Extinguishing media: Water fog, alcohol-resistant foam, CO_2, dry chemical
Conditions to avoid:
    Product is stable under normal storage conditions. The preferred working temperature is <(flash point −15 °C) to avoid flammability risks.
Materials to avoid:
    Oxidizing materials
Hazardous decomposition products:
    None known. Complete combustion will result in carbon oxides and water.

Health hazard data
Toxicology:
    Statements based on data for individual ingredients.
Ingestion:
    Single-dose oral toxicity is low. The oral LD50 for rats is >3000 mg/kg.
Eye contact:
    May cause slight transient (temporary) eye irritation. Corneal injury is unlikely.
Skin contact:
    Prolonged exposure is not likely to cause significant skin irritation. The LD50 for skin absorption in rabbits is >2000 mg/kg. Prolonged skin contact with very large amounts may cause drowsiness.
Inhalation:
    Single exposure to vapors is not likely to be hazardous. Excessive exposure may cause irritation to upper respiratory tract. Signs and symptoms of excessive exposure may be anesthetic or narcotic effects.
Systemic effects:
    Repeated excessive exposure may cause liver and kidney effects.
Occupational exposure limits:
    ACGIH TLV is 100 ppm and 150 ppm STEL for the main component.
Labeling:
    The product does not require classification of the European Communities (Council Directive 73/173/EEC).

Personal protection information
Ventilation:
    Good general ventilation should be sufficient for most conditions. Local exhaust ventilation may be necessary for some operations.
Respiratory protection:
    When airborne exposure guidelines and/or comfort levels may be exceeded, use an approved air-purifying respirator.
Eye contact:
    Use safety glasses.
Protective clothing:
    For brief contact, no precautions other than clean, body-covering clothing should be needed. Use impervious gloves when prolonged or frequently repeated contact could occur.

Ecology data
Statements based on main component.
Ecotoxicity:
    Static acute LC50 for fish is <10,000 mg/L. Material is practically nontoxic to fish on an acute basis (LC50 >100 mg/L). Static acute LC50 for daphnids (*Daphnia magna*) is 1919 mg/L. Material is practically nontoxic to aquatic invertebrates on a static acute basis (LC50 >1 mg/L).
Partitioning:
    Appreciable volatilization from water to air is not expected in the environment. Log octanol/water partition coefficient (log K_{ow}) is estimated using the Pomona-MedChem structural fragment method to be −0.064. Partitioning from water to *n*-octanol is low (log K_{ow} < 0). No bioconcentration is expected due to the high water solubility.
Degradation:
    Index of biodegradability (BOD28/ThOD, %) is 32. The bacteria toxicity threshold value (*Pseudomonos putida*) is 4168 mg/L. The assessment figure or value for bacteria toxicity according to the German assessment for water-endangering substances (Bewertung wassergefährdender Stoffe, 1979) is 2.4.

Notice: The information herein is presented in good faith, but no warranty, express or implied, is given nor is freedom from any patent owned by the Dow Chemical Company or by others to be inferred. Note: Dowanol PX-16S is a trademark of the Dow Chemical Company.

Table 5.5 Data sheet for a defluxing agent, listing health, safety, and environmental information (continued)

Emission reduction/recycling/disposal
Emission reduction:
  Air:
    Low vapor pressure allows good emission control by cooling. For drying of wasted parts, vacuum drying is to be preferred over hot air drying, since condensation and recycling of product is easier and more effective.
  Water:
    Dowanol PX-16S allows a water-free process. However,, in cases where Dowanol PX-16S-containing water is produced, the water can be purified using reverse osmosis.
  Recycling:
    Soiled product can be flash distilled under reduced pressure (<5 mbar recommended). However, product should not be distilled to dryness and guidelines of MSDS should be followed.
  Disposal:
    For final disposal, burning under carefully controlled conditions is recommended. Local, state, or national legislation has to be respected.

Notice: The information herein is presented in good faith, but no warranty, express or implied, is given nor is freedom from any patent owned by the Dow Chemical Company or by others to be inferred. Note: Dowanol PX-16S is a trademark of the Dow Chemical Company. Courtesy of Dow Chemical Company Ltd.

Appendix 5.1: Thermodynamic Equilibrium and the Boundary Conditions for Spontaneous Chemical Reaction

The thermodynamic function that provides a measure of the driving force of a chemical (including metallurgical) reaction is the Gibbs free energy, which is defined as:

$$G = E + PV - TS \quad \text{(Eq 5A.1)}$$

where E is the internal energy, S is the entropy, T is the absolute temperature, P is the pressure, and V is the volume of the materials system. The definition and physical meaning of internal energy and entropy will be explained below. As will be seen, an important property of the Gibbs free energy function is that it is always a minimum at equilibrium; the extent of its departure from the minimum value provides a measure of the tendency of a reaction to proceed spontaneously—that is, of the driving force for the reaction.

The First Law of Thermodynamics and Internal Energy

The subject of thermodynamics addresses energy changes in systems. In thermodynamics, the term "system" is used to describe a set of materials capable of undergoing a change—as, for example, through a chemical reaction.

The First Law of Thermodynamics is a statement of the Principle of Conservation of Energy. The various statements of this law are bound up with the differentiation of various types of energy, and in particular with the concept of internal energy. The internal energy of a system may be considered as the aggregate of the kinetic energies and energies of interaction (i.e., potential energies) of the atoms and molecules of which the constituent materials are composed. When the system is isolated from its surroundings, so that no exchange of energy can take place, then its internal energy remains fixed. However, if mechanical work can be done on the system but no heat is exchanged with its surroundings (i.e., the system is adiabatically isolated)—for example, by an impeller stirring a liquid or gas in an insulated container—its internal energy will change by an incremental amount equal to the work performed:

$$dE = -dW \text{ (adiabatic)}$$

the minus sign denoting that work is done on the system to raise its internal energy. This expression provides a thermodynamic definition of internal energy.

The internal energy, E, of a system depends only on the state of the system (defined in terms of macroscopic or thermodynamic properties, such as the pressure and temperature of the system). For this reason, internal energy is termed a function of state.

Where the work is done in changing the volume of a chemical system by an increment dV through the application of external pressure P, then:

$$dW = PdV$$
$$dE = -PdV \quad \text{(Eq 5A.2)}$$

In practice, most systems are not totally insulated from their surroundings, so that thermal energy may be exchanged between them.

If an increment of work dW is done on the system and an increment of heat dQ is exchanged with the surroundings, then the internal energy dE will change by the amount:

$$dE = dQ - dW \quad \text{(Eq 5A.3)}$$

This equation is a mathematical expression of the First Law of Thermodynamics, which, for a chemical system, may be written:

$$dE = dQ - PdV \quad \text{(Eq 5A.4)}$$

Entropy and the Second Law of Thermodynamics

Internal energy alone cannot determine the equilibrium state of a system. Although when a system reaches a state of equilibrium, the internal energy achieves a fixed value, this may not be a minimum. For example, the internal energy will increase when a solid melts at constant temperature and pressure through the absorption of latent heat. For this reason, in addition to the internal energy, it is necessary to stipulate the value of another state function of the system—namely, entropy—which, together with the internal energy, measures the extent to which the system is removed from equilibrium.

The concept of entropy arises in connection with the conversion of heat into mechanical work and vice versa. The Second Law of Thermodynamics defines the conditions under which this conversion from one form of energy to the other can occur. The Kelvin-Planck statement of this law relates to a device that can perform work by extracting heat from a particular source and performing an equivalent amount of work, without any other energy exchange with the surroundings. It follows that a reciprocating engine that operates by extracting heat from one source must reject some of this heat to a sink at a lower temperature. If the operating cycle of the engine is reversible, such that work can be performed to pump heat from the sink back to the source, it is possible to show that in accordance with the Second Law, the integrated ratio dQ/T over one complete cycle is zero:

$$\oint_R \frac{dQ}{T} = 0$$

The circle through the integral sign denotes that the integration is to be carried out over the complete cycle, and the letter R is a reminder that the equation applies only if the cycle is reversible. This result is known as Clausius' theorem.

If the integration is carried out over only part of the cycle, say between two states 1 and 2, then the integrated ratio dQ/T is not zero, but equals the difference between the values of a thermodynamic function at the two states:

$$\int_{R\,1}^{2} \frac{dQ}{T} = S_2 - S_1$$

This thermodynamic function of state is called entropy. If the two states are infinitesimally near, then the relationship can be written:

$$\left(\frac{dQ_R}{T}\right)_R = dS \qquad \text{(Eq 5A.5)}$$

Subscript R indicates that this equation only holds if the heat increment dQ is transferred reversibly. This equation provides a mathematical expression of the Second Law of Thermodynamics.

A consequence of the fact that entropy is a function only of state, a system that has changed from state 1 to state 2 always has entropy S_2, which differs from that of the initial state S_1 by $S_{1,2} = S_2 - S_1$, irrespective of the means used to drive the system. Thus, for example, the system may have been set in motion and some of the kinetic energy converted into heat in overcoming frictional forces, thereby raising its temperature to a value that takes the system from state 1 to state 2. In this irreversible process, the energy was not supplied to the system as heat, so that:

$$\int_{I\,1}^{2} \frac{dQ}{T} = 0$$

The letter I denotes that the process is irreversible.

However, the entropy change $S_{1,2}$ is still the same as that obtained by a reversible change between states 1 and 2, because it depends only on these states and not on the process connecting them, that is, here too $S_{1,2} = S_2 - S_1$. Thus, in all irreversible processes, the entropy change is greater than $\left(\int \frac{dQ}{T}\right)$, where dQ is the heat absorbed at each incremental step in the irreversible change. This result can be generalized to the statement that in a spontaneous irreversible change, the entropy of an isolated system will increase, and when in equilibrium, it will remain constant.

Considering the system and its surroundings together (i.e., the universe), any kind of process can be represented in entropy terms by the relationship:

$$dS \text{ (universe)} > 0$$

Therefore, from a thermodynamic viewpoint, which is macroscopic, entropy can be understood as the propensity of a system to undergo a change, such as a chemical reaction. A clearer physical picture of entropy can be obtained at the microscopic level, where a system may be regarded as an ensemble of atoms or molecules. On this basis, it can be shown that entropy provides a measure of the degree of atomic or molecular disorder that exists in the system, and this will always tend to increase. This concept is consistent with the observations that all metals are intersoluble, albeit in some cases only to a small extent, and that all liquid metals will wet the clean surfaces of solid metals.

Dependence of Gibbs Free Energy on Pressure

Having defined the thermodynamic functions of internal energy, E, and entropy, S, and explained their physical significance, it is possible to demonstrate the significance of the Gibbs free energy function, G, to determine the temperatures and pressures under which chemical reactions are thermodynamically favorable, as well as the direction of the reactions.

In incremental form, Eq 5A.1 can be written:

$$dG = dE + PdV + VdP - TdS - SdT$$

Substituting for dE and TdS from Eq 5A.4 and 5A.5 (all chemical/metallurgical processes being reversible) gives:

$$dG = dQ - PdV + PdV + VdP - dQ - SdT$$
$$= VdP - SdT$$

For a reversible process at constant temperature (isothermal) and constant pressure (isobaric), that is, when the system is in equilibrium:

$$dG = 0$$

where G is constant and has a minimum value. This is an important result for metallurgical reactions, because these can be considered as taking place usually at constant temperature and pressure.

More generally, at constant pressure, $dP = 0$, and then:

$$dG = -SdT \quad \text{(Eq 5A.6)}$$

and at constant temperature, $dT = 0$, so that:

$$dG = VdP \quad \text{(Eq 5A.7)}$$

If the system is an ideal gas, the Gas Law

$$PV = nRT$$

applies, where n is the number of moles of gas and R is the gas constant. Then,

$$dG = nRTdP/P$$

at constant temperature, so that the Gibbs free energy change resulting from a change from state 1 to state 2 at constant temperature is:

$$G_2 - G_1 = nRT \ln \frac{P_2}{P_1}$$

The Gibbs free energy, like any other measure of energy, must have some reference point. By convention, a zero value of G is assigned to the stable form of elements at 25 °C (77 °F) and 1 atm of pressure. Then the Gibbs free energy change of a gas at constant temperature from its value G^o at atmospheric pressure, which is defined as its standard state value, is given by the expression:

$$G - G^o = nRT \ln P \quad \text{(Eq 5A.8)}$$

where P is the pressure corresponding to the free energy state G, expressed in atmospheres. Although Eq 5A.8 is strictly valid for ideal gases, it is also approximately applicable to real gases and can be used for them at pressures close to normal atmospheric pressure (1 atm).

In the case of solids, the molar volumes are small compared with those of gases, so that the change in the Gibbs free energy of solids resulting from small pressure excursions ΔP, such that $\Delta P \ll 1$ atm (100 kPa), at constant temperature is small and, to a first approximation, may be neglected in reactions involving solids and gases. It is also assumed that the solubility of the gaseous species in the solid phases is negligible at the temperatures of interest, as is largely the case in practice.

It is now possible to determine the pressure dependence of the Gibbs free energy of the reagents that participate in a chemical reaction. Consider a reaction involving four gases, A, B, C, and D, and two solids, X and Y, all at constant temperature T, as follows:

$$xX + aA + bB \leftrightarrow yY + cC + dD \quad \text{(Eq 5A.9)}$$

where a, b, c, d, x, and y are the number of moles of each of the reagents. The gaseous reagents are assumed to behave as though they are ideal gases.

The Gibbs free energies $G(X)$ and $G(Y)$ of the solid constituents at moderate pressures are approximately equal to their values at atmospheric pressure, as explained above. Therefore:

Free energy of x moles of solid $X = xG(X)$
$\quad = xG^o(X)$
Free energy of y moles of solid $Y = yG(Y)$
$\quad = yG^o(Y)$

The Gibbs free energies of the gaseous constituents are:

For a moles of gas A: $aG(A) = aG^o(A) + aRT \ln P(A)$
For b moles of gas B: $bG(B) = bG^o(B) + bRT \ln P(B)$
For c moles of gas C: $cG(C) = cG^o(C) + cRT \ln P(C)$
For d moles of gas D: $dG(D) = dG^o(D) + dRT \ln P(D)$

where $G(A)$, $G(B)$, etc. are the Gibbs free energies of one mole of the reagents A, B, etc. at pressures $P(A)$, $P(B)$, etc., and $G^o(A)$, $G^o(B)$, etc. are the corresponding values at 1 atm.

The free energy change for the reaction is, from Eq 5A.8:

$$\Delta G = \Delta G \text{ (products)} - \Delta G \text{ (reactants)}$$
$$= cG^o(C) + dG^o(D) - aG^o(A) - bG^o(B)$$

$$+ RT \ln \frac{[P(C)]^c [P(D)]^d}{[P(A)]^a [P(B)]^b}$$

$$= \Delta G^o + RT \ln \frac{[P(C)]^c [P(D)]^d}{[P(A)]^a [P(B)]^b}$$

The Gibbs free energy changes of the solid reagents can be neglected, for the reasons given above.

Under equilibrium conditions, temperature and the respective pressures $P(A)$, $P(B)$, etc. are constant, and:

$$\Delta G = 0$$

Hence, the Gibbs free energy change when the gaseous reactants A and B in their standard states are transformed to the products C and D in their standard states may be expressed in terms of the partial pressures of the respective reactants in equilibrium, thus:

$$\Delta G^o = -RT \ln \frac{[P(C)]^c [P(D)]^d}{[P(A)]^a [P(B)]^b} \quad (\text{Eq 5A.10})$$

Since the Gibbs free energy change ΔG^o, for a particular reaction at a fixed temperature and at atmospheric pressure has a fixed value, so too does the argument of the logarithm. This constant is called the equilibrium constant, K_P, because it can be used to determine the equilibrium state that a reacting system will attain:

$$K_P = \frac{[P(C)]^c [P(D)]^d}{[P(A)]^a [P(B)]^b} \quad (\text{Eq 5A.11})$$

The subscript P denotes that the equilibrium constant is specified in terms of pressure. Equation 5A.10 becomes:

$$\Delta G^o = -RT \ln K_P$$

For an oxidizing reaction described by the equation

$$x M + \left(\frac{y}{2}\right) O_2 \leftrightarrow M_x O_y$$

there is one gaseous constituent and two solids, so that the equilibrium constant is simply

$$K_p = \frac{1}{(P_{O_2}^M)^{y/2}}$$

where $P_{O_2}^M$ is the partial pressure of oxygen required to effect the oxidation reaction, or the dissociation pressure of the oxide, and

$$\Delta G^o = RT \ln P_{O_2}^M \quad (\text{Eq 5A.12})$$

per mole of oxygen participating in the reaction. That is, the driving force needed to oxidize a metal, as expressed by the Gibbs free energy change, is directly related to the oxygen partial pressure of the atmosphere according to Eq 5A.12.

References

Akselsen, O.M., 1992. Review—Advances in Brazing of Ceramics, *J. Mat. Sci.*, Vol 27, p 1989-2000

Aluminum Association, 1990. *Aluminum Brazing Handbook*, 4th ed., p 35

Bannos, T.S., 1984. The Effect of Atmosphere Composition on Braze Flow, *Heat Treat.*, Vol 16 (No. 4), p 26-31

Baskin, M., 1992. How to Select a Brazing Flux, *Weld. Des. Fabr.*, Vol 65 (No. 3), p 63-66

Cooke, W.E., Wright, T.E., and Hirschfield, J.A., 1978. "Furnace Brazing of Aluminum with a Non-corrosive Flux," SAE Technical Paper Series No. 780300, Society of Automotive Engineers

Ellis, B.N., 1991. Water Soluble Fluxes, Their Reliability and Their Usefulness as a Means of Eliminating CFC-113 Usage, *Solder. Circuit Mount Technol.*, Vol 8 (No. 6), p 16-23

Jones, J.B. and Thomas, J.G., 1956. Ultrasonic Soldering of Aluminum, *Proc. Symp. Solder*, ASTM, p 15-29

Jordan, M.F. and Milner, D.R., 1951. The Removal of Oxide From Aluminum by Brazing Fluxes, *J. Inst. Met.*, Vol 85, p 33-40

Kim, S.T. and Kim, C.H., 1992. Interfacial Reaction Product and Its Effect on the Strength of Copper to Alumina Eutectic, *J. Mater. Sci.*, Vol 27, p 2061-2066

Kuzumaki, T., Ariga, T., and Miyamoto, Y., 1990. Effect of Additional Elements in Ag-Cu Based Filler Metal on Brazing of Aluminum Nitride to Metals, *ISIJ Int.*, Vol 30 (No. 12), p 1135-1141

Lea, C., 1991. After CFC's—Making the Economic and Technical Choice, *Circuit World*, Vol 18 (No. 1), p 28-33

Loehman, R.E., 1988. Joining and Bonding Mechanisms in Nitrogen Ceramics, *Proc. Conf. Int. Meet. Adv. Mater.*, Vol 8, *Metal-Ceramic Joints*, Tokyo, June 2-3, Materials Research Society, p 49-59

Lugscheider, E. and Tillmann, W., 1991. Development of New Active Filler Metals for Joining Silicon-Carbide and -Nitride, Paper 11, *Proc. Conf. Br. Assoc. Brazing and Soldering*, 6th Int. Conf., Stratford-upon-Avon, Sept 3-5

Mei, Z. and Morris, J.W., 1992. Characterization of Sn-Bi Solder Joints, *J. Electron. Mater.*, Vol 21 (No. 6), p 599-607

Milner, D.R., 1958. A Survey of the Scientific Principles Related to Wetting and Spreading, *Br. Weld. J.*, Vol 6, p 90-105

Minegishi, T., Sakurai, T., and Morozumi, S., 1991. Electric Field-Assisted and Field-Depressed Segregation of Reactive Metals to the Bond Interface in Braze Alloy Joining, *J. Mater. Sci.*, Vol 26, p 5473-5480

Mizuhara, H. and Mally, K., 1985. Ceramic-to-Metal Joining With Active Brazing Filler Metal, *Weld. J.*, Vol 64 (No. 10), p 27-32

Morrell, R., 1985. *Handbook of Properties of Technical and Engineering Ceramics*, Her Majesty's Stationery Office, London, p 168

Naka, M. and Okamoto, I., 1989. Ultrasonic Soldering and Brazing of Ceramics to Metal in Metal/Ceramic Joints, *Proc. Conf. Int. Adv. Mater.*, Vol 8, *Metal-Ceramic Joints*, Tokyo, 2-3 June, Materials Research Society, p 79-84

Naka, M., Kubo, M., and Okamoto, I., 1989. Brazing of Si_3N_4 to Metals with Al Filler (Report II), *Trans. Jpn. Weld. Res. Inst.*, Vol 2 (No. 18), p 33-36

Naka, M. and Taniguchi, H., 1992. Ultrasonic Soldering of Alumina to Copper Using Sn-Pb Solders, *Trans. Jpn. Weld. Res. Inst.*, Vol 20 (No. 2), p 113-118

Nakao, Y., Nishimoto, K., and Saida, K., 1989. Bonding of Si_3N_4 to Metals with Active Filler Metals, *Trans. Jpn. Weld. Soc.*, Vol 20 (No. 1), p 66-76

Nicholas, M.G., 1988. Reactive Brazing of Ceramics, *Proc. Conf. Int. Meet. Adv. Mater.*, Vol 8, *Metal-Ceramic Joints*, Tokyo, June 2-3, Materials Research Society, p 49-59

Nicholas, M.G., 1989. Reactive Metal Brazing, *Proc. Conf. Joining Ceramics, Glass and Metal*, Bad Nauheim, Germany, p 3-16

Nicholas, M.G., 1990. *Joining of Ceramics*, The Institute of Ceramics, Chapman and Hall, London

Nicholas, M.G. and Peteves, S.D., 1991. Reactive Joining of Silicon Nitride Ceramics, Paper 10, *Proc. Conf. Br. Assoc. Brazing and Soldering*, 6th Int. Conf., Stratford-upon-Avon, Sept 3-5

Sax, N.I. and Lewis Sr., R.J., 1989. *Dangerous Properties of Industrial Materials*, 7th ed., Van Nostrand Reinhold

Schaffer, H. *et al.*, 1962. How to Ultrasonically Seal Hermetic Ceramic Transistor Packages, *Ceram. Ind.*, Vol 79 (No. 6), p 50-64

Schultze, W. and Schoer, H., 1973. Fluxless Brazing of Aluminum Using Protective Gas, *Weld. J.*, Vol 52 (No. 10), p 644-651

Singleton, O.R. 1970. A Look at the Brazing of Aluminium—Particularly Fluxless Brazing, *Weld. J.*, Vol 49(No. 11), p 843-849

Staniek, G. and Wefers, K., 1991. Chemical Surface Treatment of Aluminum Powder Alloys, *Aluminum*, Vol 67 (No. 2), p 160-166

Stubbington, C.A., 1988. Materials Trends in Military Airframes, *Met. Mater.*, Vol 4 (No. 7), p 424-431

Sugiyama, Y., 1989. Brazing of Aluminum Alloys, *Weld. Int.*, Vol 3 (No. 8), p 700-710

Terrill, J.T. *et al.*, 1971. Understanding the Mechanisms of Aluminum Brazing, *Weld. J.*, Vol 50 (No. 12), p 833-839

The Aluminium Association, 1985, Aluminium Soldering Handbook, Fourth ed.

Turbini, L.J. *et al.*, 1991. Characterizing the Corrosion Properties of Flux Residues, Part 1: Test Method Development and Failure Mode Identification, *Solder. Surface Mount Technol.*, Vol 8 (No. 6), p 24-28

VAW 1982. "VAW Process for Fluxless Brazing of Aluminum," technical bulletin, Vereinegte Aluminium-Werke Akteingesellschaft, Bonn

Weise, W., Malikowski, W., and Krappitz, H., 1989. Wetting and Strength Properties of Ceramic to Metal Joints Brazed with Active Filler Metals Depending on Brazing Conditions and Joint Geometry, *Proc. Conf. Joining Ceramics, Glass and Metal*, Bad Nauheim, Germany, p 33-42

Wicks, C.E. and Block, F.E., 1963. "Thermodynamic Properties of 65 Elements—Their Oxides, Halides, Carbides and Nitrides," U.S. Bureau of Mines, Bull. 605, U.S. Government Printing Office

Xian, A. and Si, Z., 1991. Wetting of Tin-Based Active Solder on Sialon Ceramic, *J. Mater. Sci. Lett.*, Vol 10 (No. 22), p 1315-1317

Chapter 6

Assessment of Joint Quality

6.1 Introduction

Joints are frequently the weak link in an assembly. Even when the measured strength of the joints, as prepared, exceeds that of the joined components, a joint can become a source of weakness during service through corrosion, fatigue, and degradation by diffusion. The diffusion of constituents can lead to intermetallic embrittlement, the formation and proliferation of Kirkendall voids, and other microstructural changes.

Evaluation of the quality of joints and their durability under a range of conditions that are representative of the service environment is vitally important in the manufacture of reliable products. An effective assessment strategy must therefore be built into any program concerned with the design and development of new joining processes and configurations, with mechanical properties used as the main yardstick for establishing the tolerance of a joining process. The term "mechanical properties" embraces a wide spectrum of stress configurations applied under static or dynamic conditions, and the test used must be made relevant to the stress regime that applies in the practical situation. These can be very diverse: compare, for example, a brazed cutting tool subjected to high shear stresses in a corrosive environment with a soldered joint in surface mount circuitry that experiences cyclic fatigue at temperatures fairly close to its melting point.

Because the mechanical integrity of a joint is critically dependent on achieving good wetting by the filler, the testing of joints has been grouped together with assessment of the actual joining process by measurement of wettability and the quality of coatings applied to surfaces to aid wetting and spreading of the filler. The assessment procedures should be performed in a certain order, so that at each stage a particular type of weakness is screened out. By using such an approach, the testing is carried out in an efficient manner, with the more costly and laborious tests reserved for actual joined assemblies that come close to attaining the required integrity. For the same reason, the initial tests are not conducted on manufactured components of commercial value but on idealized test-pieces composed of offcuts or reject material that has little value added through manufacture.

This first stage of the assessment can be used to ascertain whether there are any fundamental problems with the joining process—for example, a materials incompatibility manifested by the formation of brittle intermetallic phases or voids. It will not establish whether the process is entirely suitable for the manufactured product. This information can only be obtained through a detailed assessment carried out on the actual assemblies in the later stages of the test sequence.

A sequence recommended for assessing joint quality within the framework of qualifying a new bonding process is shown in Fig. 6.1 and is outlined below:

1. Where metallizations are applied to the components, an assessment should be made of their quality prior to attempting any joining process; a defective metallization can lead to substandard joints. Simple tests often are suffi-

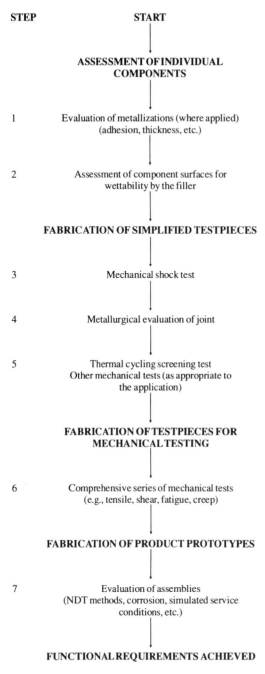

Fig. 6.1 Schematic flow chart for effectively assessing a joining process.

ciently discriminating to enable poor metallizations to be identified. Suitable procedures are described in section 6.2.

2. The next step is to determine, by means of a standard wetting test, whether the filler adequately wets the bare or metallized components (see section 6.3). The degree of wetting required will depend on the application. In some cases, a relatively high wetting angle may be preferred in order to limit the spread of the filler outside the joint, as explained in section 1.3.2.

3. Sacrificial testpieces should be prepared using the candidate materials and joining processes, and then tested to check that the assemblies are actually bonded and do not suffer from any catastrophic weakness. Poor joints often escape notice by visual inspection and gentle handling. A simple mechanical shock test, such as dropping an assembly from a prescribed height onto a hard surface, is widely used in industry as an initial test because it is simple to perform and highly discriminating. This test, which is described in several national standards, is reasonably representative of service conditions pertaining to products such as domestic and mobile telephone sets and personal stereo systems. This is a somewhat crude approach, but has been found to be highly effective for rapidly revealing gross inadequacies and inherently unsound joints that contain weak interfaces or are poorly bonded.

4. If the assembly survives step 3, it should then be sectioned in two, perpendicular to the joint. One of the cut parts can be used for a metallographic assessment of the joint quality (see section 6.4).

5. The other cut part is subjected to a thermal cycling screening test, such as that described in section 6.5. Because this test is purposely made more severe than one designed to simulate service conditions, the reduction in the stress incorporated during the joining operation by paring down the assembly should be more than compensated for by the extra severity of the temperature variation. If there is any doubt about this, the thermal cycling should also be carried out on full-size assemblies.

6. Simplified mechanical tests, such as tensile, shear, creep, and fatigue tests, are then justified, following which the joints can be deemed satisfactory at a first level of confidence. Several commonly used mechanical tests are discussed in section 6.5. If the joining system is well developed, such as tin-lead solder used in the joining of copper components, there may be sufficient data available in the literature to permit this and the previous steps to be omitted. The empirical results must be treated with caution, because the conditions of the simpli-

fied test might not be able to replicate those of the service regime. For example, the intrinsic tensile/shear strength of a soldered joint to an electronics component is often more than adequate for the application, but the joints fail in service as a result of complex processes that are not readily replicated using idealized testpieces. The soldered joints in this case may be subjected simultaneously to temperature and stress fluctuations and sometimes also to vibration while under a noncritical background stress [Frear, Jones, and Kinsman 1991, p 191-237; Coombs 1988].

7. The second level of confidence is attained following more detailed evaluation of subassemblies—that is, using real components—which establishes whether the joints meet specific application requirements. This involves simulating service conditions, where possible, using large numbers of assemblies to provide statistically reliable data. A number of tests have been devised to qualify joined assemblies, but mostly with respect to one service parameter at a time. There are special tests for assessing the resilience of joined assemblies to low-cycle fatigue, thermal shock, thermal soak, power cycling, humidity, corrosive environments, and so on. These tests are described in the technical literature, and several have entered the National Standards that govern manufacturing. A commonly cited set of standards used for electronics is MIL-STD-883. Nondestructive testing (NDT) methods can be profitably introduced at this stage to screen for incipient defects (namely, cracks and voids) that will impinge on the mean service life of the assembled product. A number of suitable NDT methods are described in section 6.6.

After the incorporation of the subassemblies into the final product, it is standard practice to check the functional characteristics. Only if these are found to be substandard is a more searching examination undertaken, which might include a reevaluation of the joints.

Having established the adequacy of the joining process and implemented it in production, continuing quality assessment is advisable. This should involve nondestructive testing of actual components and assemblies, but it is also recommended that more thorough testing along the lines outlined above be carried out occasionally on appropriate sacrificial testpieces that are included in the full-scale processing operations. This practice should help to forewarn against processing steps changing in a manner that causes the joints to become unsatisfactory.

Selected tests and evaluation methods will now be described briefly and their significance considered. Many of the tests outlined here are based on those specified in National Standards. They have the advantage of being purposefully designed, are discriminating, and should be followed wherever possible. Because differences exist among the various National Standards and among tests for different materials, and because individual standard tests are subject to occasional revision, the tests will not be described explicitly.

6.2 Evaluation of Metallization Quality

Metallizations often need to be applied to joint surfaces, especially in situations where the components are not wetted by the selected filler metal. Many metallizations must be multilayer to meet functional requirements, for the reasons given in section 4.2.2.

These metallizations should be subjected to some form of assessment, because their quality can affect joint integrity. In an extreme case where the composition of a layer happens to be at variance with that specified, this discrepancy is likely to manifest itself in the formation of "peculiar" phases and microstructures that will affect the properties of the joint. These effects normally become apparent during metallographic and mechanical property evaluation. Assuming that the composition of the metallization layers is correct, the following features then remain to be assessed:

- Adhesion between the various layers
- Cleanliness and wettability of the layers
- Thickness and uniformity of the layers

Gross deficiencies in the quality of metallization layers are usually easy to detect. Common examples include poor adhesion to the underlying material and inadequate cleanliness of the surface due to the presence of organic and inorganic films, including oxides. For checking that metallizations are capable of meeting a basic standard of integrity, one or more of the tests described below are recommended prior to the joining operation, as they are relatively quick to perform, simple to in-

6.2.1 Tests for Adhesion

A simple pull-off test for the adhesion of coatings involves applying an adhesive tape to the metallization, peeling it away, and then inspecting for evidence of the metallization parting from the substrate. The test is only qualitative, but it is able to disclose whether the adhesion of a coating is grossly inadequate.

The blister test is another easy to apply adhesion test that is suitable for coatings thicker than 1 µm (40 µin.). This involves either rapidly heating or cooling the component from room temperature to about ± 200 °C (± 360 °F), as appropriate. Both forms of the test rely on thermal expansion mismatch between the component and the metallization to produce interfacial shear stresses. Poorly adhered coatings will readily delaminate. Heating to an elevated temperature will also reveal the presence of trapped gas or organic residues, as blisters will appear if these contaminants are present.

The adhesion of coatings on brittle components can be tested at a rudimentary level by shattering a sacrificial testpiece. Provided that the metallization is well adhered, the fracture line through the metallization will match the crack propagation path through the component and will not be deflected along the interface. If the components are ductile, adhesion of the coatings can be tested by bending them around a small radius (see section 6.5.3); poorly adhered coatings will tend to delaminate from the component.

These tests are fairly crude and unquantifiable, but they possess the merit of being quick and easy to perform. More sophisticated tests that can provide quantitative data on adhesion are available. These are described in the literature [Mittal 1978; Valli 1986], and several are summarized in Fig. 6.2.

Adhesion testing of coatings is surrounded by controversy; no "standard" test procedure is universally accepted. Test methods have been designed to address this deficiency, but only time will tell as to whether any of these will become widely adopted. One test procedure that has been put forward for consideration is based on the inboard wire peel test shown schematically in Fig. 6.2, conducted under highly specified conditions [Williams 1988].

6.2.2 Tests for Surface Cleanliness and Wettability

For the purposes of making a joint, the metallization must be sufficiently clean to ensure wetting by the molten filler. The attainment of this condition can be assessed by conducting a spreading test (see section 6.3.1) and comparing the results with a standard spread specimen. If the molten filler metal fails to wet or spread sufficiently far, the metallization may be deemed to be unsatisfactory. More quantitative data can be obtained by using a wetting balance if the component is of simple symmetry (see section 6.3.2).

Visual inspection is not reliable for ascertaining cleanliness, because it is highly subjective. The surface appearance of metallizations can be deceptive, as color and reflectivity are dependent on a variety of factors, such as grain size and type (whether equiaxed or columnar), surface texture, and any films present on the surface.

6.2.3 Tests for Thickness and Thickness Uniformity

The thickness and uniformity of layers can be examined in metallographic section. Due to the thinness of many metallizations, taper sections should be used to provide the best chance of revealing all the layers present. Then, layers as thin as 0.1 µm (4 µin.) can be measured using a standard optical microscope (see section 6.4.1). Many coatings tend to be pure metals. These are relatively easy to apply; consequently, chemical etches can be effective in resolving adjacent layers in multilayer coatings. Figure 6.3 shows an etched metallographic section through an electroformed component. The three layers of metal comprising the electroform can be resolved, together with the two-layer metallization applied to one face of the component in readiness for the joining operation.

Instruments are available that can provide a nondestructive measure of the thickness and composition of metallizations at a particular spot. Those that are widely used in the electroplating industry exploit X-ray fluorescence and the backscatter of beta particles (electrons) [Liebhafsky *et al.* 1960, p 146-159; Tertian and Claisse 1982, p 278-306; Latter 1975].

Assessment of Joint Quality

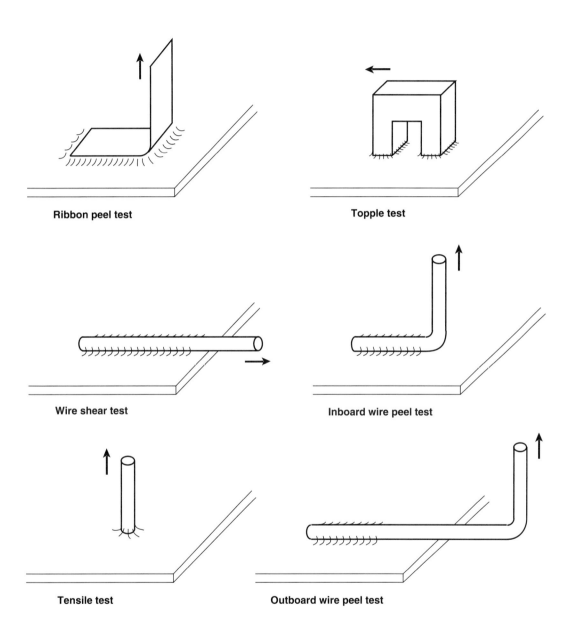

Fig. 6.2 Testpiece configurations for various adhesion tests applied to metallizations. Arrows indicate the direction of applied force.

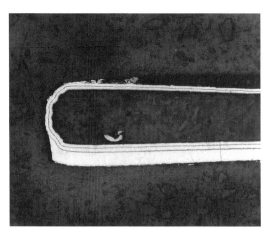

Fig. 6.3 Microsection through a three-layer electroformed component, metallized on one side with a two-layer coating. 540×

6.3 Assessment of Wetting

The formation of a strong joint is contingent on the ability of the molten filler to wet the joint surfaces over their entire area. Because wetting tends to be inhibited by surface oxides and contaminants, wettability measurements can be used to assess the effectiveness of a flux, the sensitivity of surfaces to exposure to various atmospheres, and shelf life in general. The evaluation of shelf life tends to involve artificially accelerated testing to obtain data within a reasonable timescale. Because soldering and brazing processes are dynamic in nature—that is, they are sensitive to the time during which the filler is molten—two aspects of wetting need to be considered, namely:

- The rate at which substrates and components are wetted by a filler
- The degree of wetting (and spreading) that is obtained at the end of the process cycle, when the filler solidifies

A number of evaluation procedures and tests have been developed to measure these characteristics. The following tests are used for determining the rate of wetting; these procedures are normally carried out in an ambient air atmosphere using fluxed testpieces:

- Wetting force measurements as a function of time (using a wetting balance)
- Split globule test
- Rotary dip method
- Meniscus rise as a function of time

The extent of wetting achieved is measured by:

- Area of spread measurements
- Sessile drop test
- Edge dip method
- Meniscus rise end value determination

Both classes of test are used for solders, but only the second, concerned with the extent of wetting, is widely applied to brazes. This is because industrial brazing operations are usually much less rapid than, for example, wave soldering. In wave soldering, the joints must be formed within the few seconds it takes for the printed circuit board (PCB) to pass through the solder wave. There, part of the soldering cycle is taken up by heating the joints to a temperature where wetting by the solder can occur. Thus, the thermal demand characteristics of soldered components must be considered in conjunction with wettability. This subject is treated in soldering textbooks such as Manko [1992, p 237-238] and Klein Wassink [1989, Chapter 3].

The principles of the wetting tests listed above are summarized in the paragraphs that follow.

Wetting force measurements. The testpiece is suspended from an arm of a sensitive balance and then partly immersed to a defined depth in a bath of molten solder that is held at constant temperature. The force exerted by the solder on the testpiece is recorded as a function of time. Further details are given in section 6.3.2.

Split globule test. A horizontally mounted test wire is fluxed and then lowered into a globule of molten solder mounted on a heated plinth at a controlled temperature, as shown in Fig. 6.4. This causes the globule to be bisected until the wire is wetted. When wetting occurs, the two parts of the globule coalesce—a condition that can be readily observed. The time delay between the splitting of the solder globule by the wire and the encasement of the wire in the reformed globule provides a measure of solderability. The test can be carried out manually based on visual observation or automatically using a wetting balance. Evidence has been presented to show that the split globule test is not sufficiently reliable to recommend its use for quality control inspection of items intended for manufacture by soldering [Roberts 1983].

Assessment of Joint Quality

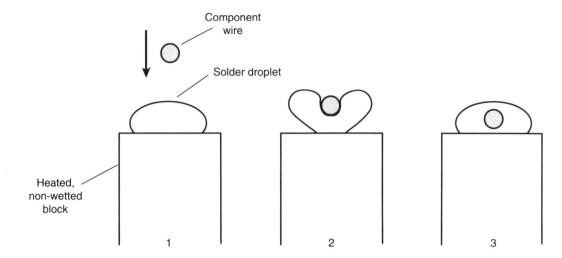

Fig. 6.4 Side view of a split globule solderability test. 1, start of test; 2, commencement of time-to-wet measurement; 3, termination of time-to-wet measurement

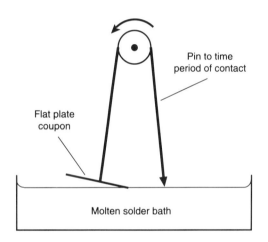

Fig. 6.5 Schematic of the rotary dip solderability test

The rotary dip method is designed to reproduce the wave soldering process. In this test, a flat coupon that has been fluxed is made to describe a circular path about a horizontal axis, which takes it through a bath of molten solder held at a specific temperature, as shown schematically in Fig. 6.5. The nominal time of contact of the specimen with the solder is measured with a needle that follows the motion of the specimen; its tip is equidistant from the axis as the center of the lower surface of the testpiece. The time of contact is held constant, commonly 2 s, and the maximum depth of immersion of the lower face of the specimen in the solder is fixed at typically 2 mm (0.08 in.). The specimen is examined for evidence of wetting after being dipped. The test is used for a simple pass/fail verdict and is somewhat arbitrary.

Meniscus rise as a function of time. In this test, a length of wire is dipped vertically into a bath of molten solder held at a predetermined temperature. As the wire is wetted, the meniscus of the solder is pulled up along the wire. The rise of the circle of contact with the wire is measured at specific intervals of time after the wire first makes contact with the solder, using an apparatus of the form depicted in Fig. 6.6. From the height of the solder rise, it is possible to calculate the contact angle, θ. The value of θ after the fixed time of the test provides a measure of wettability. In this test method, there is no sharp boundary between pass and fail, and the result tends to be somewhat imprecise.

Area of spread measurement. A fixed volume of solder or braze is melted on a flat coupon of the substrate held at a fixed temperature. A flux is generally added if the test is carried out in air. The degree of spread obtained after a set period of time is measured quantitatively. Details are given in section 6.3.1 below.

The sessile drop test is a variant of the area of spread assessment, in which the contact angle of the molten solder or braze is measured by direct observation. The principle of the method is illustrated in Fig. 6.7. The contact angle is usually measured after spreading but before solidification of the molten filler, while its surface is still smooth. Apparatus has been developed to allow dynamic measurement of contact angle in the

Principles of Soldering and Brazing

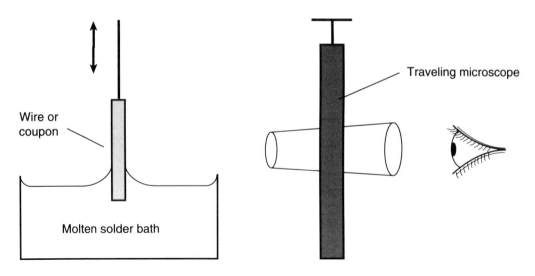

Fig. 6.6 Setup used for measuring solderability by observing the solder meniscus rise

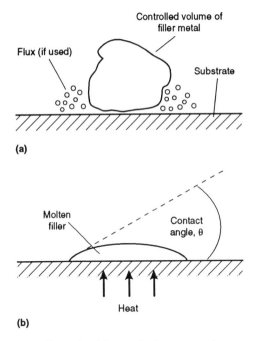

Fig. 6.7 Principle of the sessile drop test used to assess wettability. (a) A controlled volume of filler metal is melted onto the substrate under controlled conditions. (b) The contact angle is measured with a calibrated viewfinder.

ual assessment of the nature and extent of the filler coating on the specimen.

Meniscus rise end value determination. This test is the same as the meniscus rise determination outlined above, except that here the measured parameter is the maximum height rise that is achieved by the filler. This test and the edge dip method are not normally used to assess brazing processes because of the practical difficulties arising from the elevated temperatures involved.

Arguably the two most widely applied methods of assessing wettability are spreading tests and dynamic force measurements made on a wetting balance. Each of these test methods is described in greater detail below. When used in tandem, they furnish a comprehensive description of wetting behavior. This enables the effects of process conditions and material characteristics on soldering and brazing operations to be evaluated clearly [Jacobson and Humpston 1990]. Other test methods tend to provide less detailed information on wettability, without offering other clear-cut benefits, and will not be discussed here. The results from the various test methods can usually be correlated. For a brief review of their relative status, see Thwaites [1981].

course of solder spreading [Matienzo and Schaffer 1991].

Edge dip method. A flat specimen, coated with flux, is partly dipped edgewise into a bath of molten filler for a defined period and then withdrawn. This is a qualitative test that relies on vis-

6.3.1 Spreading Tests

One of the simplest and most direct methods that has been devised for assessing wetting by liquid metals on solid substrates involves measuring the area of spreading by the molten metal. This type of basic spreading test measures wetting by a molten filler of a solid over the interface of contact and not the enhancement that occurs in narrow joints through the action of capillary forces. This distinction is detailed in sections 1.3.2 to 1.3.4.

Because wetting balance and other methods of measuring wetting have not been developed for use at high temperatures, spread testing is relied on heavily for brazing assessments [Cibula 1958]. The procedure that has been widely adopted is to melt a filler metal pellet of known volume in a specified atmosphere, with or without fluxes, and to allow it to spread over the surface of the testpiece for a fixed period of time under controlled conditions. These conditions, where possible, should be representative of the intended application, because the spreading of the filler metal is usually sensitive to component-specific variables (e.g., surface finish) and to process variables (e.g., time at temperature).

The area of spreading of the filler metal is measured; this provides a relative index of wettability for comparative purposes. This index is the "spread ratio," which is defined as:

$$\text{Spread ratio} (S_r) = \frac{\text{Total plan area wetted by the molten metal}}{\text{Original plan area of a metal pellet (of a specific geometry)}}$$

Because the geometry of the solidified filler metal is seldom perfectly circular, image analysis should be used to provide an accurate measure of the total wetted area. The technique of image analysis, which is outlined in section 6.4.1, is capable of providing an assessment in almost real-time, so that with suitable instrumentation the spread area can be monitored as a function of time on a single testpiece. A sequence of spread tests made to evaluate small changes to the composition of a filler alloy is shown in Fig. 7.21.

An alternative means of determining relative spreading involves quantitatively defining a "spread factor" in terms of the volume, V, of filler metal used, and the maximum height, h, of the solidified pool:

$$\text{Spread factor} (S_f) = \frac{(6V/\pi)^{1/3} - h}{(6V/\pi)^{1/3}} \times 100$$

where $(6V/\pi)^{1/3} = D$, the diameter of a sphere corresponding to volume V of the metal pellet before the spreading test. This term can be calculated from the density of the metal pellet and its mass.

The normal method for measuring the height (h) uses a micrometer. However, this approach introduces errors because of the necessity to subtract the thickness of the substrate, which is likely to be much greater than h, from the measured height. A more accurate method of measuring h is to determine the peak height from metallographic sections or profilometer traces. Both of these methods also enable simultaneous measurements to be made of the contact angle between the resolidified pellet and the substrate.

Assuming that the initial pellet of filler can be approximated to a sphere and the resolidified filler to a spherical cap of radius R on the surface of the substrate, the spread ratio (S_r) and spread factor (S_f) can be expressed in terms of the angle of contact (θ), as follows:

$$S_r = \frac{4 \cot^2 \theta/2}{(1 + 3 \cot^2 \theta/2)^{2/3}}$$

$$S_f = 1 - \frac{1}{(1 + 3 \cot^2 \theta/2)^{1/3}}$$

The contact angle θ can be written in terms of the radius, A, and height, h, of the resolidified pool of filler, thus:

$$\sin \theta = \frac{2}{(A/h) + (h/A)}$$

The mathematical derivations are given in the appendix to this chapter (Eq 6A.1, 6A.2, and 6A.3, respectively). Some numerical values relating spread ratio and spread factor to the wetting angle are given in Table 6.1. Obviously, the situation represented by these expressions is an idealized one, and the assumptions made to derive them become less realistic with increasing solder spread. Nevertheless, these relationships do provide a reasonably close concordance with measured values. Some results of spreading tests are given in Fig. 1.9, which provides spread ratio data for a range of solder alloys melted on thin chromium metallizations covered with a flash of gold as a function of the process temperature.

The spread ratio becomes progressively more sensitive as the contact angle declines and wetting improves, whereas the spread factor varies almost linearly with contact angle, from a value of 1 at $\theta = 0°$ to 0 at $\theta = 180°$, as shown in Fig. 6.8. Spread ratio, therefore, provides a better differentiation between small differences in measured contact angle when the values of the latter are small. How-

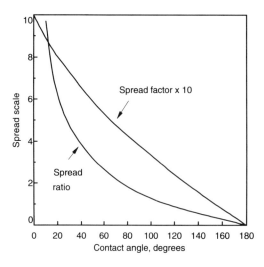

Fig. 6.8 Relationships among spread ratio, spread factor, and contact angle

Table 6.1 Calculated values of spread ratio and spread factor corresponding to selected contact angles

Values are derived from the expressions given in Appendix 6.1.

Contact angle, degrees	Spread ratio (S_r)	Spread factor ($S_f \times 10$)
180	0	0
90	1.59	3.70
40	3.67	6.52
10	9.74	8.63
0	Infinite	10.00

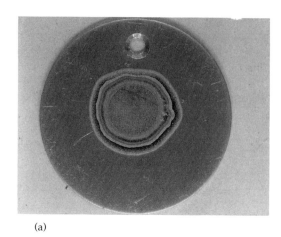

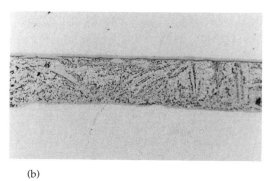

Fig. 6.9 (a) Solder spread sample on a gold-plated substrate. The solder is Pb-60Sn, and the substrate is copper plated with 5 μm (200 μin.) of nickel and then 5 μm (200 μin.) of gold. Three distinct microstructural bands are visible. (b) Micrograph of a joint made using the same substrate and solder described in (a). The joint has a regular microstructure, and all of the gold coating has dissolved in the solder.

ever, the converse is true when the contact angle is greater than 90°.

The results of spreading tests over a single surface must be treated with some caution when attempting to relate them to the wetting and filling of joints. Because a joint comprises a pair of facing solid surfaces, with which the filler metal can react, the capillary forces can govern the spreading characteristics; there are hydrostatic forces to consider as well. The relevance of the spreading test is also questionable when the joints are made using foil preforms. For this type of configuration, it is not necessary for the filler metal to spread significantly in order to fill the joint, so that a low spread in the test does not necessarily mean that a joint formed under similar conditions will be poor. Indeed, a high degree of spreading can be detrimental to joint filling, as the filler metal can flow out of the joint, resulting in voids and unwanted coverage of other parts of the component. Nevertheless, it is often but erroneously assumed that low spreading in the test necessarily implies poor bonding and weak joints.

An additional problem in attempting to correlate spreading test results with joint quality is that the filler metal reacts and spreads over a single surface in a conventional spreading test to produce a somewhat different microstructure from that of an actual joint. This is demonstrated in Fig. 6.9. Therefore, the wetting and spreading characteristics might be different in the two cases.

The spreading test is relatively simple to perform and well suited to pass/fail quality control applications. However, because of the different geometries of spreading test samples and actual joints, spreading test data usually must be supple-

Assessment of Joint Quality

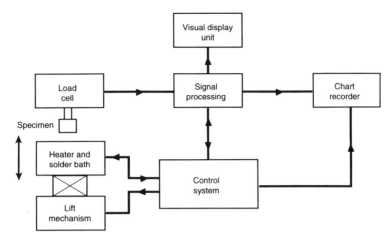

Fig. 6.10 Block diagram of a wetting balance

Fig. 6.11 Commercially available wetting balance

mented by other measurements when assessing soldering and brazing processes.

6.3.2 Wetting Balance Testing

The wetting balance solderability test is designed to furnish quantitative data on the wetting of a substrate by a molten metal under a closely specified set of conditions, including the atmosphere and flux that are used. The method has been adopted as a standard test for measuring the solderability of electronic component leads and substrates. In principle, the test can be applied to brazes equally well, assuming that the instrument is suitably adapted for operation at high temperatures.

A block diagram and a photograph of a typical wetting balance are shown in Fig. 6.10 and 6.11, respectively. The apparatus comprises:

- A load cell and a signal processing system that furnishes a measurement of load versus time and provides automatic taring of specimen weight
- A temperature-controlled solder bath
- A bath lift mechanism with speed and positional control
- A chart recorder or computer to display the force/time curve

The substrate under test is held in a specimen holder, which is itself suspended from the load cell. The bath containing the molten filler metal is raised at a preselected speed to immerse the testpiece to a given depth. The bath is held in this position for a preset dwell time and is then returned to its rest position.

The resolved vertical forces acting on the specimen are recorded as a function of time over the entire test cycle. Figure 6.12 shows the typical form of the trace that is recorded, together with the corresponding position of the specimen relative to the solder bath at each stage.

The wetting balance provides a measurement of the vertical component of the force exerted on the testpiece as it is lowered into a reservoir of the molten solder or braze, as a function of time. This force is theoretically equal to the sum of the vertical component of the surface tension force, F_γ, between the filler and the testpiece and the buoyancy of the testpiece, F_B. Figure 6.13 shows an equilibrium situation appropriate to partial wetting. The resolved force in the vertical direction, F_R, is the parameter measured in the test. The variation of

Principles of Soldering and Brazing

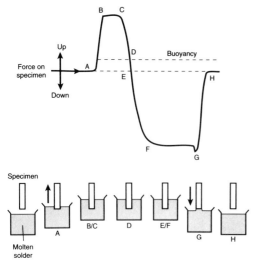

Fig. 6.12 Typical trace of the wetting force during a solderability test cycle, with the corresponding position of the specimen relative to the solder bath

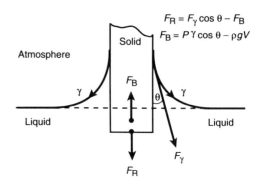

Fig. 6.13 Forces diagram for a solid plate partially immersed in a liquid. P, specimen periphery length; γ, liquid surface tension; θ, contact angle; ρ, liquid density; $g = 9.81$ m$_2$/s; V, immersed volume of specimen

this force as a function of time provides information on the dynamics of the wetting process, which the other methods of evaluating wettability described do not, apart from the dynamic measurement of contact angle in an enhanced form of the sessile drop test referred to above [Matienzo and Schaffer 1991].

Typical wetting balance force/time traces are given in Fig. 6.14. These measurements were made using a GEC Meniscograph solderability tester. The graphs show the effect of different cleaning methods on the wetting behavior of lead-tin eutectic solder, heated to 235 °C (455 °F), on mild steel fluxed with a mildly activated rosin flux. In Fig. 6.14(a), the steel coupon is in the as-received condition, and wetting by the solder is consequently very poor. Abrasive cleaning of the steel improves wetting, as shown in Fig. 6.14(b), but chemical cleaning is necessary in order to meet the quality acceptance criteria, which are indicated by the box in the lower left corner of the graph (Fig. 6.14c). Acceptance criteria for electronic components as specified in National Standards are a wetting force that exceeds two-thirds of the theoretical maximum force, achieved in a time representative of the soldering process to be used. For wave soldering, this is set at 2.1 s.

A wetting balance test can be performed rapidly and the results are quantitative, inasmuch as reproducible numerical data can be obtained for a well-defined set of sample and instrumental parameters and operating procedures, as explained in Barranger [1989] and Lea [1991]. Furthermore, the change of wetting as a function of time can be monitored. The surface tension of the molten filler can be calculated from data obtained on the wetting balance using nonwetted ($\theta = 180°$) substrates such as polytetrafluoroethylene (PTFE) or ceramic coupons, using the equation:

$$F_R = P\gamma \cos\theta - \rho g V \text{ (see Fig. 6.13)}$$

From this value and the measured wetting force, the angle of contact between the molten filler and the testpiece can be calculated.

Attempts have been made to correlate wetting balance data with the results given by other methods used for assessing wetting [Thwaites 1981; Wooldridge 1988]. Moreover, adaptations have been made to the wetting balance for solderability testing on specific types of components, and in particular for surface-mounted electronic devices (SMDs) [Gunter 1986; Yoshida, Warwick, and Hawkins 1987; Klang and Nylen 1989], and also in controlled atmospheres, including vacuum [Gunter and Jacobson 1990].

There is not necessarily a close correlation between test measurements and the wetting characteristics of actual joints. One of the reasons is that the configuration of the test does not strictly mirror that of an actual joint [Jacobson and Humpston 1990]. In particular, the volume of the molten filler relative to the volume of the components, including surface metallizations that react with it, may be significantly different in the two cases. For example, the extremely fast dissolution rates of

Assessment of Joint Quality

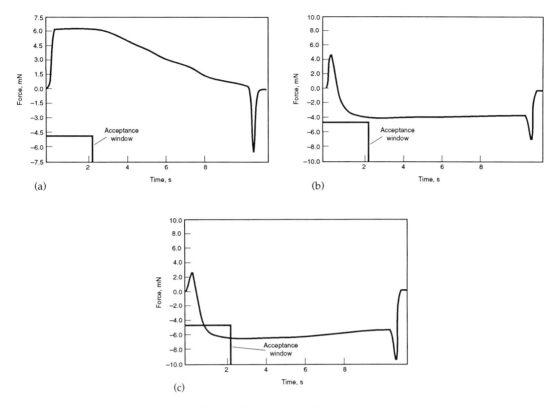

Fig. 6.14 Wetting behavior of mild steel, by lead-tin eutectic solder, measured on a wetting balance. Temperature: 235 °C (455 °F). (a) As-received condition. The pass condition, which is a wetting force of –4.5 mN achieved within 2.1 s of immersion, is not achieved in this case. (b) Following abrasion of the coupon surfaces. The pass condition is not achieved in this case. (c) Following chemical cleaning. The component satisfies the acceptance criterion.

gold coatings in solder, reported for samples dipped into the reservoirs of solder in a wetting balance, of typically 1 µm/s (40 µin./s) do not apply to joints soldered using foil preforms of much smaller volume [Bader 1969]. Thus, gold coatings that completely dissolve in the solder bath are often observed to partly survive in actual joints after the same time and temperature of the bonding operation.

Another difference that may need to be considered is a nonequivalence of the temperature gradient between the testpieces and the solder in the two cases. In the wetting balance, the solder is preheated to the test temperature, while the testpiece is at room temperature until immersion. In joints made by most methods other than wave soldering, the situation is largely reversed, with the solder being heated via the components being joined. In many materials systems, the filler metal will react with the testpiece differently as the temperature changes. This is a critical aspect because, as explained in section 1.3.2, metallurgical reaction and not surface tension is usually the dominant driving force for wetting in liquid-solid metal systems. Because such a metallurgical reaction occurs, to an extent that depends on the materials involved and their temperature, the wetted interface changes its composition and geometry with time. This, in turn, means that the measured wetting force tends to vary in the course of a test, and only in exceptional cases is the wetting force equation strictly valid, depending as it does on the classical model of wetting. Considerations such as these limit the value of wetting balance tests for absolute measurements. However, as a means of obtaining comparative data, and in particular for quality control assurance purposes (pass/fail determination), this method is most useful [Thwaites 1981].

6.3.3 Accelerated Aging of Components

The wetting characteristics of component surfaces may degrade over a period of time. This can have serious implications for the yield of fabri-

Table 6.2 Typical conditions used in accelerated aging assessments

Test name	Duration	Temperature °C	°F	Humidity, %	Atmosphere
Steam aging	1-4 h	100	212	100	Air
Damp heat	4-56 days	40	105	90	Air
Dry heat	2-96 h	155	310	5	Air
Mild environment	21 days	21	70	30	"Town air"
Severe environment	24 h	21	70	30	$H_2S/SO_2/NO_2$
Climatic test	8 cycles of 12 h in "severe environment" followed by 12 h in "dry environment"				

cated assemblies. The usual source of degradation is reaction with the environment, which may result in straightforward oxidation, the formation of corrosion products, or metallurgical changes affecting the surface region of the component. These effects are collectively described as aging. Accelerated aging tests are designed to simulate an extended period of natural aging. The wettability of the components being assessed is measured before and after aging, with the aim of screening out components having a wettability that will degrade during storage prior to use.

Numerous tests have been developed to mimic natural aging, some of which have been incorporated into standards. Each test involves exposing components to hostile conditions for a set period of time. The conditions used in typical tests are given in Table 6.2.

Each test variable has significance, but also poses problems, as follows:

- *Duration.* For convenience, the duration of the aging test is normally made much shorter than the maximum period of storage. This, then, constitutes an accelerated test.
- *Temperature.* Elevated temperature increases the rate of chemical reactions and thus corrosion of the sample. However, there is also the danger that this condition will make the test unrepresentative, because raising the temperature will promote metallurgical reactions, which may not occur at ambient temperature, that proceed by solid-state diffusion.
- *Humidity* acts to spread corrosion products and other localized surface contamination across the surface of the testpiece. It also sets up local electrochemical cells, enabling galvanic corrosion to proceed. Humidity may also affect the composition of the atmosphere immediately adjacent to the testpiece surfaces in a manner that is difficult to control reliably.
- *The atmosphere* that simulates the ambient environment tends to be corrosive to some ex-

tent. However, controlling the composition of mixed-gas atmospheres is notoriously difficult, particularly as it is often minor constituents that play the critical role in corrosion processes.

Given the complex interrelationship among these variables, it is not surprising that no "universal" accelerated aging test for solderability has been developed. Nevertheless, by trial and error it is possible to devise an aging test that will provide meaningful results for particular situations involving closely specified components. For example, the decrease in the average wetting time of copper wire coated with tin between 0.5 and 8 µm (20 and 315 µin.) thick during storage for 6 and 12 months in an open laboratory at a site in West London is reasonably approximated by the "damp heat" aging test for 4 and 6 h, respectively [Ackroyd 1975]. However, even small changes in the components, including in the thickness and microstructure of surface coatings, can invalidate the results of an aging test. Likewise, the method used to assess the wettability of aged substrates may also affect the correlation between natural and accelerated aging [Ackroyd 1975].

6.4 Microscopic Examination of Joints

Microscopic examination is widely used to directly evaluate the manner in which a molten filler metal reacts with a solid substrate and to identify defects in joints, such as voids and cracks. Microscopic analysis can also be performed on specimens that have failed in mechanical testing to identify the source of weakness.

Three principal types of microscopic examination are widely employed in the evaluation of soldered and brazed joints. They may be used both on

Assessment of Joint Quality

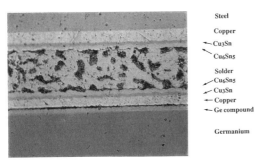

Fig. 6.15 Micrograph of a soldered joint, made using lead-tin eutectic solder, between copper-metallized steel and germanium components following aging at elevated temperature. A crack is clearly visible at the interface between the copper and the germanium-containing intermetallic layer. 420×

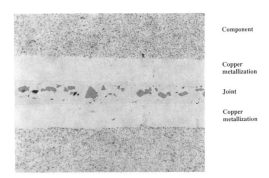

Fig. 6.16 Micrograph of a joint, made using Al-12Si braze, between components that have been metallized with copper. The joint microstructure comprises silicon in a matrix of the intermetallic compound $CuAl_2$, which accounts for its embrittlement. 420×

joints as bonded and after failure or degradation. These techniques are:

- Metallographic examination
- Scanning acoustic microscopy
- Scanning electron microscopy

To a large extent, the three methods are complementary in that they provide different information on joints.

6.4.1 Metallographic Examination

Examination of metallographically polished and etched joints under an optical microscope can be used to identify heterogeneous features, such as intermetallic phases, porosity, cracks, filler/substrate interfaces, and lack of fusion. Photographic guides are available that catalog the structure of soldered joints and the reactions with common component materials [ITRI 708 (undated)]. Fractured specimens can be used to pinpoint the source of failure, whether intermetallic phases, weak interfaces, or void clusters. Examination of edge fillets can provide an indication of the filler/substrate contact angle at the onset of solidification. However, metallographic inspection of joint sections does not provide any direct information on the kinetics of joint formation.

An example of the use of metallographic examination in the evaluation of joints can be given with reference to Fig. 6.15, which is a micrograph of a tin-lead soldered joint between components of steel and germanium, following soak testing at elevated temperature. The components were metallized by electroplating with copper to provide solderable surfaces, and, as expected, copper-tin intermetallic phases are observed at the interfaces of the joint. As made, the joints possessed the requisite mechanical properties, but these were not maintained following the thermal soak test. The micrograph shows that the origin of the weakness is a crack on the germanium side of the joint that runs between the Cu_3Sn interfacial layer and a germanium-containing intermetallic phase that has formed during the heat treatment.

In another case, Fig. 6.16 shows a brazed joint made with aluminum-silicon eutectic alloy between components of an engineering material that was also metallized with copper. The metallographic section reveals that the joint contains particles of silicon embedded in a matrix of the intermetallic compound $CuAl_2$. As both of these phases are hard and brittle, the joint is likely to have similar mechanical characteristics, which would be inappropriate for the intended application.

Metallographic examination is often regarded as involving little more than straightforward observation. It does, in fact, require considerable skill to carry out effectively. For example, great care must be taken to avoid introducing artifacts of the sample preparation, such as smearing of soft solders, polishing out of hard phases, embedding of polishing debris, and inducing incipient melting of low-melting-point phases, all of which can easily give rise to unrepresentative joint microstructures [Bjerregaard *et al.* 1992]. Many chemical etches used to reveal phases can also cause staining, opening the possibility of misleading interpretation.

The correct evaluation of even perfectly polished and etched specimens requires additional

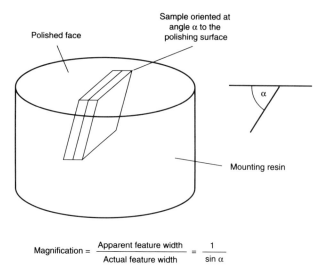

Fig. 6.17 Metallographic taper section, used to magnify the width of a thin layer to facilitate resolution

expertise, particularly when multicomponent filler metal/substrate combinations are involved. Because of the thinness of many soldered and brazed joints, taper sections (see Fig. 6.17) are often used to effectively magnify the width of the joint gap and surface metallizations. If quantitative analysis has to be made of the dimensions of any features in a taper section, the orientation of the section must be known. Special metallographic procedures must be followed when preparing sections of nonmetallic materials, particularly ceramics, which tend to be harder, more brittle, and more friable than metals [Lay 1991].

With the aid of modern image analysis techniques, features of joints, such as the volume fractions of phases present and their anisotropy, can be quantitatively measured [Pover 1987]. Image analysis involves directing a television camera at a metallographically prepared specimen through a microscope. The image is digitized by a computer, and the data obtained are processed to enhance specific features. For example, shade correction can be applied to compensate for nonuniform sample illumination. Edge contrast of features can be increased, and spurious detail, produced by artifacts such as scratches or staining, can be removed. These procedures are applied in order to isolate or differentiate features of interest, which can then be quantitatively measured. Typical dimensional measurements that may be made on individual particles using an image analysis system are illustrated in Fig. 6.18. Statistical measurements on an ensemble of particles, voids, or other features can also be made, enabling their size distribution and volumetric density to be calculated. The data can be used to establish whether particles or voids lie in a plane within a joint and thereby constitute a line of possible weakness.

The principal limitation of metallographic examination is that only single sections of an entire joint area are viewed. There is therefore a risk that the section examined might not be representative of the joint as a whole. For example, center-to-edge variations in microstructure are particularly common due to compositional changes that occur during flow of a molten filler. This limitation can be reduced by evaluating complete sections, extending from one edge to the other, and multiple sections through the joint.

6.4.2 Scanning Acoustic Microscopy

Scanning acoustic microscopy (SAM) is a relatively new technique that is capable of nondestructively assessing the distribution of voids, cracks, inclusions, and other hard phases over the area of essentially parallel-sided joints [Burton and Thaker 1985; Matijasevic, Wang, and Lee 1990; Kauppinen and Kivilahti 1991]. It is a high-resolution version of ultrasonic nondestructive testing (see section 6.6.5).

The technique involves focusing an acoustic wave, generated from a piezoelectric transducer, via a sapphire lens, onto the specimen and scanning it in a raster fashion as shown in Fig. 6.19. Changes in the reflected acoustic signals from boundaries between features having different

Assessment of Joint Quality

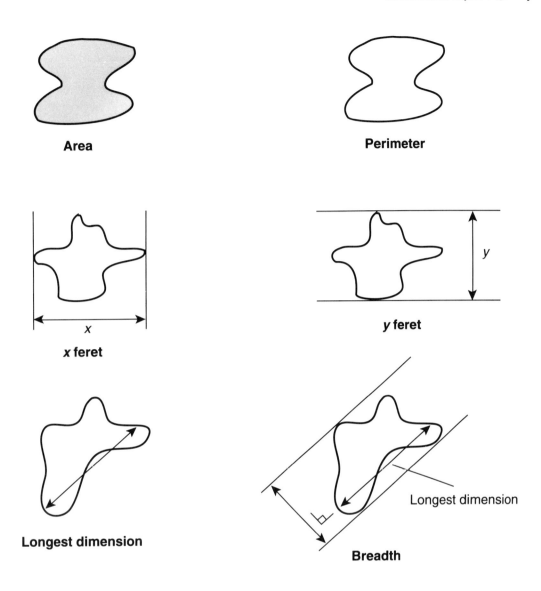

Fig. 6.18 Typical feature-specific measurements that can be made using image analysis

acoustic properties are recorded and mapped to produce the image. A fluid, usually deionized water, is used to couple the acoustic energy between the lens and the sample. The detected signal is transmitted to a synchronous display and recorded.

The operational frequency range for acoustic microscopy is generally 20 MHz to 2 GHz. Frequencies at the higher end of the range (above 800 MHz) provide the best resolution. However, increasing the resolution reduces the penetration depth, owing to a progressive absorption by the material of acoustic energy with increasing frequency. For subsurface imaging by this technique, the optimum resolution transverse to the acoustic beam is given by the expression:

$$\text{Resolution (microns)} = \frac{0.7 \times V_s}{f}$$

where V_s is the velocity of sound in the base material, assuming that the joints are relatively thin, and f is the frequency of the acoustic wave in Megahertz.

Thus, at 2 GHz, the resolution can be on the order of 1 to 3 µm (40 to 120 µin.), with a penetration

Principles of Soldering and Brazing

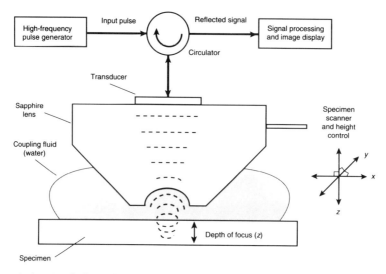

Fig. 6.19 Schematic showing the basic elements of a scanning acoustic microscope

depth of only about 10 μm (400 μin.) (depending on the material and its condition), while at 50 MHz detail can be resolved down to approximately 40 μm (1600 μin.), but now the beam can probe to a depth of as much as 5 mm (0.2 in.) [Burton and Thaker 1985]. An acoustic image of a brazed joint in a 50 mm (2 in.) diam assembly is shown in Fig. 6.20.

The interpretation of a scanning acoustic micrograph of a joint can present difficulties—for example, determining whether a certain feature corresponds to a crack, void, or intermetallic particle. The need to mechanically scan the probe in very close proximity to the surface of the components restricts the geometries that can be examined. Moreover, the limited depth of sample from which clear images can be obtained means that one of the components must be thin. The correspondence that can be obtained between the images of a voided joint produced by SAM and by X-radiography is illustrated in Fig. 4.12 and 4.13.

6.4.3 Scanning Electron Microscopy

Scanning electron microscopy (SEM) can be used to identify the elemental composition of phases present in joint sections and thereby to correlate the microstructure with the alloy phase diagram. In an SEM system, electrons are accelerated by a voltage, normally up to 30 kV, and passed through focusing coils to produce a fine beam that impinges on the surface of the specimen. Scanning coils cause the electron spot to be swept across the specimen in a raster fashion. The incident electrons excite electron and X-ray emissions from atoms in the specimen. These are collected separately and the modulations in the respective signals are mapped to produce images on fluorescent screens.

The events following the impingement of the electron beam are indicated in Fig. 6.21. The beam penetrates the specimen to a depth L, which depends on the accelerating voltage and on the density and atomic weight of the materials being probed, but is usually on the order of 1 to 5 μm (40 to 200 μin.). As the beam penetrates the material, it spreads from the incident spot diameter d to a cross section D. All of the emitted X-rays and electrons derive from volume V, which is on the order of at least 1 μm^3 (6 μin.^{3}). The secondary electrons, which are generated between 10 and 20 nm (100 and 200 Å) beneath the surface, are selected and imaged to provide topographical information, such as the presence of voids and cracks. On the other hand, the X-rays, which are derived from the entire volume, V, are collected at an energy-dispersive analyzer, which furnishes quantitative measurements of elemental composition and its variation across the scanned surface to a resolution that can detect 0.5 at.% from areas larger than about 1 μm^2 (16 μin.^{2}). Energy-dispersive X-ray analysis (EDAX) can be used to detect X-rays from all elements with an atomic number greater than 10.

Assessment of Joint Quality

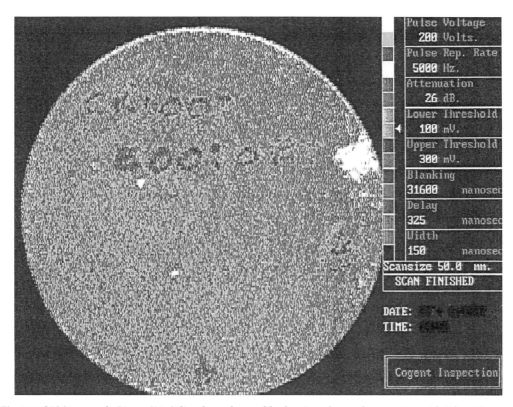

Fig. 6.20 SAM image of a 50 mm (2 in.) diam brazed assembly showing a large edge-opening void at the right

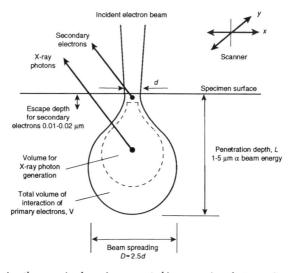

Fig. 6.21 Schematic showing the energized species generated in a scanning electron microscope

A representative SEM system is shown schematically in Fig. 6.22. A compact review of SEM, outlining its modes of operation and capabilities, is given by Richards and Footner [1989]. Additional information on electron microscopy at an introductory level may be found in Kossowsky

Principles of Soldering and Brazing

[1983]. An SEM image of a fatigue crack in a lead-tin soldered joint between a leadless ceramic chip carrier and the copper termination on a PCB is reproduced in Fig. 6.23. The large depth of focus that is a feature of this technique can often help those skilled in the art to elucidate the mode of failure and sources of weakness in samples that have failed by either mechanical stress or corrosion.

An example of the application of the X-ray mapping technique can be illustrated with reference to Fig. 6.24. Figure 6.24(a) is a backscattered electron image obtained in an SEM of a brazed joint in an aluminum engineering alloy made using a filler metal containing aluminum, silicon, and copper. The SEM image has just sufficient contrast to indicate that the reaction zone contains three phases, two of which are relatively massive

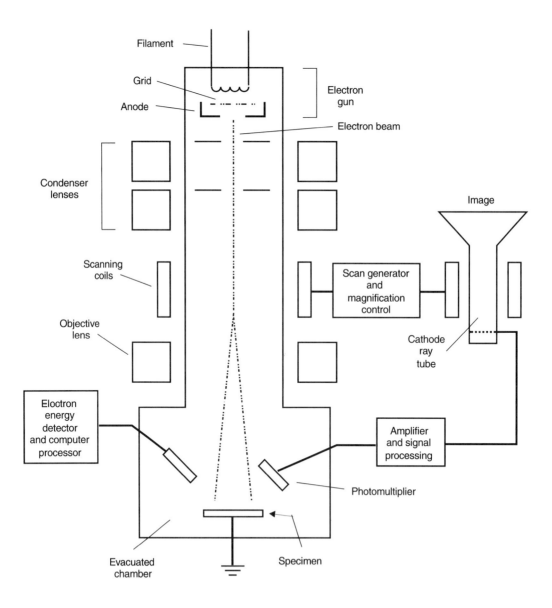

Fig. 6.22 Schematic of a scanning electron microscope

Assessment of Joint Quality

in appearance (one having a lacelike morphology) and the third comprising small particles of a darker phase. EDAX maps of the same region of the sample, made for the elements aluminum, silicon, and copper, are given in Fig. 6.24(b) to (d), respectively. From these it is clear that aluminum and copper are present throughout the joint region, with the lacelike phase containing more copper than the "dense" phase. Neither of the massive phases contains silicon, which is concentrated solely in the darker phase. With reference to the relevant ternary phase diagram it is possible to deduce that the dense phase is Al_2Cu, that the lacy phase is AlCu, and that the dark phase is probably a solid solution of aluminum and copper in silicon.

Fig. 6.23 SEM image of a soldered joint between a leadless chip carrier and a PCB, showing a fatigue crack through the filler metal

6.5 Mechanical Testing of Joints

The most usual requirement of soldered and brazed joints is to provide bonds between components that are sufficiently strong and durable for meeting the design life of the assembly. For this reason, the strength of joints is frequently measured in various stressing modes—typically ten-

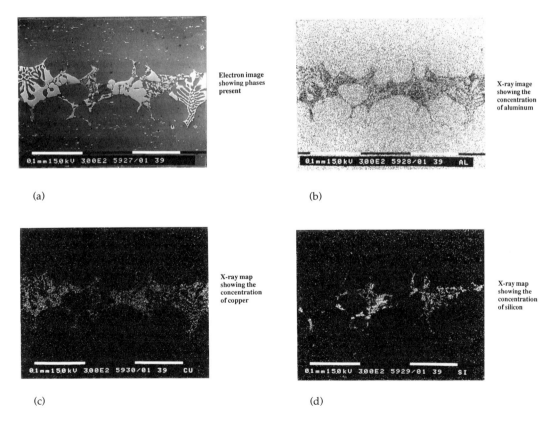

Fig. 6.24 SEM image (a) and X-ray maps for concentrations of aluminum (b), copper (c), and silicon (d) for a joint, made using Al-Cu-Si braze, to an aluminum engineering alloy. Courtesy of TNO—Industrial Research, Apeldoorn, The Netherlands

sile, shear, peel, and cyclic fatigue, depending on the intended application. Mechanical testing should also encompass the effects of corrosion on the mechanical properties, as appropriate.

Mechanical testing on its own will not provide information on the sources of weakness in a joint, such as the presence of voids, brittle interfacial phases, or the lack of fusion of interfaces. It therefore must be complemented by other methods of assessment.

The mechanical properties of joints normally have to meet a certain minimum standard. The set of properties that are of importance tend to be application specific and will depend on the stress, temperature, and chemical conditions that the assembly will experience in service. This environmental regime is usually complex and interactive, and a testing schedule for quality control and assessment purposes can only approximate the set of conditions that apply in practice and thus will be necessarily selective. For practical and economic reasons, the testing is often limited to scaled-down specimens of simplified geometry. Despite the apparent remoteness of the idealized testpieces from products in the field, it is frequently possible to develop test procedures that can offer a high degree of confidence regarding the adequacy of the product for the intended application environment.

In defining a worthwhile testing schedule, it is necessary to identify and rank the limiting conditions under which the assembly has to operate, according to their influence on mechanical integrity. Thus, in the case of a brazed bicycle frame, which has a load-bearing function and is subjected to weather exposure, the joints must possess a modicum of strength and be resistant to failure by fatigue and corrosion. The strength of this structure can be assessed by a simple tensile test, its fatigue life approximated by a standard fatigue test, and its susceptibility to corrosion gauged by an accelerated salt-spray test. Satisfactory performance in all three tests is needed to provide confidence of fitness-for-purpose. If an assembly is likely to experience a combination of "distressing" influences, such as stress in a corrosive environment, then it is necessary to include tests for the appropriate condition—here, stress corrosion.

For a different example, consider a surface-mounted electronic component soldered to a circuit board. In this instance, the most critical mechanical parameter is usually the fatigue life, which is generally limited by the thermal expansion mismatch stresses that are cyclically induced whenever the circuitry is operated. Mechanical strength *per se* is less important in this case, because the component is lightweight in relation to the joint area, and low loads are superimposed.

Measurement of joint strength provides the most direct indication of whether a soldered or brazed joint has achieved the rudimentary level of mechanical integrity necessary to satisfy the function of an assembly. Assessment of mechanical properties usually involves destructive tests in which representative or simplified assemblies are stressed until failure occurs. There is a benefit in carrying out the testing on a given parent/filler combination using different stressing modes inasmuch as this is more likely to furnish a clearer picture of the mechanism of failure and the critical flaws that are responsible than if a single test configuration is employed [Chilton, Smart, and Wronski 1983].

Several difficulties are associated with the mechanical testing of joined assemblies for quality control and assuance purposes. These are addressed in the paragraphs that follow.

Devising mechanical tests that will furnish usable data. Assuming that the conditions which operate during the service life of an assembly are known, tests that reproduce these conditions often are too complicated to be practical. In such circumstances it is common practice to carry out simplified tests, designed to provide a high degree of semiquantitative discrimination of joints. One sort of simplification that is widely applied is to test under resolved stresses, for example, pure tensile or pure shear. For this purpose joints of simplified geometry are used, notably butt and lap joint configurations (Fig. 6.25). Resolution of the actual applied stresses that develop in the testpieces is usually difficult, because a test intended to measure tensile strength, for example, may inadvertently subject the joint to shear or peel forces.

An additional difficulty is encountered in attempting to relate these measurements on idealized testpieces to the different joint geometries used in assemblies of practical interest, which include scarf, step, and various types of strap joints, as shown in Fig. 6.26, and many of the configurations found in the microelectronics industry. In these situations, the most prudent approach is to modify the tests so as to reproduce the actual joints as closely as possible, although matching a realistic stress regime presents a more difficult problem. Indeed, for this reason electronic circuitry is tested in a very different manner. By way of example, Marshall [1988] and others [Frear, Jones, and Kinsman 1991, p 197-201] have devised ther-

Assessment of Joint Quality

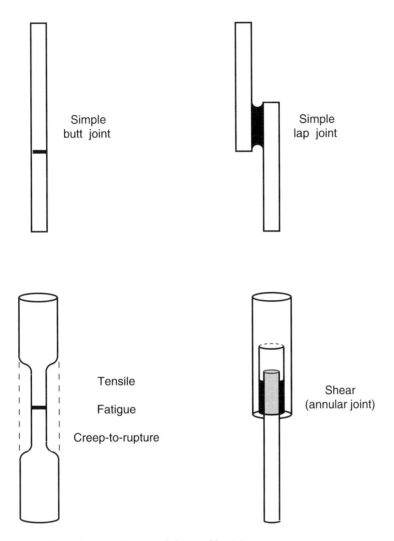

Fig. 6.25 Testpiece configurations based on simple butt and lap joints

momechanical fatigue testing procedures for simplified joints in test specimens that reflect microelectronic configurations. These are described in section 6.5.5.

The test conditions will generally have a significant influence on the results obtained. Associated parameters that need to be closely defined include strain rate, test temperature, test environment, and size and shape of any fillets.

Establishing the magnitude of the stresses to which the joint will be subjected in service. In order to ensure that the joint is "fit for purpose," the bonded assembly must be tested under conditions of stress that are at least as severe as those to which it is subjected over its service life. This presupposes knowledge of the form and distribution of the stresses that are actually exerted on the joint during use and of the associated environmental conditions. These factors and their variability during service can have a critical effect on mechanical properties. By contrast, most standard mechanical tests are carried out under a resolved stress, usually uniaxial tensile or shear, coupled with a normal air environment. How often are joints in engineering components kept within such a constrained service regime?

More usually, the stress distribution in joints is exceedingly complex and highly variable, and the atmosphere may be corrosive. An example is a carbide cutting tip brazed in a steel tool (see Fig. 6.27). This component is subjected to a number of hostile conditions in service, such as several dif-

Principles of Soldering and Brazing

ferent stress modes (including vibration), the corrosive effects of cutting fluids, and thermal shock.

Replicating the thermal history. Prolonged service at elevated temperature can cause metallurgical changes to occur in joints, which can affect their mechanical integrity. Intermetallic compounds frequently form by solid-state diffusion reactions between the components and the solidified filler alloy. These compounds may themselves be brittle or have weak interfaces with other phases in the joint. A commonly encountered compound of this type is the Cu_6Sn_5 intermetallic phase that forms as an interfacial layer between tin-base solders and copper alloys at temperatures as low as 50 °C (120 °F) (see section 3.2.2). This phase normally forms in association with the Cu_3Sn compound, which is relatively strong and ductile. Also associated with the development of intermetallic phases is the formation of Kirkendall porosity, which can generate cracks, as occurs when aluminum conductors are joined to gold contacts [Footner, Richards, and Yates 1987].

Solid-state diffusion can also cause grain growth and microstructure coarsening, which can lead to weakening of the filler. This phenomenon is especially conspicuous in soldered joints seeing service at ambient temperatures and above, which are high in relation to their melting temperatures. This point is discussed in section 2.2.3.3 and in Frear, Jones, and Kinsman [1991, p 1-27]. Because heating excursions during service can influence the mechanical properties of joints, it is im-

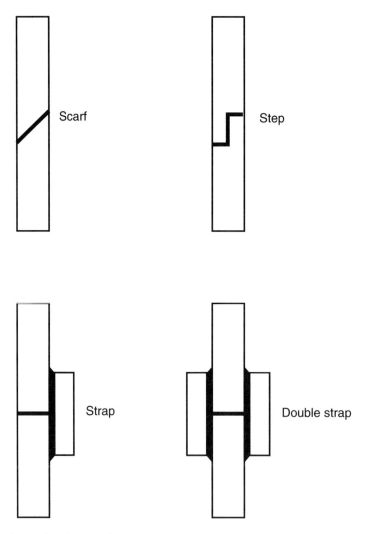

Fig. 6.26 Practical examples of joint configurations encountered in mechanical applications

portant to evaluate the test joints both before and after an aging treatment that represents a "worst case" condition for the assembly in the field.

Reproducing the conditions under which the manufactured products are joined. The conditions under which the testpieces are prepared, including the peak joining temperature, the time that the filler is molten, and the cooling rate from the peak temperature, can influence their performance in the test [Saxton, West and Barrett 1971]. If the conditions under which the test joints are prepared do not closely reflect those used in manufacture, the test results may not be applicable to the performance of the fabricated assembly being modeled. Because these and the test conditions referred to below are not completely specified for most of the published data, test measurements frequently differ widely for a given filler/component combination, even though each of the laboratories involved can readily reproduce its own set of data.

Influence of fillets. Mechanical testing can be complicated by the presence of fillets at the periphery of a joint that form naturally during the bonding operation. These can have a significant effect on the measured joint properties. Well-formed fillets of filler metal can enhance the measured tensile, shear, and peel strengths by as much as an order of magnitude, the value depending on the geometry of the joint (see Fig. 6.28) and the mode of stressing. This improvement can be attributed to the role of fillets in reducing stress concentrations at the edges of joints, as discussed in section 4.4.3.1. Because the formation of fillets of reproducible geometry is difficult to achieve, testpieces for mechanical testing are often designed to exclude fillets, either by preventing their formation through the use of nonwettable surfaces outside the edge of the joint or by removing any that happen to form. Although this practice makes the measurements more readily reproducible, it modifies joint strengths to an extent whereby they may not be representative of most practical situations.

Despite these complexities, a program of "proof" testing, involving more than one stressing mode, is recommended. Provided that this program is carefully designed and the limitations appreciated, it can be effective in promoting quality assurance, as attested by the impressive safety record in materials performance achieved by the aviation industry. Ultimate confidence in the mechanical integrity of a joint is attained when the

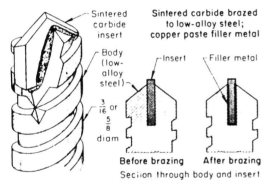

Fig. 6.27 Typical masonry drill bit comprising a sintered carbide insert brazed into a steel shank

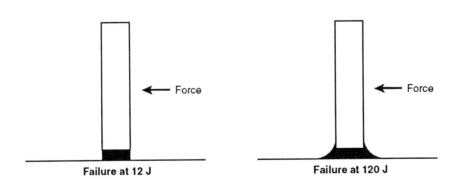

Fig. 6.28 Impact test on brazed T-joints, clearly demonstrating the role of fillets in enhancing joint strength. Substrate: mild steel. Braze: Ag-Cu-Cd-Zn

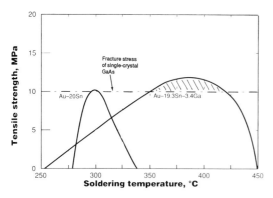

Fig. 6.29 Tensile strength of soldered joints made between GaAs samples, coated with 0.5 µm (20 µin.) of evaporated gold, showing the effect of adding 3.4 wt% Ga to the Au-20Sn solder. The fracture stress of single-crystal GaAs is approximately 10 MPa (1.5 ksi). Thus, over a range of process temperatures the joint strength exceeds that of GaAs.

strength of the joint exceeds that of the adjacent components (i.e., failure occurs through the components), assuming, of course, that the joining process does not itself degrade the components nor the joint geometry impose high stress concentrations on the components (see section 4.4.3.1). Figure 6.29 shows the strength of gallium arsenide (GaAs) die-bonded assemblies as a function of joining temperature. Joints are judged to be acceptable when failure occurs in the GaAs or the substrate, rather than through the soldered joint.

Mechanical property assessments often have been made on the bulk filler metals, but these are of questionable value in relation to joints, because:

- Alloying within joints changes the composition of the filler.
- New phases formed by alloying in joints tend to segregate in a manner that can enhance embrittlement.
- The cooling regimes are different in the two cases, so that grain size and texture will also differ.
- The stress distribution in joints often differs radically from that produced in standard testpieces of bulk filler metals.

Therefore, tests must be performed on joints with geometries that reflect those in the assemblies of interest.

Thermal cycling provides a useful initial sorting test for joint quality, especially when the thermal expansivities of the two joined components differ. A discriminating test, sometimes referred to as thermal shock, involves shuttling the testpieces rapidly between a furnace set at a chosen upper temperature and a liquid nitrogen bath (–196 °C, or –320 °F), with the dwell time in the two environments being about 5 min and 1 min, respectively. Suitable apparatus is shown schematically in Fig. 6.30. The shorter dwell period in the liquid nitrogen bath reflects the faster thermal equilibration that can be achieved in this medium. Boiling cyclohexanol (160 °C, or 320 °F) or hot silicone oil (180 °C, or 355 °F) are sometimes used in place of the furnace [Matijasevic, Wang, and Lee 1990]. The cold bath may also be at a temperature other than –196 °C (–320 °F). Although this type of test may be overly severe in relation to actual operating conditions, it is useful as a sorting test, enabling rapid screening of joints in prototype assemblies that are inferior on one or more counts, because it involves subjecting them to conditions that encompass shear and peel stresses, mechanical fatigue, elevated-temperature soak testing, and thermal shock, while the atmosphere repeatedly changes from cold and wet to hot and dry.

Individual mechanical properties that are most commonly tested are discussed in the sections that follow.

6.5.1 Tensile Strength

The tensile strength of a planar joint is a measure of its ability to sustain a stress applied perpendicular to the joint without breaking. The measured strength will be sensitive to the geometry of the testpieces and of the joint, particularly if the latter is comparable in strength to the parent materials and if the filler has some elasticity, a point that needs to be borne in mind when comparing strength data [West *et al.* 1971; Saxton, West, and Barrett 1971; Trimmer and Kuhn 1982].

A typical tensile test for soldered and brazed joints will be briefly outlined to indicate its main features. The testpiece comprises two cylindrical bars, each 100 mm (4 in.) long and 10 mm (0.4 in.) in diameter, joined end-to-end to form a butt joint using a special jig that provides axial alignment. The joint gap is defined, as necessary, by inserting a spacer wire between the two cylindrical bars. Flux, if required, and the filler are placed within the gap, and the joint is made following a set procedure. A gage length of typically 50 mm (2 in.) is formed by turning down the bar to a diameter of 9 mm (0.35 in.) in a region that symmetrically en-

Assessment of Joint Quality

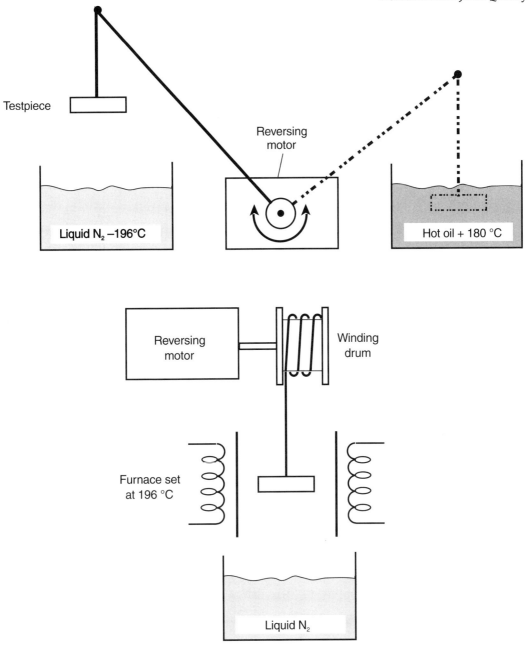

Fig. 6.30 Schematic of two methods used for subjecting samples to severe thermal cycling/shock

compasses the joint, as shown in Fig. 6.25. A rigid, close-fitting sleeve placed over the gage length can be used to ensure that the test is conducted under pure tensile loading. Parameters such as the maximum load to fracture (known as the ultimate tensile strength, or UTS), the elastic limit, and the elongation to failure are measured by pulling the joined testpiece apart at a fixed rate—for example, under constant loading of 10 MPa/s (1.5 ksi/s) or a constant rate of extension of 1 mm/s (0.04 in./s).

A typical load/extension graph for a brazed joint in a ductile material subjected to tensile testing is reproduced in Fig. 6.31. The computer used to produce the graph has calculated and marked (the solid box) the limit of proportionality of the sample, that is, its yield point. The high elongation

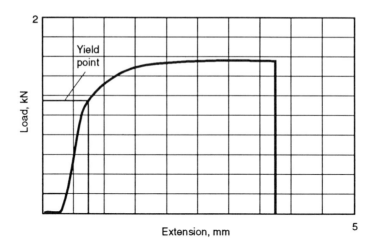

Fig. 6.31 Typical load/extension curve obtained during tensile testing of a brazed joint in relatively ductile components

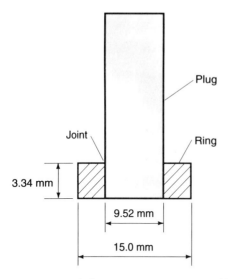

Fig. 6.32 Ring and plug testpiece geometry suitable for measurement of the shear strength of soldered joints. Source: After ITRI 656 (undated)

to failure exhibited in this example is a consequence of the ductility of the parent material.

6.5.2 Shear Strength

The shear strength of a planar joint is a measure of its ability to sustain a stress applied parallel to it without breaking. Most shear tests are designed to be carried out under tensile loading of the testpiece assembly, so that the same equipment can be used to perform both tensile and shear testing.

A testpiece geometry that is often used to test brazed joints comprises a cylindrical bar joined to an annular component in such a way that the joint is circumferential to the bar and will be loaded in shear when the assembly is subjected to a uniaxial tensile stress, as shown schematically in Fig. 6.25. The length of the annular joint is of a specified value, for example, 6 mm (0.2 in.). The joint gap can be defined by spacer wires. The bonding operation has to be carried out by infiltrating the filler metal into the joint. For soldered joints, a "ring and plug" configuration of the type shown in Fig. 6.32 is preferred, because it can be prepared relatively easily with a precision that provides reproducible test assemblies [ITRI 656 (undated), p 7]. The testpiece is pulled at a fixed rate of applied load or of extension to failure. As in the case of the tensile test, the results must specify such test-sensitive conditions as the rate of application of the stress (or of strain), the parent metals and their dimensions, the filler, the average joint gap, any flux used, and details of the joining operation.

A simple lap joint configuration is more representative of many engineering assemblies. However, a test assembly with this type of joint cannot be relied on to provide reliable measurements of shear strength. This is because a simple lap testpiece is, by virtue of its geometry, misaligned with respect to the axis of applied stress, which will cause it to bend when a tensile load is applied and will result in misleading shear strength measurements. A symmetrical double lap joint overcomes this particular problem, because yield is constrained along a single axis as it is in the ring and plug configuration. However, it is more difficult to prepare a double lap joint with precision. The stress concentrations present in lap joints are discussed in section 4.4.3.1.

6.5.3 Impact Resistance, Bending, and Peeling Strength

The bending and peeling strength of a joint is its ability to withstand stresses that act in a manner so as to tear or peel the joint apart. Most joints in service experience these types of stresses as part of their loading regime. Impact resistance is equivalent to bending strength as measured at very high strain rates, that is, under conditions of mechanical shock.

Impact, bending, and peeling tests are effective in disclosing brittle interfacial phases, voids, inclusions of flux, and so on. Specimens with such defects may give high failure strengths in tensile and shear tests but have little capacity to bear impact, bending, or peeling forces. Bending tests are usually confined to brazed butt joints. Soldered joints of this configuration are rarely of sufficient strength to support extensive bending, and thin lap-jointed assemblies tend to be used instead. For acceptance purposes, the joined testpiece must survive being bent through a specified angle over a former of a given radius, which is related to the most severe conditions likely to be experienced by the assembly represented by the testpiece. A 180° bend of a joined plate around a former having a radius 1.5 times its thickness is sometimes specified in standards as the test criterion for brazed joints in ductile materials such as mild steel.

The test is applied by bending the joined assembly onto a former. Inhomogeneity of the joint, however, can cause the specimens to take on an irregular shape if the bending is carried out without restraints, and thus the radius of the bend can vary along the joint width. This problem can be overcome by using a guided or controlled bend test. The specimen configurations used for these types of tests are illustrated in Fig. 6.33. These standard

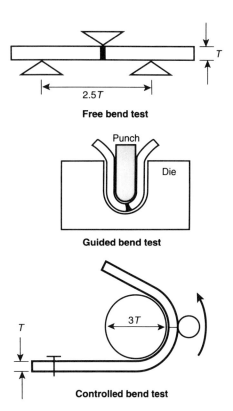

Fig. 6.33 Methods for conducting bend tests on joined testpieces of plate geometries

bend tests can only be used to either pass or fail joints.

Semiquantitative data can be obtained in a bend test by linking it to an impact test. In such a test, a standard notched specimen is held in a grip and a pendulum is made to swing from a fixed height to strike it, as shown in Fig. 6.34. The maximum deflection of the pendulum after bending or fracturing the specimen provides a measure of the energy imparted to the specimen as a result of the deformation. One version of this semiquantitative bend test involves fabricating a notched testpiece from two tapered components, with the joint formed at the butt ends of the tapers [Boughton and Sloboda 1970]. This is shown schematically in Fig. 6.34. In the subsequent impact test, the testpiece will either deform plastically under impact loading through a sufficiently large angle to allow the pendulum of the testing machine to swing clear, or fracture will occur with limited plastic deformation. The test parameter measured is the energy absorbed by the testpiece.

Peel tests are a particular type of bending test designed to measure the resistance to tearing or

Principles of Soldering and Brazing

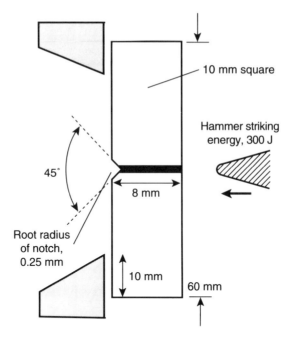

Fig. 6.34 Dimensions of typical testpieces used in a semiquantitative high-strain-rate bend test, based on the Charpy impact test (not drawn to scale)

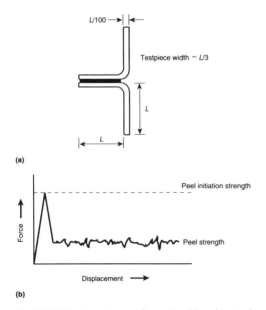

Fig. 6.35 T-lap testpiece configuration (a) and typical force/displacement curve (b) of a T-peel (Chadwick) test

peeling of a joint. In the Chadwick configuration, two L-shape strips are soldered or brazed together along one of the arms [ITRI 656 (undated), p 9]. Again, spacer wires can be used to fix the joint gap. The two unjoined ends of the resulting T-shape testpiece (see Fig. 6.35) are pulled apart in a tensile testing machine at a fixed rate of (typically 10 mm/min). Figure 6.36 shows a typical load/extension curve for a Chadwick peel test of a joint in a ductile material. The peaks in strength at each end of the joint are associated with small fillets in those regions.

Two values are furnished by this test: the stress required to initiate tearing or peeling and the stress required to propagate this mode of failure. Interpretation of the results is straightforward only if the joint is much weaker than the components. Otherwise, part of the measured stress is expended in bending the components and thus does not strictly correspond to the failure load as applied to the joint. On the other hand, if the components are of a strong metal, such as steel, and suffer severe distortion in the test, this can provide qualitative assurance that the joint has a high resistance to peeling.

Other testpiece geometries are also used to assess peel strength, but usually for reasons of ease of assembly and testing rather than in an attempt to permit better interpretation of the results (e.g., see Kurihara et al. [1992]). Relationships among peel strength, rate of specimen loading, and peel crack growth rate and the correlation among crack initiation, testpiece geometry, and the form of the

Assessment of Joint Quality

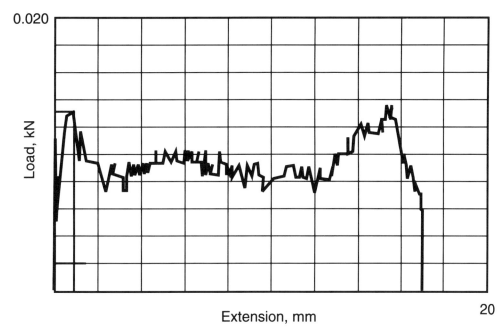

Fig. 6.36 Typical load/extension curve obtained during Chadwick peel testing of a brazed joint

Table 6.3 Breaking strengths of soldered joints to copper testpieces in different configurations

Testpiece configuration	Strength value
Butt joint, tensile loading	92 N/mm² (MPa)
Single lap, shear loading	39 N/mm² (MPa)
Ring and plug, shear loading	36 N/mm² (MPa)
T-piece, peel loading	8.8 N/mm

Source: ITRI 656 (undated), p 9

load/extension curve have been demonstrated by Dunford and Partridge [1992].

A comparison of the ultimate (breaking) strengths of joints as measured by different test configurations is given in Table 6.3. The data were obtained on copper components joined using Pb-60Sn solder [ITRI 656 (undated), p 9]. All of the tests were performed using a tensile testing machine and a strain rate of 2 mm/min (0.08 in./min). The shear strength measured using a single lap joint is, in this case, close in value to that obtained using a ring and plug assembly. On the other hand, the tensile strength obtained for a butt joint was approximately 2.5 times the measured shear strength (see section 6.5.6). The peel strength value cannot be directly related to the other data, because it is measured in different units. The value of the peel strength quoted is the load per unit width of the joint that is required to initiate tearing. However, it can be seen that, in terms of the applied load, the joints are weaker with respect to peel than to the other stressing conditions.

6.5.4 Creep Rupture Strength

Mechanical failure of materials can occur by time-dependent plastic deformation when subjected to stress. This phenomenon, known as creep, is especially common in stressed assemblies that are held at elevated temperatures for long periods of time, such as boiler systems [Dieter 1976, p 451-489]. Failure through creep is referred to as stress rupture, and the reduction in stress in an assembly through creep is termed stress relaxation.

The rate of creep accelerates toward the melting point of a metal, being governed to a first approximation by an Arrhenius relationship, that is:

$$\frac{de}{dt} = A\sigma^n \exp\left(-\frac{E}{RT}\right) \qquad \text{(Eq. 6.1)}$$

where de/dt is the steady-state strain rate, σ is applied stress, E is the activation energy for creep, R is the gas constant, T is temperature in degrees Kelvin, A is a constant, and n is a power factor.

Figure 6.37 presents creep test results obtained at different temperatures for joints to copper made

Principles of Soldering and Brazing

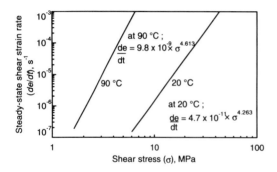

Fig. 6.37 Steady-state strain-rate creep of Bi-42Sn soldered joints to copper, as a function of applied shear stress and test temperature. Source: After Mei and Morris [1992]

with Bi-42Sn eutectic solder; these show reasonable concordance with the Arrhenius relationship. In other words, the rate of creep increases exponentially as the temperature of testing is raised and is also dependent on the applied shear stress in accordance with Eq. 6.1, as shown in Fig 6.37.

Because the temperatures at which most solders are used are relatively close to their melting point—that is, they find application at a high homologous temperature (T/T_m)—creep and stress relaxation are often important determinants of the mechanical properties of soldered joints, and also of brazed joints if these experience temperatures that approach their melting points [de Kluizenaar 1990, p 27-28]. Therefore, it is important to include creep rupture strength measurements in an assessment schedule that addresses the mechanical integrity of joints.

Creep rupture strength is defined as the stress at a particular temperature that produces a life to failure, or rupture, of a certain duration, normally 1000, 10,000 or 100,000 h. It can be determined by subjecting the joint to a constant stress (static loading) in tension or in shear, at temperatures within the range of interest. Stress relaxation, which is the relief of stress through creep, is measured by subjecting the testpiece to a fixed strain and monitoring the falloff in the corresponding stress as a function of time. The testpieces for these two types of measurements are usually similar to those used for tensile and shear testing, as described above.

In most practical situations, the creep rupture strength of soldered joints is extremely low for timescales that reflect the normal design life of most products. This means that soldered joints cannot be expected to bear appreciable loads, typically greater than 0.1 MPa (14.5 psi).

6.5.5 Fatigue Strength

When a material or an assembly is subjected to cyclic stresses, failure can occur when the stresses are a mere fraction of those needed to produce fracture under static loading, and even below the nominal yield stress. This type of failure is known as fatigue and is widely encountered in metals. Fatigue failure in joints usually occurs within the filler metal, which tends to be the softest material in an engineering assembly. The mechanism of fatigue failure of a joint will depend on the softness of the filler, which itself is a function of the homologous temperature (T/T_m).

At low homologous temperatures, a situation that is more applicable to brazed assemblies used at ambient temperatures than to soldered systems, failure occurs through the generation of one or more cracks at or near a free surface and their propagation through the joint. Under these conditions, the fatigue strength is only weakly dependent on the frequency of straining. The degradation process is affected by a variety of factors, many of which are difficult to control in a reproducible manner. For this reason, data scatter in a fatigue test is generally high, and large numbers of repeated tests (approximately six per point on the curve) are needed to obtain statistically reliable data.

Fatigue cracks can be initiated either at a surface of the assembly or internally within the material at microstructural inhomogeneities. Discontinuities in the vicinity of a surface where stresses tend to be concentrated, such as abrupt interfaces, rough surface texture, edge-opening fissures, and other irregularities that are common features of brazed and welded joints, are especially prone to nucleating fatigue cracks. Where there are no such surface defects and the joint has smooth and well-rounded fillets, the applied stress can cause internal damage that is manifested as voids, which can lead to cracks.

At high homologous temperatures, represented by soldered joints, the strain is manifested as plastic deformation (i.e., classical creep), accompanied by recrystallization, grain growth, and other microstructural changes. The mechanism of failure in this case is complex and is not fully understood at this time, but appears to involve the formation of cavities at grain boundaries, which coalesce to form gross internal voids. These grow

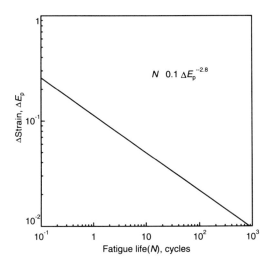

Fig. 6.38 Fatigue life (number of cycles for a 50% reduction in stress amplitude) of Bi-42Sn soldered joints to copper, as a function of strain amplitude, when tested at 65 °C (150 °F). Source: After Mei and Morris [1992]

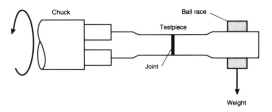

Fig. 6.39 Rotary fatigue testing machine. The stressing mode is sinusoidal and symmetrical tension/compression.

to produce fracture surfaces rather than the fatigue cracks that are observed at low homologous temperatures [Frear, Jones, and Kinsman 1991, p 135-153].

A common cause of fatigue failure in soldered joints is cyclic strain induced by repeated heating and cooling when the components have different thermal expansivities. This type of fatigue, known as low-cycle fatigue (but more accurately described as low-frequency fatigue), can be approximately modeled by the Coffin-Manson formula, which relates the number of cycles to failure to the amount of plastic deformation per strain cycle, thus:

$$N_f = C \times \Delta E_p^{-n}$$

where N_f is the number of strain cycles to failure, E_p is the plastic deformation per strain cycle, and C is a constant that is related to the ductility of the metal subjected to fatigue. This is a purely empirical relationship, and it is often found that the power factor n is close to 2 for soldered joints [de Kluizenaar 1990, p 28-30]. The validity of the Coffin-Manson relationship for joints to copper soldered with the Bi-42Sn eutectic alloy is attested by Fig. 6.38. The fatigue behavior is affected by the frequency of the strain cycling, which is not encompassed by the Coffin-Manson relationship. This and other aspects of low-cycle fatigue are treated in some detail by de Kluizenaar [1990] and by Frear, Jones, and Kinsman [1991].

As the frequency of straining increases toward the audible range (>100 Hz), the fatigue characteristics at high homologous temperatures become dominated by crack initiation and growth. This mode of stressing is restricted to a narrow range of applications, such as those employing vibrating sensors. For further information, the reader is referred to de Kluizenaar [1990, p 30-31].

A basic test for fatigue generally involves the application of a sinusoidal stress that alternates symmetrically in magnitude from tension to compression, using a machine of the form shown in Fig. 6.39. The stress can be applied either directly by mechanically loading the specimen (representative profiles are depicted in Fig. 6.40) or by thermal cycling of assemblies in which there is a mismatch in the thermal expansivities of the abutting components.

The latter type of assembly has been used to model fatigue in soldered joints to electronic components. The testpiece configurations are of relatively simple geometry and are primarily designed to place the sample joint under nominally pure shear stress. Two of these testpieces are illustrated in Fig. 6.41. The first comprises a sandwich of a flat strip of one metal (here, aluminum) to which similar-size strips of a second metal (copper) are soldered on both sides. Where necessary, the metals are suitably plated to ensure good solderability. The thermal expansivity mismatch between the metal strips provides fatigue stress when the sample is thermally cycled.

The second testpiece in Fig. 6.41 works on a similar principle, but here the shear stress is generated by the thermal expansivity mismatch between a metal pin (brass) and a polymer that encapsulates the pin. This geometry corresponds to a through-hole soldered joint to a PCB. The simplicity of the strain distributions in these test configu-

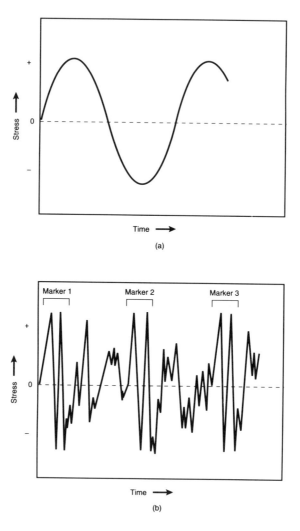

Fig. 6.40 (a) Simple sinusoidal stressing mode produced by a rotary fatigue testing machine. Complex fatigue stress profile patterns are occasionally used to replicate service environments. (b) The lower pattern is random, but incorporates periodic marker cycles to assist in analysis of the failure.

rations makes the interpretation of fatigue behavior relatively straightforward, although it may not represent the situation in practice. However, because the evolution of fatigue cracks can only be assessed by destructive examination of the testpiece, the progressive failure of the joints cannot be accurately followed under practical testing conditions [Frear, Jones, and Kinsman 1991, p 197-201].

In electrical and electronic applications, it is not uncommon for components and assemblies to be subjected to thermal cycles that arise due to heat generated by operation of the circuits. The assessment of fatigue life by *in situ* thermal cycling of this type is frequently referred to as "power cycling" and can obviously only be carried out on electrically functioning circuits. This form of testing has the merit that it is possible to closely simulate the actual operating environment and that a large number of thermal cycles can be generated relatively quickly, compared with thermal cycling involving an external heat source.

In a fatigue test the number of cycles, N, to failure for different maximum applied stresses, S, or

Assessment of Joint Quality

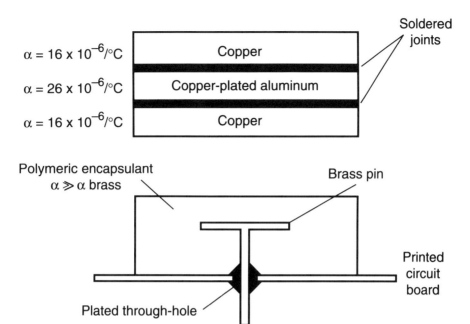

Fig. 6.41 Schematic of testpieces used to impose shear strain on soldered joints when thermally cycled. Source: After Frear, Jones, and Kinsman [1991, p 200]

for a range of ratios of maximum to minimum stress, S/S_o, is plotted for an assembly of specific dimensions and metallurgical specifications. Both types of plots are referred to as S-N fatigue curves. Representative S-N curves for butt joints in mild steel brazed with copper and a Ag-Cu-Zn-Cd filler alloy, obtained by cyclic mechanical loading in torsion, are shown in Fig. 6.42. As the magnitude of the stress is reduced, the number of cycles to failure increases. In some materials the number of cycles increases exponentially to an infinite value, as can be seen in Fig. 6.42. The stress level at which this situation is reached is defined as the endurance limit. Not all materials have an endurance limit; in such case, even a nominal stress will lead to ultimate fatigue failure.

6.5.6 Other Mechanical Tests

The tests described above represent those that are most widely used to assess mechanical integrity. Many other test procedures are described in the literature, some designed to evaluate the effects of corrosion in its various forms. Because corrosion and mechanical stress can act together in undermining joint integrity, tests have also been designed to assess resistance to failure modes such as stress corrosion, corrosion fatigue, and so on.

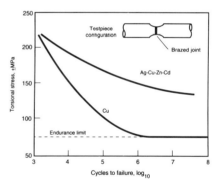

Fig. 6.42 Fatigue curves for butt joints in mild steel, made using copper and Ag-Cu-Zn-Cd brazes, subjected to sinusoidal, reversed torsional stress. The joint region is necked to ensure that failure occurs there. Source: After Brooker, Beatson, and Roberts [1975]

Of the many available tests, only those that reflect the service conditions of the assembly need to be considered, as explained above. Fortunately, it is often possible to further reduce the number of tests to be performed. For example, published data on actual joints show that the values obtained in the tensile and shear modes of testing are generally related, with the maximum shear strength being approximately twice that of the tensile strength [Trimmer and Kuhn 1982, p 7]. Therefore, either of these two tests is usually sufficient for an assessment of strength under static loading, the choice being made with regard to the service conditions. Moreover, thermal and mechanical fatigue tests conducted at low homologous temperatures are often interchangeable, because the failure modes tend to be closely related.

6.6 Nondestructive Evaluation of Joints and Subassemblies

The cost associated with component failure in an engineering system is invariably higher, usually by orders of magnitude, than the value of an individual component. Nondestructive testing provides a means of examining fabricated assemblies, as well as the constituent parts, for defects without damaging or destroying them. For applications where failure can be catastrophic, such as nuclear reactors and aeronautics, critical components and subassemblies are normally required to be submitted to 100% inspection by nondestructive methods.

The types of flaws in joints that can be revealed and measured by nondestructive means, together with the appropriate techniques for detecting and assessing them (in parentheses) are:

- Lack of fusion at joint interfaces and cracks in joints (ultrasonic, acoustic emission)
- Voids (ultrasonic, X-ray, eddy current)
- Embrittling phases (ultrasonic, acoustic emission, microhardness)
- Poor fillets (visual, dye penetrant)
- Distortion and lack of joint planarity (metrology)
- Damage to components during the joining process (ultrasonics, visual, dye penetrant)
- Modification to the "temper" of the parent materials by the joining process (hardness)

On the whole, nondestructive testing tends to be reserved for what are regarded as highly critical components and assemblies for reasons of cost and time. In order for nondestructive testing to be effective, it must be tailored to screen for particular defects in predefined sizes and distributions that are known to reduce the integrity of the joint to an unacceptable level. A commonly encountered type of defect in joints that can be screened effectively by NDT (non-destructive test) techniques is a plane of weakness produced by a layer of pores due to poor wetting by the filler. Nondestructive tests need to be qualified by cross correlations with mechanical property measurements in order to establish acceptance criteria.

Brief descriptions of the major NDT techniques used in industry for inspecting joints are summarized and compared in Forshaw [1989]. Inspection techniques for soldered joints in electronic manufacture are more developed than for most other applications of soldering and brazing; these and some of the more novel NDT techniques that are of potential value are reviewed in Goodall and Lo [1991]. An outline of the NDT techniques that are most widely applied to the examination of soldered and brazed joints is given below, in approximate order of their frequency of use.

6.6.1 Visual Inspection and Metrology

The quality of fillet formation is usually assessed by eye, perhaps with the aid of a magnifying glass. Because this method of inspection is easy to carry out, it is the most widely used, although it is purely qualitative. Attempts have been made to refine the classification of defects by comparison against photographic reference "standards" [ITRI 555 (undated); ITRI 700 (undated)]. Efforts have also been made to automate the examination procedure, for example, by employing a laser to illuminate the joint and then using a computer to evaluate the image and characterize defects [Tunick 1986; Berquist 1989]. Even so, assessment of the surface of joints will simply indicate whether good wetting and spreading have been achieved toward the observable extremities of the joint and no more than that [Lea 1991].

The alignment and distortion of components and the variability of joint gap are measured with either purely optical devices, such as traveling microscopes, or mechanical instruments, notably calipers, micrometers, and profilometers. Instruments are available to provide virtually any required accuracy, although it should be borne in mind that in general, the higher the resolution, the lower the size of the region that is sampled. Optical-fiber light pipes have been developed, largely for medical applications, which can be guided through internal passages of engineering assemblies for viewing internal surfaces of interest. The

Assessment of Joint Quality

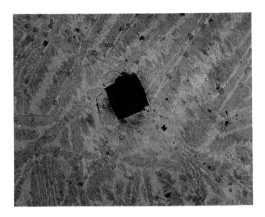

Fig. 6.43 Vickers hardness indentation made in a bulk sample of an experimental filler alloy. The extensive matrix cracking emanating from the indentation suggests that the alloy has very limited ductility. 59×

images obtained can be viewed in real time using video cameras, and they can be enlarged and enhanced using computer-based image processing techniques, as outlined in section 6.4.1.

6.6.2 Hardness Measurements

The hardness of a material is its ability to resist mechanical deformation [Tabor 1951]. For most engineering applications, and certainly as far as the assessment of joints is concerned, hardness is specifically concerned with the resistance of metals to indentation. In this context, hardness measurements involve determining the magnitude of the indentation produced by a tool of a defined geometry under a given load. Because the stress configuration produced by an indentor is complex and changes in the course of the measurement, hardness does not accurately relate to other mechanical properties. However, a hardness measurement, which can be made relatively quickly, does provide a convenient means of gauging the yield or tensile strength of a metal and is especially useful for comparative purposes and for monitoring changes in the temper of the material. It has the additional benefit of being essentially nondestructive, because the indent is usually microscopic in size.

The loose relationships between hardness and both yield and tensile strength, and the qualifications that need to be applied to these correlations, are succinctly outlined in Wyatt and Dew-Hughes [1977, p 156-158]. Conversion charts for obtaining mechanical strength values from hardness data have been compiled and are widely used. A qualitative indication of the ductility of a metal can be obtained by microscopic examination of a hardness indent: cracking in the adjacent area is an indication of brittleness, an example of which can be seen in Fig. 6.43.

There are four principal tests in use, each furnishing its own scale of hardness [Wyatt and Dew-Hughes 1977, p 152-158]. It is often possible to relate these scales, but they are dependent on the material and, in particular, on its degree of hardness.

In the Brinell hardness test, the surface of a metal is indented by a hardened steel ball, usually 10 mm (0.04 in.) in diameter (see Fig. 6.44a). The applied load is chosen such that the indentation made will be relatively small, that is, such that the ratio of the diameter of the cavity to that of the ball will be between 0.25 and 0.5. The appropriate test parameters for different materials can be found in ASTM standards. From a measurement of the diameter of the indentation, normally made with the aid of a microscope, and the value of the applied load, it is possible to calculate (or obtain from tables) the hardness value on the HB (or BHN) scale, in units of kgf/mm^2.

A similar procedure applies to the Vickers measurement of hardness, except that a square-base diamond pyramid is used as the indentor and that lower loads, in the range of 1 to 100 kgf, are applied (see Fig. 6.44b). Hardness values are specified on the HV scale, again in units of kgf/mm^2. The Vickers test is also used for microhardness measurements on individual grains of the material. In this case, the test load is in the range of 1 to 100 gf, and the indentations are of the order of microns in diameter. For reliable measurements, extremely smooth surfaces are needed, such as are provided by polished metallographic specimens. The small size of the indentor means that it is possible to measure the hardness of individual phases and produce hardness profiles across a joint. Figure 6.45 is a hardness profile made across a brazed joint in an item of 18 karat gold jewelry, which shows that the filler alloy is somewhat harder than the base material, which itself has softened adjacent to the faying surfaces as a result of local heating by a torch in the joining operation.

The Knoop test developed from the Vickers hardness test. It employs an indentor having a parallelogram rather than a square profile, which produces an indentation with a length to breadth ratio of 7:1. The applied loads range between 0.2 and 4

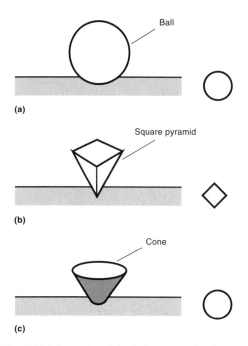

Fig. 6.44 Schematics of the indentors and indentations formed in three static indentation hardness tests. (a) Brinell. (b) Vickers. (c) Rockwell

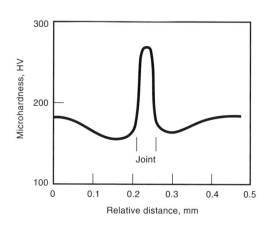

Fig. 6.45 Microhardness profile across a brazed joint. The filler alloy is harder than the parent material, which is a gold alloy in the cold-worked condition. The heating of the joint has softened the parent material on both sides of the joint gap.

kgf, and the optimum length of the indentation is of the order of 100 μm (4000 μin.). As might be expected, hardness values on the Knoop and Vickers scales correspond fairly closely. An important advantage of the Knoop indentor is that it produces satisfactory indentations in hard materials, including glasses.

The Rockwell hardness test is similar in principle to the previous two, but it uses a dial gage to furnish a direct measure of hardness, and the measurement procedure is different. The indentor takes the form of a steel ball 1.58 mm ($1/16$ in.) in diameter for testing relatively soft metals (scale B), whereas a diamond cone with a hemispherical tip is used for testing hard metals (scale C) (see Fig. 6.44c). In the Rockwell test, the dial is first set to zero with a load of 10 kgf applied to the indentor, which ensures that any slack in the instrument is removed. The test load is then applied, with a magnitude of 100 kgf for measurements on scale B and 150 kgf for measurements on scale C. On reducing the load back to the setting load of 10 kgf without disturbing the testpiece, the dial registers the Rockwell B (HRB) or Rockwell C (HRC) hardness value in dimensionless units. The dial reading is proportional to the difference in depth of the indentations corresponding to the first and second application of the 10 kgf setting load.

Each of the different tests has advantages and limitations. For example, the Brinell test is not suitable for very hard materials, because the steel ball flattens and does not produce a significant indentation. On the other hand, the Vickers test is designed to produce small indentations and is therefore suitable only for testing on smooth surfaces. The Rockwell test is especially favored for use in production environments, because it lends itself to rapid measurement and is designed to be relatively free from operator subjectivity. The 7:1 aspect ratio of the Knoop indentor makes it especially suitable for measuring the hardness of thin coatings or layers in cross section.

6.6.3 Liquid Penetrant Inspection

Liquid penetrant inspection is used for detecting surface discontinuities on nonabsorbent materials, particularly cracks that extend to the surface. It is not capable of revealing internal defects. The liquid penetrant, which is frequently a dye, provides a contrast enhancement to these irregular features, making them more readily discriminated. It is normally possible to detect cracks having opening widths down to about 1 μm (40 μin.) using this method.

The test involves first cleaning and degreasing the testpiece and then applying a low-viscosity

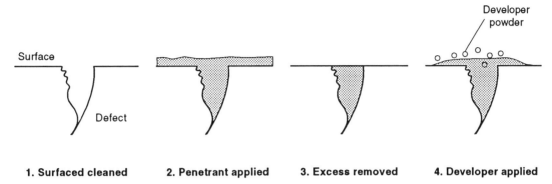

Fig. 6.46 Stages in liquid penetrant testing

liquid to its surface. Sufficient time is allowed for the liquid to soak into surface-opening flaws. The excess liquid is then removed and the surface coated with an absorbent powder. After a period of time, the liquid seeps out of the flaw and into the powder by capillary action, spreading laterally as it does so, thereby magnifying the area of the defect. Because the powder agent reveals the defects, it is referred to as the "developer." These stages are illustrated in Fig. 6.46. There are many variations of this technique; coloring agents or fluorescent dyes are frequently added to increase sensitivity. A practical guide to penetrant testing is given by Lovejoy [1991].

Liquid penetrant testing is simple to use, versatile, and inexpensive. Specimen geometry is not a limiting factor, provided that there is reasonable access to the relevant surfaces. However, it tends to be operator dependent and is only able to detect the existence of surface-opening defects. The size and shape of the defects usually cannot be determined by this technique. In joint assessment, penetrant testing is mostly useful for revealing edge voids and cracks.

Magnetic particle inspection may be considered to be a variant of liquid penetrant inspection, although it is also sensitive to features immediately below the surface. This technique is limited to ferromagnetic materials. A magnetic field is applied to the area to be inspected. The resulting magnetization will be perturbed by inhomogeneities, which will behave like local magnets and attract fine magnetizable particles. The particles are normally suspended in an ink or dye to provide visual contrast.

6.6.4 X-Radiography

The X-ray spectrum comprises electromagnetic radiation of high frequency in the range of 10^{16} to 10^{21} Hz. The high frequency and energy of X-rays enable them to penetrate materials with a degree of absorption that is specific to the particular material and that is also a function of the X-ray frequency. In general, the higher the frequency or energy of the radiation (the energy E being related to the frequency v by the relationship $E = hv$, where h is Planck's constant), the deeper it penetrates. Exposure curves relating optimum equipment settings for particular materials, as a function of their thickness, normally must be determined by the operator prior to carrying out routine analysis.

The ability of X-rays to penetrate materials enables them to reveal internal features, including defects, which absorb the radiation to a different extent than does the surrounding material. It is this characteristic that provides the contrast in the X-ray image. Accordingly, voids, inclusions of heterogeneous material, and cracks parallel to the X-rays will be more conspicuous to this technique than cracks and interfaces that are perpendicularly oriented. The sensitivity and resolution provided by X-radiography also depend on the thickness of the testpiece, its X-ray absorption characteristics at the wavelength (or intensity) of the radiation, and the X-ray optical configuration used. To a first approximation, for a defect such as a crack or void to be resolved reliably by X-radiography, its width parallel to the X-ray beam must be more than 5% of the thickness of the material under examination. Radiographs revealing voids in joints are shown in Fig. 1.10 and 4.12.

X-rays are generated by applying a high voltage (typically several hundred kilovolts) between a heated filament cathode and a target anode contained in an evacuated tube. Figure 6.47 shows a schematic representation of a conventional X-ray tube. A broad beam of electrons generated at the cathode is accelerated by the electric field and bombards the target at a focal spot, producing a beam of X-rays that emerges from the sealed tube through a window.

The filament is commonly a helical wire, which causes the X-rays generated at the target to diverge from a line focus. The size of the focus can be reduced by placing the filament in a cathode structure known as a focusing cup. By tapering the face of the target in such a manner that there is an acute angle between the target surface and the tube axis, the projected X-ray beam is made to emerge with a small, square cross section, down to about 1 mm (0.04 in.) in width.

The clarity of a radiograph obtained with a conventional X-ray tube, even under optimum conditions, is limited by the degree of geometrical unsharpness. The unsharpness, U_g, is directly dependent on the focal spot size (diameter), F, via the relationship:

$$U_g = \frac{F \times t}{D_o} \quad \text{(Eq 6.2)}$$

where D_o is the distance from the focal plane to the object, and t is the distance from the object to the film. This is shown in Fig. 6.48.

To minimize the geometrical unsharpness, the ratio t/D_o must be made as small as possible. In other words, the focal plane-to-object distance must be large, while the separation of the film from the object must be small. This condition is usually best achieved by placing the radiographic film in contact with the back surface of the object.

Increasing use is being made of X-radiographic systems that employ a television camera system in place of a film. The use of a television camera enables the X-ray image of the object to be viewed in real time. In a typical system, the testpiece is placed on a turntable that can also be moved in position. The combination of movement of the testpiece, by remote control, coupled with real-time viewing permits a comprehensive examination to be made rapidly. The nature of visual perception is such that an area which differs marginally in contrast from its surroundings is more easily detected when in motion, so that defects are more readily noticed. Additionally, the controlled movement enables defects to be viewed at an optimum angle.

An important innovation in X-radiography, and one which has had a major impact on NDT, is the development of systems incorporating microfocus sources. Small focal spot sizes can be achieved at the X-ray target by electromagnetically focusing the incident electron beam, exploiting a technique that is widely used in electron microscope technology. The focal spot size can be as small as 1 µm (40 µin.) in diameter.

A microfocus X-ray system offers three distinct advantages. First, it uses fine rod anode sources that can be inserted into hollow assemblies, such as tubular structures, permitting single-wall exposures to be obtained in situations where conventional X-ray systems could only provide double- or multiple-wall radiographs, owing to their bulky tube heads. This simplifies the projection geometry of the radiography and increases the relative sensitivity of the radiograph to defects, with respect to the absorbing material of the assembly.

Second, it is possible to obtain geometrically magnified images of high definition by distancing the recording medium, either a film or camera, from the testpiece. The magnification obtainable from an idealized point source, M, is given by the expression:

$$M = \frac{t + D_o}{D_o}$$

This is illustrated in Fig. 6.49.

A diminution of the focal spot size means that the unsharpness, given by Eq 6.2 above, is correspondingly reduced. A consequential benefit of spatially separating the image sensor from the testpiece is a reduction in the scattered ray fraction, generated by the testpiece itself, that contributes to the image. This results in a corresponding improvement in sensitivity to flaws. By simply moving the testpiece along the axis of the X-ray beam, it is possible to continuously zoom in on detail.

The third advantage of microfocus systems is the ability to produce X-ray beams with a closely defined directionality. For example, by suitable shaping of the X-ray target, the radiation can be projected with a defined polar distribution, which is ideal for selectively inspecting circumferential joints in tubular assemblies. A selection of beam profiles that can be obtained from a microfocus source is shown in Fig. 6.50.

Assessment of Joint Quality

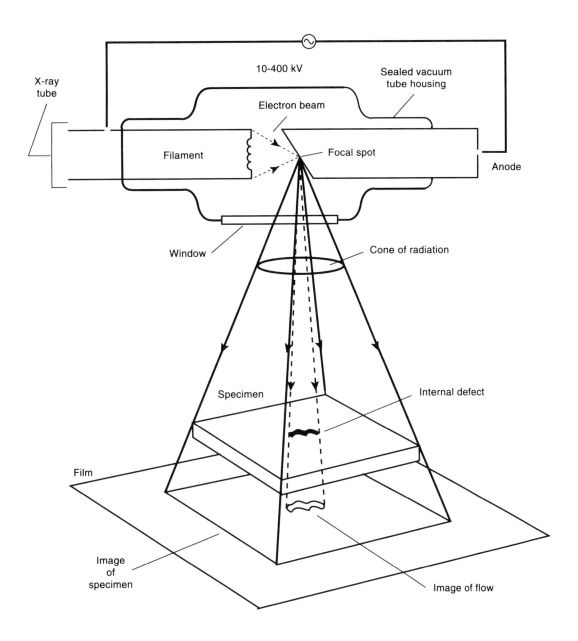

Fig. 6.47 Schematic showing the principle of radiographic nondestructive testing

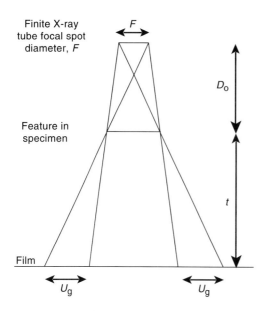

Fig. 6.48 Origin of geometric unsharpness in conventional X-radiography due to the finite size of the focal spot

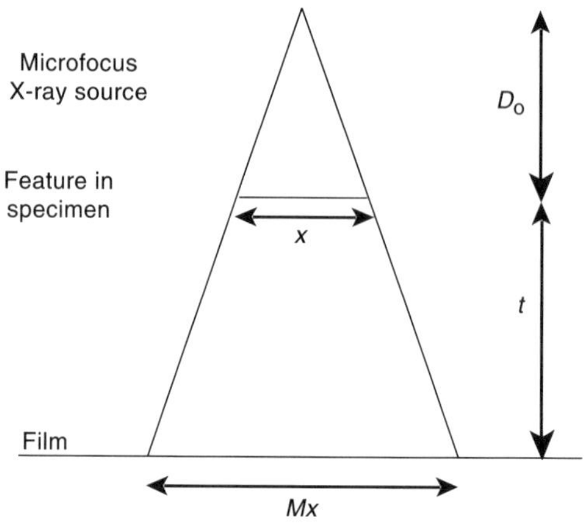

Fig. 6.49 Representation of geometric magnification in microfocus X-radiography

6.6.5 Ultrasonic Inspection

Ultrasound is defined as pressure waves that have frequencies higher than sound waves and that cannot be heard—in practice, in the range of 0.5 MHz to 5 GHz. The particular characteristic of ultrasound that is exploited in NDT is its ability to travel through solid materials, while obeying the same laws of reflection and refraction as light. At its most simple, an ultrasonic beam may be thought of as a torch, which shows up defects inside materials as reflections. Because ultrasound travels more slowly than light, by a factor of typically $1/20,000$ in metals, it is relatively easy to identify defects and measure their depth by recording the echoes produced from incident pulses launched into the material. A pulse of ultrasound travels with a fixed velocity through a given material; the time delay for the echo to be detected provides a measure of the distance traversed. Echoes produced by reflection at different surfaces and interfaces will be temporally resolved, and the corresponding distances can then be calculated. This is the basic principle of pulse-echo ultrasonic inspection, and its use in the internal inspection of multilayer chip capacitors is described by Bradley [1981], who also correlates his results with those provided by acoustic microscopy and destructive examination. Gilmore et al. [1979] describe the theory and equipment used to ultrasonically assess brazed and diffusion-bonded silicon/substrate interfaces. Ensminger [1988, p 330-332] provides a brief overview of ultrasonic inspection of soldered joints.

The time delay, T, of the echo obtained from the back wall of the testpiece of thickness D can be used to calibrate the depth, d, of a defect, the latter being given in a homogenous material by:

$$d = \frac{t \times D}{T}$$

where t is the time delay of the echo produced at the defect. The corresponding test configuration and pulse trace are shown in Fig. 6.51. The pulses are represented as being launched in a direction normal to the surface of the testpiece.

This simple pulse-echo method is best suited for use on planar joints. A well-filled joint will allow the ultrasonic pulse to propagate through it with little attenuation, while voids or cracks encountered by the pulse will give rise to an increased signal from the joint and a decrease in the back-wall echo, as shown in Fig. 6.52. Obviously, if the joined materials are not of the same composition, then refraction and other effects resulting from the change in the ultrasonic velocity between them must be considered. Such complexities lie outside the scope of this book; they are treated in specialist texts on ultrasonic nondestructive test-

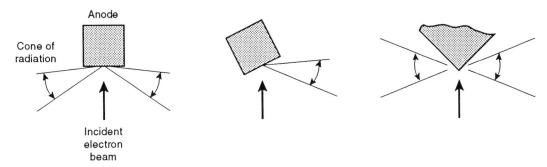

Fig. 6.50 Typical X-ray beam profiles from microfocus sources

ing. An indication of the nature of the problem involved and the approaches that can be used to overcome it are briefly treated by Gilmore *et al.* [1979].

Reflections of ultrasound will be strongest at interfaces between materials that differ greatly in their elastic properties, in particular between solids and gases. Ultrasonic methods of inspection are therefore particularly sensitive to cracks that lie transverse to the direction of propagation of the ultrasonic pulse. If a defect is located at an angle to the incident beam, the echo that is received at the surface will be correspondingly weaker. However, it is possible to launch ultrasonic pulses at an angle to the free surface, using a suitably oriented probe. In this connection, defects in joints tend to occur in predictable directions, because they generally lie in the plane of the joint, and the orientation of the joint with respect to the outer surfaces of an assembly is generally known *a priori*. This is not often the case with intrinsic defects in materials; therefore, inspecting defects in joints is usually easier than it is in components.

In commercial instruments, ultrasound is generated by a piezoelectric transducer, mounted as a probe that is coupled to the surface of the testpiece via a liquid or pasty coupling agent. Two or more probes tend to be used, one to transmit the pulse and the others to detect echoes. The higher the ultrasonic frequency, the higher the resolution but the stronger the signal attenuation due to absorption by the materials through which it travels, as noted in section 6.4.2. For this reason, a frequency of 1 to 10 MHz provides a reasonable compromise for the inspection of metallic components and assemblies by the pulse-echo method.

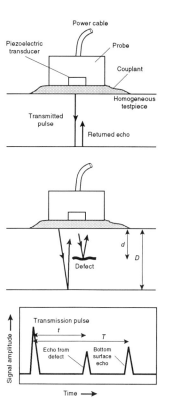

Fig. 6.51 Basic principle of pulse-echo ultrasonic testing and signal interpretation

Ultrasonic signals, like other forms of wave energy, can resolve features down to approximately the size of a single wavelength. Accordingly, ultrasound of 10 MHz frequency should be capable of detecting cracks down to 0.5 mm (0.02 in.) in length in steel. Even to approach these lev-

Principles of Soldering and Brazing

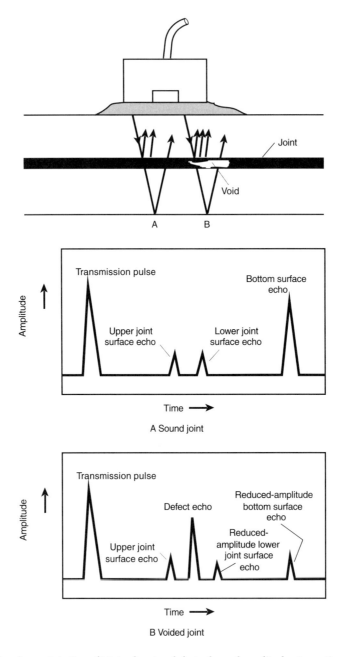

Fig. 6.52 Pulse-echo ultrasonic testing of joints, showing defect echo and amplitude attenuation of back-wall echoes if the joint is not homogeneous

els of resolution requires better signal-to-noise discrimination than is available from simple pulse-echo methods and the filtering out of various interfering effects. This in turn demands sophisticated procedures for collecting and analyzing signals from ultrasonic probes. Such methods are under development, but the complexity of the instrumentation that is presently needed tends to make them difficult to use and time consuming from the viewpoint of industrial applications. Moreover, in order to interpret the mass of data acquired, reductive algorithms need to be applied, which are usually valid only for simple geometries.

A resolution of 0.5 mm (0.02 in.) is clearly inadequate for the inspection of defects in joints that may themselves be of comparable size or even smaller—for example, those made to surface mount electronic components. To improve on this level of discrimination, higher frequencies are required. The SAM technique has been developed to operate in the frequency range of 20 MHz to 2 GHz and thus offers the finest level of resolution of the ultrasonic test methods, although the depth of penetration is limited to below 10 mm (0.4 in.). This technique is described in section 6.4.2.

6.6.6 Technique Selection

Among the numerous NDT techniques—in addition to visual inspection and metrology, which are almost universally applied—three tend to be heavily relied on for joint inspection. Dye penetrant testing is extensively used for identifying surface-opening cracks and recesses. Standard ultrasonic and X-ray techniques are mostly used for surveying the interior of joined assemblies. Interpretation of the data is often difficult, and serious defects are occasionally missed. The relatively high absorption of X-rays by materials, especially metals, means that X-radiography can only be used to inspect limited thicknesses, typically no more than 5 mm (0.2 in.). Virtually all of the techniques are sensitive to defect shape and orientation. Ultrasound is most sensitive to cracks perpendicular to the direction of propagation, whereas X-rays are least sensitive to defects in this orientation. These points should be borne in mind when selecting nondestructive techniques for testing particular components and assemblies.

As a discipline, nondestructive testing suffers from the existence of a two-tier activity. At one end of the spectrum is the research, which covers a plethora of ingenious concepts and techniques, while at the other extreme lies the practice of NDT techniques by industry for quality control and assurance purposes. Many of the fruits of the research are not being tapped in the testing of real components because they are difficult or prohibitively expensive to implement. For this reason, the balance of emphasis here has been given to the more widely applied and versatile techniques.

6.7 *Evaluation of Fabricated Products*

The final step in the qualification of a product is the testing of complete, fully functioning assemblies. As might be expected, efforts at this stage focus on functional characteristics, and for products for which reliability is at a premium, this will include carrying out condition-monitoring routines. For example, a radio will be switched on and tuned with minor "screwdriver" adjustments made to optimize its response. In a communications-quality model, its performance over time will be monitored. The product will also be visually inspected for the quality of finish. Only if performance is unsatisfactory will it be subjected to a more searching examination, which may, if necessary, extend to the reexamination of joints. Normally, it is assumed that if the joints of the components and subassemblies have been validated previously, they should be of adequate quality to sustain the later stages of manufacturing.

Appendix 6.1: Relationships Among Spread Ratio, Spread Factor, and Contact Angle of Droplets

Expressions describing the spread of a molten metal droplet are derived under the following set of idealized conditions:

- The original metal pellet is in the form of a spherical bead of radius a (and diameter $D = 2a$).
- The droplet resolidifies after spreading on the substrate as a spherical cap of radius R and height h, its interface with the substrate having a diameter of $2A$, as shown in Fig. 6A.1.
- The volume of the original pellet is equal to the volume of the resolidified droplet. This means that any volatilization of the molten droplet and reaction with the substrate do not measurably affect its volume.

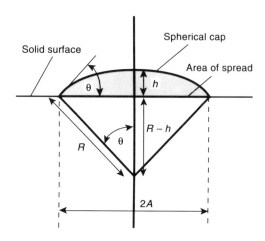

Fig. 6A.1 Spherical cap geometry

The volume of the spherical cap is:

$$V = \frac{1}{6}\pi h (h^2 + 3A^2)$$

Spread Ratio and Contact Angle

The spread ratio, S_r, is defined as:

$$S_r = \frac{\text{Plan area of spread on the substrate surface}}{\text{Plan area of the original spherical pellet}}$$

A and a are related by the conservation of the volume of the droplet, that is:

$$V = \frac{4}{3}\pi a^3 = \frac{1}{6}\pi h (h^2 + 3A^2)$$

Therefore:

$$a = \tfrac{1}{2}[h(h^2 + 3A^2)]^{1/3}$$

and

$$S_r = \frac{4A^2}{[h(h^2 + 3A^2)]^{2/3}}$$

From the geometry (Fig. 6A.1), $A = R \sin\theta$ and $h = R(1 - \cos\theta)$:

$$S_r = \frac{4A^2/h^2}{(1 + 3A^2/h^2)^{2/3}}$$
$$= \frac{4\cot^2\theta/2}{(1 + 3\cot^2\theta/2)^{2/3}} \quad 0° \le \theta \le 180° \quad \text{(Eq 6A.1)}$$

Spread Factor and Contact Angle

The spread factor is defined by the following formula:

$$S_f = \frac{D-h}{D} = \frac{(h^3 + 3A^2h)^{1/3} - h}{(h^3 + 3A^2h)^{1/3}}$$
$$= 1 - \frac{1}{(1 + 3A^2/h^2)^{1/3}}$$
$$= 1 - \frac{1}{(1 + 3\cot^2\theta/2)^{1/3}} \quad 0° \le \theta \le 180°$$

(Eq 6A.2)

Contact Angle and Dimensions of the Solidified Pool of Filler

From Fig. 6A.1, it can be seen that:

$$A^2 + (R-h)^2 = R^2$$

according to the Pythagorean theorem. Rearranging this equation:

$$R = \frac{A^2 + h^2}{2h}$$

Therefore:

$$\sin\theta = \frac{2}{(A/h) + (h/A)} \quad \text{(Eq 6A.3)}$$

References

Ackroyd, M.L., 1975. A Survey of Accelerated Aging Techniques for Solderable Substrates, Publ. No. 531, International Tin Research Institute

Bader, W.G., 1969. Dissolution of Au, Ag, Pt, Cu and Ni in a Molten Tin-Lead Solder, *Weld. J.*, Vol 48, p 551s-557s

Barranger, J., 1989. Critical Parameters of Measurement Using the Wetting Balance, *Solder. Surf. Mount Technol.*, Vol 1 (No. 2), p 11-13

Berquist, K., 1989. Automatic Versus Visual Inspection: Which Is Truly Superior?, *Solder. Surf. Mount Technol.*, Vol 3 (No. 6), p 36-37

Bjerregaard et al., L., 1992. *Metalog Guide*, Struers A/S, Rødovre, Denmark

Boughton, J.D. and Sloboda, M.H., 1970. Embrittling Effects of Trace Quantities of Aluminum and Phosphorus on Brazed Joints in Steel, *Weld. Met. Fabr.*, Vol 8, p 335-339

Bradley, F.N., 1981. "Ultrasonic Scanning of Multilayer Ceramic Chip Capacitors," Tech. Info. Sheet 1-7, AVX Corp.

Brooker, H.R., Beatson, E.V., and Roberts, P.M., 1975. *Industrial Brazing*, 2nd Ed., Butterworths, London

Burton, N.J. and Thaker, D.M., 1985. Weld Bond Evaluation Using SAM, *Met. Mater.*, Vol 1 (No. 7), p 435-436

Chilton, A.C., Smart, D.J., and Wronski, A.S., 1983. An Assessment of the Mechanical Properties of a Silver-Free Brazing Alloy, *Brazing Soldering*, Vol 5 (No. 3), p 7-9

Cibula, M.A., 1958. The Soundness of High Temperature Brazed Joints in Heat-Resisting Alloys, *Br. Weld. J.*, Vol 5, p 185-201

Coombs, C.F., Jr., 1988. *Printed Circuits Handbook*, 3rd ed., McGraw-Hill

de Kluizenaar, E.E., 1990. Reliability of Soldered Joints: A Description of the State of the Art, *Solder. Surf. Mount Technol.*, Vol 4 (No. 2), p 27-38

Dieter, G.E., 1976. *Mechanical Metallurgy*, McGraw-Hill

Dunford, D.V. and Partridge, P.G., 1992. Strength and Fracture Behavior of Diffusion Bonded Joints in Al-Li (8090) Alloy. Part III: Peel Strength, *J. Mater. Sci.*, Vol 27 (No. 11), p 5769-5776

Ensminger, D., 1988. *Ultrasonics: Fundamentals, Technology, Applications*, 2nd Ed., Marcel Dekker

Footner, P.K., Richards, B.P., and Yates, R.B., 1987. Purple Plague: Eliminated or Just Forgotten?, *Qual. Reliab. Eng. Int.*, Vol 3, p 177-184

Forshaw, M., 1989. Application of Non-destructive Testing to Inspection of Soldered Joints, *Circuit World*, Vol 15 (No. 3), p 14-17

Gilmore, R.S. et al., 1979. High Frequency Testing of Bonds: Application to Silicon Power Devices, *Mater. Eval.*, Vol 37 (No. 1), p 65-72

Goodall, A.J.E. and Lo, E.K., 1991. A Review of Inspection Techniques Applicable to PCB Manufacturing and Assembly, Particularly With Respect to SMT, *Adv. Manuf. Eng.*, Vol 3 (No. 1), p 45-53

Gunter, I.A., 1986. The Solderability Testing of Surface Mount Devices Using the GEC Meniscograph Solderability Tester, *Circuit World*, Vol 13 (No. 10), p 8

Gunter, I.A. and Jacobson, D.M., 1990. The GEC Meniscograph Solderability Tester: Adaptation to Vacuum Soldering, *GEC Rev.*, Vol 6 (No. 2), p 86-89

ITRI 555. "Photographic Guide to Soldering Quality," Publ. No. 555, International Tin Research Institute (undated)

ITRI 656. "Solder Alloy Data: Mechanical Properties of Solders and Soldered Joints," Publ. No. 656, International Tin Research Institute (undated)

ITRI 700. "Soldering Surface Mount Devices," Publ. No. 700, International Tin Research Institute (undated)

ITRI 708. "Metallurgy of Soldered Joints in Electronics," Publ. No. 708, International Tin Research Institute (undated)

Jacobson, D.M. and Humpston, G., 1990. Assessment of Fluxless Solderability With Reference to Three Tin-Based Solders, *GEC J. Res.*, Vol 8 (No. 1), p 21-31

Kauppinen, P. and Kivilahti, J., 1991. Evaluation of Structural Defects in Brazed Ceramic-to-Metal Joints With C-Mode Scanning Acoustic Microscopy, *Nondestr. Test. Eval. Int.*, Vol 24 (No. 4), p 187-190

Klang, K.G. and Nylen, M., 1989. A Proposal for a Standard Solderability Testing Method for SMDs, *Solder. Surf. Mount Technol.*, Vol 2 (No. 3), p 33-38

Kossowsky, R., 1983. Designing an Analytical Microscopy Laboratory, *J. Met.*, Vol 3, p 47-53

Kurihara, Y. et al., 1992. Bonding Mechanism Between Aluminum Nitride Substrate and Ag-Cu-Ti Solder, *IEEE Trans. Components Hybrids Manuf. Technol.*, Vol 15 (No. 3), p 361-368

Latter, T.D.T., 1975. Measuring Coating Thickness by the Beta-Backscatter Technique, *Br. J. Non-Destr. Test.*, Vol 10, p 145-152

Lay, L.A., 1991. Metallographic Procedures for Advanced Ceramics, *Met. Mater.*, Vol 7 (No. 9), p 543-547

Lea, C., 1991. Evidence That Visual Inspection Criteria for Soldered Joints Are No Indication of Reliability, *Solder. Surf. Mount Technol.*, Vol 9 (No. 10), p 19-24

Lea, C., 1991. Quantitative Solderability Measurement of Electronic Components. Part 5: Wetting Balance Instrumental Parameters and Procedures, *Solder. Surf. Mount Technol.*, Vol 7, p 10-13

Liebhafsky, H.A. et al., 1960. *X-Ray Absorption and Emission in Analytical Chemistry*, John Wiley & Sons

Lovejoy, D., 1991. *Penetrant Testing—A Practical Guide*, Chapman & Hall, London

Marshall, J.L., 1988. Characterization of Solder Fatigue in Electronic Packaging, *Brazing Soldering*, Vol 15 (No. 3), p 4-9

Matienzo, L.J. and Schaffer, R.R., 1991. Wetting Behavior of Eutectic Tin/Lead Solder and Fluxes on Copper Surfaces, *J. Mater. Sci.*, Vol 26, p 787-791

Matijasevic, G.S., Wang, C.Y., and Lee, C.C., 1990. Void Free Bonding of Large Silicon Dice Using Gold-Tin Alloys, *IEEE Trans. Components Hybrids Manuf. Technol.*, Vol 13 (No. 4), p 1128-1134

Mei, Z. and Morris, J.W., Jr., 1992. Characterization of Eutectic Sn-Bi Solder Joints, *J. Electron. Mater.*, Vol 21 (No. 6), p 599-607

Mittal, K.L., 1978. *Adhesion Measurement of Thin Films and Bulk Coatings*, ASTM

Pover, P.S., 1987. Image Analysis in Materials Science—An Introduction, *Met. Mater.*, Vol 3 (No. 12), p 717-720

Richards, B.P. and Footner, P.K., 1989. Scanning Electron Microscopy of Semiconductor Devices, *Met. Mater.*, Vol 5 (No. 2), p 75-80

Roberts, D.F.T., 1983. The Globule Test—Is It Reliable?, *Brazing Soldering*, Vol 4 (No. 1), p 28-30

Saxton, H.J., West, A.J., and Barrett, C.R., 1971. Deformation and Failure of Brazed Joints—Macroscopic Considerations, *Metall. Trans.*, Vol 2 (No. 4), p 999-1007

Saxton, H.J., West, A.J., and Barrett, C.R., 1971. The Effect of Cooling Rate on the Strength of Brazed Joints, *Metall. Trans.*, Vol 2 (No. 4), p 1019-1028

Tabor, D., 1951. *The Hardness of Metals*, Clarendon Press, Oxford

Tertian, R. and Claisse, F., 1982. *Principles of Quantitative X-ray Fluorescence Analysis*, Hehden & Sons, London

Thwaites, C.J., 1981. Solderability and Some Factors Affecting It, *Brazing Soldering*, Vol 1 (No. 3), p 15-18

Trimmer, R.M. and Kuhn, A.T., 1982. The Strength of Silver-Brazed Stainless Steel Joints—A Review, *Brazing Soldering*, Vol 2 (No. 1), p 6-13

Trimmer, R.M. and Kuhn, A.T., 1982. The Tensile Strength of Small-Scale Silver Brazed Joints, *Brazing Soldering,* Vol 2 (No. 3), p 5-7.

Tunick, D., 1986. Laser-Soldering System Closes Inspection Loop, *Electron. Des.*, Vol 18 (No. 9), p 29-34

Valli, J., 1986. A Review of Adhesion Test Methods for Thin Hard Coatings, *J. Vac. Sci. Technol.*, Vol 4A (No. 6), p 3007-3013

West, A.J. *et al.*, 1971. Deformation and Failure of Brazed Joints—Microscopic Considerations, *Metall. Trans.*, Vol 2 (No. 4), p 1009-1017

Williams, K.J., 1988. Adhesion Test Methods for Solderable Thick Film Conductors, *Brazing Soldering*, Vol 15 (No. 3), p 10-18

Wooldridge, J.R., 1988. Lessons Learned During a Year of Production Solderability Testing With a Wetting Balance, *Brazing Soldering*, Vol 15 (No. 4), p 24-27

Wyatt, O.H. and Dew-Hughes, D., 1977. *Metals, Ceramics and Polymers*, Cambridge University Press

Yoshida, H., Warwick, M.E., and Hawkins, S.P., 1987. The Assessment of the Solderability of Surface Mounted Devices Using the Wetting Balance, *Brazing Soldering*, Vol 12 (No. 1), p 21-29

Chapter 7

Characterization and Process Development in Soldering and Brazing: Selected Case Studies

7.1 Introduction

In this chapter, selected examples are used to provide a flavor of different types of soldering and brazing investigation and process development activities. These case studies utilize the principles that were outlined in the preceding chapters and draw on some of the material that is covered there. They are arranged under two separate headings: the identification of factors that affect joint integrity, which can give rise to problems, and the development of new or improved joining solutions. In the interest of clarity, these case studies have been simplified.

7.2 Illustration of Factors That Can Influence Joint Integrity

A number of factors can determine the effectiveness of a joining process and the performance of joints in service. Case studies, which address the characterization of soldered and brazed joints, will be used to illustrate this point. These examples have been culled from the technical literature and abridged to focus on specific features. Fuller information can be found in the references cited.

7.2.1 Effect of Impurity Elements on the Properties of Filler Metals and Joints

During the development of a joining process in the laboratory, conditions tend to be strictly controlled, including the purity of the materials used. In production, standards are usually more relaxed. However, relatively low levels of impurities in joints, either present in the filler alloy or infiltrated from the parent material during the joining process, can have an adverse effect on properties.

7.2.1.1 Formation of Embrittling Interfacial Phases in Brazed Joints

The embrittlement of brazed joints by impurity elements is spectacularly illustrated by the degrading effect of trace quantities of aluminum and phosphorus on joints made between mild steel and brass with a widely used Ag-Cu-Zn-Cd brazing alloy. This effect is not observed in similar joints made between all steel or all brass components.

Boughton and Sloboda [1970] made a study to quantify this problem and to establish its cause. A series of V-notch Charpy impact testpieces were prepared by brazing together two equal lengths of steel bar with beveled ends, at which the joint was made. The braze contained controlled fractions of either aluminum or phosphorus, and the joining operation was carried out under fixed conditions (temperature, time, fluxing, cooling rate, etc.). The testpieces were then cleaned and subjected to

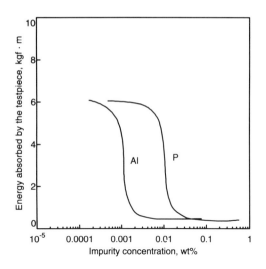

Fig. 7.1 Effect of impurity elements on the impact strength of joints made in mild steel using an Ag-Cu-Zn-Cd filler alloy. Source: After Boughton and Sloboda [1970]

an impact bending test (see section 6.5.3). From the data a graph was plotted of the energy absorbed by the testpiece versus impurity concentration; this is reproduced in Fig. 7.1, which shows that for both aluminum and phosphorus there is a critical concentration threshold above which the joints are completely embrittled and the impact bending strength plummets by an order of magnitude.

Although the critical impurity concentration is substantially higher than the maximum permitted in the brazing alloy by the standards specification, the problem can arise in practice, because aluminum and phosphorus are sometimes present in brass. On wetting of brass by the molten braze, these elements will dissolve into the molten filler alloy. Elemental analysis obtained in an electron microscopic examination of the fractured surfaces showed that the cause of the weakness is the formation of brittle iron-base phases containing aluminum and phosphorus at the braze/steel interface [Boughton and Sloboda 1970]. This finding accounted for the problem being evident only in brazed joints between brass and steel. It can be prevented by tightening the impurity limits in brass components when used in such assemblies.

7.2.1.2 *Generation of Voids That Compromise the Strength of Soldered Joints*

Brass components can prove similarly troublesome when joined with antimony-containing solders, although in this case the cause is entirely different from the one cited above. Antimony-containing solders are widely used in many of the less exacting engineering applications; the antimony substitutes for some of the tin, which is a significantly more expensive element. Although lead-antimony-tin solders are suitable for joining copper and steel, joints made to brass can be highly variable in strength (by up to 50% of the mean value) [Stone *et al.* 1983; Tomlinson and Bryan 1986].

It might be supposed that the cause of the problem is the dissolution or diffusion of zinc into the solder to form embrittling antimony-zinc intermetallic compounds at the brass/solder interface. However, no intermetallic phases containing either antimony or zinc were detected, and this prompted a more detailed study by Thwaites and McDowall [1984]. These workers prepared a series of ring and plug assemblies, as described in section 6.5.2, comprising brass and copper substrates joined with Pb-60Sn solder or Pb-40Sn-2Sb, which were subjected to shear, creep, and cyclic fatigue stresses in a tensile testing machine at ambient temperature. Their measurements confirmed the variability in shear strength and stress-to-rupture in brass joints made with the antimony-containing solder.

Metallographic examination of sectioned joints revealed a high density of irregularly shaped micropores, typically 10 μm (400 μin.) in diameter or less, at the brass/solder interface [Tomlinson and Cooper 1986]. Clearly, such voids constitute a source of weakness. Experiments made by Thwaites and McDowall [1984] with homogeneous bulk ingots of the solder containing small additions of zinc showed that the likely origin of the voids is an abnormally high solidification shrinkage that only occurs when the solder contains both antimony and zinc. This finding is consistent with the voids being located predominantly at the joint interfaces to brass components, where the zinc concentration is likely to be highest. Differences in the levels of zinc at the surfaces of the brass components, coupled with only minor changes in the processing conditions, can account for the wide variation in measured joint strengths.

7.2.1.3 *Adverse Effect on Wetting and Spreading Characteristics*

Impurities present in filler metals can have a deleterious effect on joint formation, as occurs for over a dozen common elements in lead-tin solders when in concentrations as low as 5 ppm [Schmitt-Thomas and Becker 1988; Becker 1991]. Joint formation may be inhibited or even prevented.

Table 7.1 Lowest impurity concentrations producing detrimental effects, in terms of wetting, in Pb-60Sn solder

Impurity element	Impurity concentration, %	Detrimental effect
Aluminum	<0.0005	Oxidation
Antimony	1.0	Reduced spreading
Arsenic	0.2	Reduced spreading
Bismuth	0.5	Oxidation
Cadmium	0.15	Reduced spreading
Copper	0.29	Increased melting range
Phosphorus	0.01	Dewetting
Sulfur	0.0015	Increased melting range
Zinc	0.003	Oxidation

Source: After Ackroyd, Mackay, and Thwaites [1975]

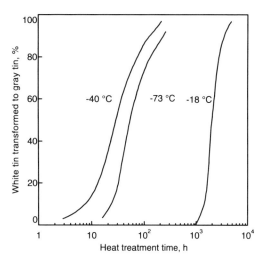

Fig. 7.2 Allotropic transformation of white tin into gray tin as a function of time and temperature. Source: After Bornemann [1956]

A detailed study of the effect of different elements on the wetting properties of eutectic lead-tin solder has been made by Ackroyd, Mackay, and Thwaites [1975]. They assessed the solderability of copper, brass, and steel coupons using area-of-spread tests, rotary dip tests, and wetting balance measurements (see section 6.3). High-purity Pb-60Sn solder was deliberately adulterated with low concentrations (<1%) of aluminum, arsenic, bismuth, cadmium, copper, phosphorus, sulfur, antimony, and zinc and used under an ambient atmosphere with selected active and nonactivated fluxes.

The measurements obtained by the various solderability assessment methods were broadly consistent, with the wetting balance showing greatest sensitivity. Of the substrate materials, brass exhibited a consistent enhancement in spread area with increasing impurity addition, irrespective of the element. By contrast, the solderability of copper and steel almost invariably deteriorated in the presence of subpercentage levels of the impurity and then continued to decline or tended to a steady value. The detrimental characteristics of the elemental additions and the critical concentrations that produce these deleterious effects are summarized in Table 7.1. There was no minimum level of aluminum that did not adversely affect solderability; the lowest concentration investigated was 0.0005%.

Impurities do not always impair wettability. On the contrary, bismuth and beryllium can have a significant beneficial effect on the wetting and flow characteristics of aluminum- and silicon-containing brazes [Schultze and Schoer 1973]. This feature is mentioned in section 5.2.3.3.

7.2.1.4 Beneficial Effect in Preventing "Tin Pest" in Solders

An important area where impurity elements have a positive role in solders is in the suppression of "tin pest" in tin-lead solders. Pure tin can exist in two allotropic forms: white tin, which is the common metallic form, and gray tin, to which white tin can transform below 13 °C (55 °F). This transformation creates a problem that has long been recognized and characterized and for which appropriate remedies are well known. Therefore, this issue is not presented as a case study in the sense of the preceding examples, but is considered here because it is closely linked in its subject matter.

Owing to the 26% increase in volume that accompanies the phase change from white to gray tin, the solid metal disintegrates into a crumbly mass having essentially no strength. The transformation from white to gray tin is clearly disastrous for the mechanical properties of soldered joints. Tin pest can affect electrical systems used in subzero environments, for example, in high-flying aircraft and in soldered containers and conduits of refrigerants.

The transformation does not occur spontaneously, but is always preceded by an incubation period, which may be as much as several years if the tin is exceptionally pure. The maximum rate of conversion occurs at about –40 °C (–40 °F). Figure 7.2 shows the extent of transformation as a function of time at various temperatures. The incubation period is much shorter if white tin is mechanically worked at low temperatures or is inoculated with either gray tin or other elements and compounds having similar diamondlike crystal structures, such as silicon and ZnSb [MacIntosh 1968]. Copper, too, accelerates the process [Rogers and Fydell 1953], whereas lead retards it slightly [Williams 1956].

Relatively small additions of certain elements have been shown to suppress the allotropic transformation of tin. The addition of 0.15% Sb will prevent it in tin-base solders. Bismuth is even more effective; a 0.05% addition is all that is necessary [Bornemann 1956]. Existing solder specifications accommodate sufficient quantities of impurity elements to prevent tin pest. However, in recent years there has been a trend toward high-purity solders in order to improve certain characteristics in wave soldering machines. In the latest draft solder specifications, the maximum level of antimony permitted has been reduced from 0.5 to 0.12%, so that tin pest could occur in soldered assemblies unless other impurities, notably bismuth, remain at appropriate levels [EP&P 1992].

7.2.2 Corrosion of Soldered and Brazed Joints in Tap Water

In many countries the use of heavy metal elements, notably cadmium and lead, in filler alloys and plumbing fittings used for drinking water appliances is now banned on health grounds. The reason is that lead and cadmium can be continuously leached out by tap water and contaminate the supply [Feller *et al.* 1984]. The native oxide, carbonate, and hydroxycarbonate films that form on the surfaces of the joints are porous and do not provide an effective barrier. Questions have also been raised about the ability of alternative filler alloys to function satisfactorily in the corrosive environment of tap water.

7.2.2.1 Water Installations of Copper

Copper installations for conveying tap water are commonly joined using silver-copper-zinc and silver-copper-phosphorus brazing alloys and soft solders of tin containing small percentages of silver and copper. Pipework for conveying water might be considered to be a relatively benign environment as the temperature is maintained within a fairly narrow range (5 to 90 °C, or 40 to 195 °F), the mechanical stresses applied to joints are usually low, and water intended for domestic use is chemically neutral, having a pH value maintained close to 7. Nevertheless, tap water frequently contains significant volumes of dissolved gases and salts, which confer a degree of electrical conductivity, thereby enabling galvanic-type corrosion to occur between different metals in contact (see section 2.2.2.1). As the electrode potential of filler alloys will tend to be negative with respect to the copper pipework, it is the joints that corrode preferentially.

A study was made by Nielsen [1984] to establish the ranking order of the four filler alloy systems mentioned above with regard to their corrosion resistance. Lengths of copper pipework were joined by standard methods using representative alloys of each of these systems and were then inserted into a hot water circuit. The variables of the test conditions included water composition (characteristic of three different geographical locations), temperature (60 to 87 °C, or 140 to 190 °F), and flow rate (0.3 to 0.5 m/s, or 1 to 1.6 ft/s). After approximately a year of the test, the couplings were cut through and the joints examined in metallographic section. The results were as follows:

- *Silver-copper-zinc brazing alloys* were found to be extensively corroded, with an attrition of the filler of about 5 mm (0.2 in.) per year, due to the dezincification of the filler metal. This form of corrosion, which is illustrated schematically in Fig. 7.3(a), is to be expected, because the zinc phase in the braze has a large and negative electrode potential (i.e., is cathodic) with respect to the other phases in the filler, and especially to the copper pipe. The leaching of the zinc-rich phase out of the joint loosens the silver- and copper-rich phases, which are then washed away by flowing water.

- *Silver-copper-phosphorus brazing alloys*, which are self-fluxing (see section 5.3.4), exhibited no corrosion at all. However, the copper pipework immediately adjacent to the joints was moderately corroded, at 0.2 to 0.3 mm (0.008 to 0.012 in.) per year (see Fig. 7.3b), which suggests that the solidified filler is significantly cathodic with respect to copper. Although copper phosphide is strongly cathodic to copper, corrosion from this source is unlikely in view of the fact that the phosphorus becomes incorporated in the oxide slag during the brazing cycle and the residual silver-copper alloy is only slightly cathodic to copper. A more likely cause of the observed corrosion is that the braze fillet is covered with

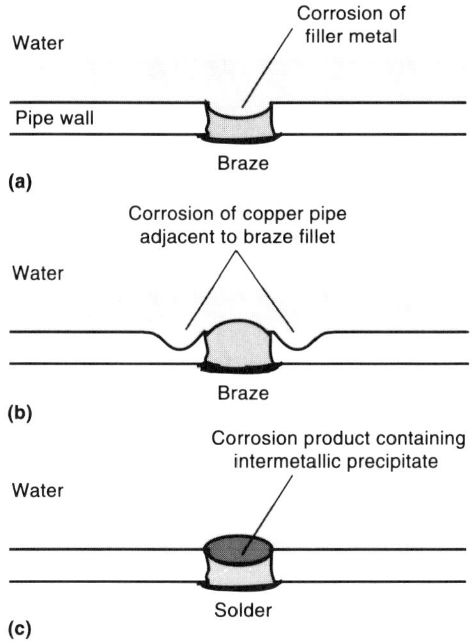

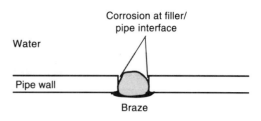

Fig. 7.3 Schematic illustrations showing the corrosion of butt joints between copper pipes conveying tap water. (a) Ag-Cu-Zn braze. (b) Ag-Cu-P braze. (c) Tin-base solder

Fig. 7.4 Schematic illustration showing the corrosion of butt joints between stainless steel pipes conveying tap water, joined with an Ag-Cu-Zn braze

a tenacious layer of the slag, which consists primarily of cuprous oxide and is electrically conductive and also strongly cathodic to copper (i.e., the copper is anodic toward the slag).

- *Silver-tin and copper-tin solders* suffered corrosion of the tin matrix at a rate of between 0.1 and 0.4 mm (0.004 and 0.016 in.) per year. The residues remaining on the surfaces of the soldered joints contained the more resistant silver-tin and copper-tin intermetallic phases, which have the effect of slowing down the rate of corrosion. A schematic illustration of a corroding soldered joint is shown in Fig. 7.3(c).

Nielsen's study [Nielsen 1984] showed that there are no major differences between the types of corrosion that occur in different tap waters; only the rates of attack vary. The tin-rich soft solders are recommended for joining copper pipework except where high mechanical strengths are required, in which case phosphorus-containing brazes are preferred and the joints should be cleaned scrupulously of oxide slag.

7.2.2.2 Water Installations of Stainless Steel

Copper alloys used for plumbing are seldom more than 99% pure. The impurity elements are not removed so as to minimize manufacturing costs, and some elements (e.g., lead and bismuth) are deliberately added to improve machinability and formability [BBI 1991]. Because of the concern over leaching of heavy metals into drinking water, stainless steel and plastics are now often recommended in place of copper pipework.

Investigations have been undertaken to characterize and understand the corrosion of brazed joints made to stainless steel water pipes. These have similarly involved metallographic examination of joints formed with standard silver-copper-zinc brazes that have been subjected progressively to the corroding effect of tap water.

On exposure to water of steel pipes brazed in this manner, the joints fail catastrophically within a few months at the braze/steel interface, as illustrated schematically in Fig. 7.4 [Jarman, Myles, and Booker 1973]. The mechanisms responsible have been established and are as follows. First, dezincification of the braze fillet occurs, as described above, which enables water to reach the braze/steel interface within the joint. At this interface there is a thin zone that consists essentially of copper, iron, and some nickel, which is generated in the brazing operation [Kuhn, Rawlings, and May 1984]. Being less noble than both the braze and the stainless steel, this zone corrodes rapidly, particularly as its width is small [Jarman, Linekar, and Booker 1973].

Joints in stainless steel that are resistant to interfacial corrosion by tap water can be made using Ag-Cu-Zn-base brazes, if these contain more than about 1% Ni [Jarman, Myles, and Booker 1973]. Then, as described in section 2.2.1.5, this element concentrates at the joint periphery, where it attenuates the electrode potential difference between the braze and the stainless steel.

Principles of Soldering and Brazing

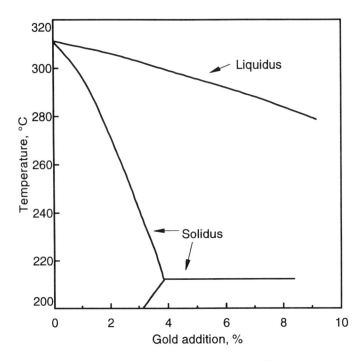

Fig. 7.5 Section through the Au-Ag-Pb-Sn quaternary system showing the effect on melting range of adding gold to Ag-97.5Pb-1Sn solder. Source: After Evans and Prince [1982]

Table 7.2 Effect of gold on common tin-base eutectic solders

Solder composition	% gold needed to form $AuSn_4$ as the primary phase	% gold that dissolves in solder at 50 °C (90 °F) superheat	Change in melting point of solder by dissolving 1% Au	
			°C	°F
In-48Sn	$AuIn_2$ formed	<1	−1	−2
Bi-43Sn	0.8	4	−2	−4
Pb-62Sn	8	13	−12	−22
Sb-95Sn	10	30	−15	−27
Ag-96Sn	10	30	−15	−27

7.2.3 Effect of Filler/Component Reactions on Properties and Performance of Joints

The metallurgical reactions between a solder or braze and the materials of the components being joined can substantially alter the characteristics of the filler, which may in turn have unexpected consequences for the resulting joints and thereby create problems in practical situations. By anticipating the effects of these reactions, problems can be forestalled. Three examples, drawn from the published literature, are described below.

7.2.3.1 Changes Introduced to the Melting Point of Soft Solders

As emphasized throughout this book, alloying inevitably takes place between the filler and the components (including applied metallizations), which affects the melting point of the filler. This point should always be borne in mind when designing a soldering or brazing process. The extent to which the alloying raises or depresses the melting point can be established by inspecting the relevant phase diagram or by using the empirical methods set out in section 3.3. These data will also furnish values for the dissolution of material from

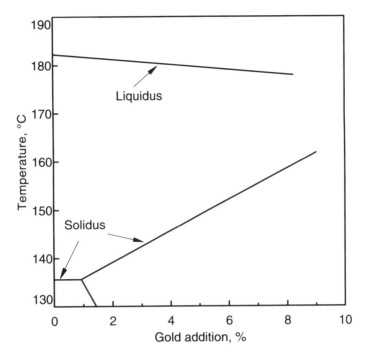

Fig. 7.6 Section through the Au-In-Pb-Sn quaternary system showing the effect on melting range of adding gold to In-18Pb-70Sn solder. This composition is not a ternary eutectic as is often claimed.

the components that will occur during the joining cycle, as a function of the peak temperature; details are given in section 3.2.1.

The change in the melting point of selected binary tin-base solders, when these have taken up 1% Au, is given in Table 7.2. The effect is particularly marked in the case of the Ag-97.5Pb-1Sn solder, where the solidus temperature is reduced from 309 to 217 °C (588 to 423 °F) when 4% Au is added. For a layer of the solder 50 μm (2000 μin.) thick, this radical depression of the melting point corresponds to a mere 1 μm (40 μin.) of gold coating being dissolved from each side of the joint.

An example of a change in the contrary direction is the dissolution of gold in In-18Pb-70Sn solder (melting point = 136 °C, or 277 °F). A 5% Au addition raises the solidus temperature by 14 °C (25 °F) to 150 °C (302 °F). Sections through these quaternary systems as a function of gold concentration are given in Fig. 7.5 and 7.6, respectively.

7.2.3.2 Metallurgical Reactions That Vary With Solder Composition

Indium-containing solders are often recommended for joining to gold-coated components, because gold is less soluble in these alloys than in tin-lead solders. This is largely due to the formation of a continuous layer of the intermetallic compound $AuIn_2$ at the solder/gold interface, which effectively suppresses further reaction between these metals (see section 2.2.3.2 and Fig. 2.33, as well as section 7.3.1.3 below).

However, indium solders containing lead do not necessarily operate in the same manner. Heating joints containing these solder alloys over extended timescales promotes continuing interaction by solid-state diffusion, albeit at a more modest rate. In the presence of lead, the interfacial layer takes the form of separate grains of the $AuIn_2$ compound embedded entirely in primary lead, and the lead provides an easy diffusion path between the solder and the gold coating and so permits the reaction to continue [Yost, Ganyard, and Karnowsky 1976].

In their study, Yost, Ganyard, and Karnowsky dipped a series of solid gold coupons into In-50Pb solder at 250 °C (480 °F) for 5 s, then transferred the coated coupons to ovens held at different temperatures and left them there for various periods of time. Following this treatment, the specimens were examined in metallographic section by optical microscopy.

Table 7.3 Mechanical properties of soldered joints to copper made with lead-tin and bismuth-tin eutectic solders

Solder alloy	Shear strength		Creep stress failure for a shear strain rate of 10^{-6} mm/mm/s(a)		Fatigue life at 10% strain at 10 min/cycle(a) cycles
	MPa	ksi	MPa	ksi	
Bi-Sn	24	3.5	4.1	0.6	45
Pb-Sn	18	2.6	1.4	0.2	150

Source: After Tomlinson and Collier [1987]. (a) Data from Mei and Morris [1992]

Table 7.4 Shear strength of soldered joints to brass made with lead-tin and bismuth-tin eutectic solders

Solder alloy	Shear strength	
	MPa	ksi
Bi-Sn	17	2.5
Pb-Sn	19	2.8

Source: After Tomlinson and Collier [1987]

7.2.3.3 Influence of Solder Composition on the Mechanical Properties of Soldered Joints

The substitution of one filler alloy by another with a lower melting point normally has a detrimental effect on the strength of joints because, at the same operating temperature, the lower-melting-point alloy has a higher homologous temperature (operating temperature/melting temperature) and is therefore usually weaker.

Studies comparing the mechanical properties of joints to copper made using bismuth-tin eutectic solder with those employing Pb-60Sn have been made by Tomlinson and Collier [1987] and by Mei and Morris [1992]. The latter also examined three other low-melting-point solders. Whereas Tomlinson and his coworkers prepared conventional single-lap shear testpieces [Tomlinson and Bryan 1986], Mei and Morris used single- and double-lap shear specimens prepared from copper-clad epoxy-fiberglass sheet, patterned to leave an array of 3 × 3 rectangles of copper, each approximately 1.2 × 2.0 mm (0.05 × 0.08 in.), which were then soldered together. The solders were applied in the form of powder in flux-loaded pastes to the copper squares, and the joints were produced by reflow. This type of testpiece was adopted to simulate joints to printed circuit boards (PCBs). Notwithstanding the different testpieces used by the two research teams, their results were broadly compatible. The soldered assemblies were tested under applied stresses that subjected the joints to shear, creep, and low-cycle fatigue.

Contrary to the normal pattern of behavior, soldered joints to copper made with the lower-melting-point bismuth-tin solder were found to be substantially more resilient in the shear and creep tests than were the lead-tin joints. However, this is not the complete picture, for as Table 7.3 shows, the fatigue resistance of bismuth-tin joints was inferior.

The relatively strong and brittle characteristics of the bismuth-tin solder can be ascribed to the greater hardness of the bismuth phase in the eutectic lamellar microstructure of this solder as compared with the lead phase in the lead-tin solder. However, as pointed out in section 2.2.1.10, it is possible to make the eutectic bismuth-tin alloy ductile by preparing it with a fine equiaxed grain structure, which can be achieved by very rapid cooling from the melt.

The parent materials can have a marked effect on the characteristics of joints made with different filler alloys. For example, as shown by the shear strength data given in Table 7.4, joints made to brass using lead-tin solder are stronger than identical joints made with bismuth-tin solder [Tomlinson and Collier 1987]. The difference in this instance is caused by the zinc in the brass, which modifies the thickness and morphology of the Cu_6Sn_5 intermetallic layer that forms between the solder and the brass in a manner which renders the joints more susceptible to fracture.

7.3 Practical Examples of Process Design

This section offers a selection of examples representative of process design assignments that have been successfully addressed by the authors.

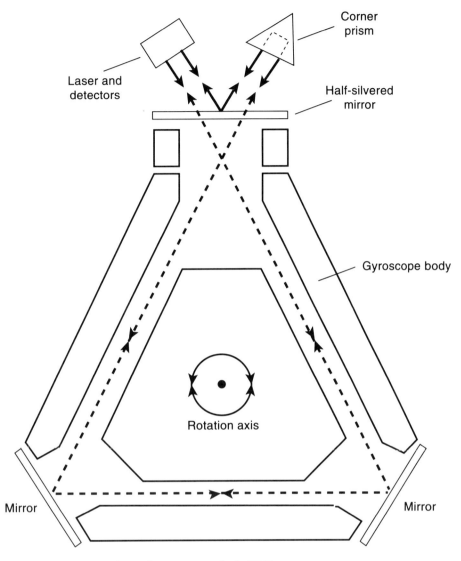

Fig. 7.7 Schematic illustration of a ring laser gyroscope body (RGL)

7.3.1 Mitigating Thermal Expansion Mismatch in Bonded Assemblies

7.3.1.1 *Background*

When materials with different coefficients of thermal expansion (CTE) must be joined, a range of solutions is available, several of which are described in section 4.3. The particular approach that is developed and implemented will be influenced by constraints such as the space available, consideration of the weight of the product, the maximum stress that must be borne, the required thermal conductance, and, of course, the cost tolerance of the product.

A ring laser gyroscope (RLG) will be used to illustrate this type of problem. This instrument is used in aircraft to detect and measure angular movements by sensing the frequency change of a laser beam that is made to traverse a space of triangular trajectory simultaneously in two opposite directions (see Fig. 7.7). The two beams have identical path lengths when the RLG is at rest, which can be detected by combining them and observing the interference fringe pattern produced. When the instrument is subject to movement that includes a component of rotation about an axis normal to the plane of the optics, one beam will have a longer optical path length than the other. This will cause a shift in the interference fringe pattern that is proportional to the rate of rotation. The frequency shift therefore provides a measure of the angular velocity [Macek and Davis 1963].

7.3.1.2 Formulating the Problem

For operational reliability and freedom from drift, the RLG must be constructed rigidly, with the semiconductor laser strongly bonded to the housing of the gyroscope. The housing of the RLG is an aluminum alloy (CTE = 20×10^{-6}/°C), because of the requirement for mechanical robustness coupled with low weight; the laser body, however, is a brittle glass-ceramic with a much lower, and indeed close to zero, coefficient of thermal expansion. Therefore, the mismatch stresses that would exist between these two components, were they to be directly bonded, need to be distributed away from the glass-ceramic in order to prevent its catastrophic failure. This can be achieved by using an intermediate plate of a material with a thermal expansivity that is closely matched to the ceramic, so that the majority of the stress due to the thermal expansion mismatch can be absorbed by the more robust components. The material selected for the expansion matching plate was a proprietary iron-nickel-base alloy, of the type referred to in section 7.3.4.2. It was found possible to select a material with a coefficient of thermal expansion of 2×10^{-6}/°C, which would be maintained over the operating temperature range of the RLG—namely, –25 to 85 °C (–13 to 185 °F).

All of the selected materials were impossible to wet directly with a filler alloy without the use of activators in the filler or fluxes and, in view of the critical function of this instrument in aircraft, these options (with their attendant risks of contamination and degradation) had to be ruled out. Therefore, suitable metallizations needed to be applied to the various joint surfaces.

The aluminum alloy specified for the gyro housing was a precipitation hardening alloy, in the peak aged condition, which conferred a known and stable coefficient of thermal expansion mismatch over the operating temperature range of the RLG. Thus, an additional consideration governing the processing temperature and time that could be permitted for the joining operation was that these had to be compatible with the conditions that enabled the aluminum alloy to be optimally precipitation hardened. The aging characteristics of

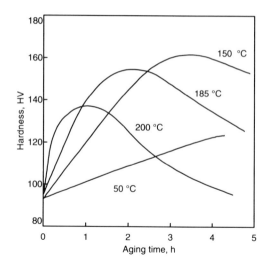

Fig. 7.8 Aging behavior of the aluminum alloy used for an RLG

the aluminum alloy were determined by hardness measurements; the results are given in Fig. 7.8.

7.3.1.3 Solution

With the introduction of the expansion matching plate, there were now two joints in the assembly. For the convenience of enabling subsequent replacement of the laser without undoing the entire assembly, a step soldering process was used, involving the use of two separate solders having well-separated melting points. Because the close expansion match of the intermediate plate to the glass-ceramic only held true up to 150 °C (300 °F), the joint between these components had to be made just below this temperature but above the maximum operating temperature of the RLG.

This was achieved by using the In-48Sn (melting point = 120 °C, or 248 °F) eutectic solder together with gold metallizations applied to the joint surfaces. As described in section 2.2.3.2 with reference to Fig. 2.33, the dissolution of gold by molten indium is sharply self-limiting due to the formation of the stable $AuIn_2$ phase at the interface. Although the converse is true for the reaction between molten tin and gold, tests conducted by the authors established that the indium-tin solder reacts with gold in a similar manner to pure indium [Jacobson and Humpston 1990].

This is to be expected from consideration of the gold-indium-tin ternary phase diagram. An isothermal section of this diagram at 25 °C (77 °F) is given in Fig. 7.9, which shows that the addition of small quantities of gold to the In-48Sn eutectic

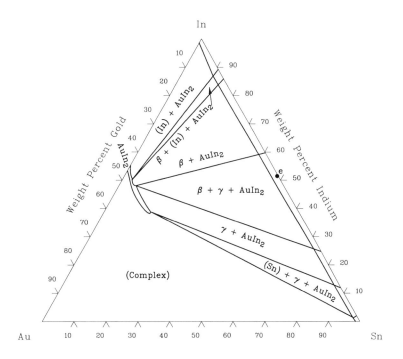

Fig. 7.9 Isothermal section at 25 °C (77 °F) of the Au-In-Sn phase diagram. Source: After Prince, Raynor, and Evans [1990]

composition alloy results in equilibrium between the beta and gamma phases of the solder and AuIn$_2$. Because indium, tin, and gold do not diffuse readily through this intermetallic compound, further dissolution of the gold coating effectively ceases. Consequently, the gold metallizations could be made extremely thin and therefore without a significant cost penalty.

In practice, a gold-loaded frit (see section 4.2.2) was fired onto the glass-ceramic surface and an electroplated gold coating applied to the expansion matching plate over intermediate layers of copper and nickel. The gold layer could have been electroplated directly onto the nickel layer. However, for the reasons given below, it was beneficial to also include the layer of copper, while retaining the nickel. It is difficult to electroplate a consistently adherent layer of gold directly onto the surface of the iron-nickel alloy plate. Although thin gold layers can be applied directly to iron-nickel alloys using vapor deposition processes, such as sputtering, this option was ruled out by the customer, who did not possess this type of facility.

In order to provide a sufficiently wide temperature interval between the two soldering operations, the Sn-5Sb solder (melting range = 235 - 245 °C, or 455 - 473 °F) was used to form the first joint between the aluminum alloy electrode and the expansion matching plate. This solder forms stronger joints to copper than to nickel surfaces, because interfacial copper-tin phases have superior mechanical properties, as noted in the previous example [Vianco, Hosking, and Rejent 1990]—hence the desirability to plate the nickel with copper. The nickel layer is still necessary, because it serves the function of providing better adhesion than can be obtained by plating copper directly onto iron-nickel alloys.

Accordingly, the following metallization sequences were used for achieving wettable surfaces on the metal components: a copper on nickel strike electroplate was applied to both faces of the expansion matching plate and also to the aluminum alloy body. A thin coating of a sensitizing metal, in this case zinc, was first applied to the aluminum alloy component in order to achieve strong adhesion of the electroplated metallizations. The complete sequence of layers applied to the component surfaces is shown schematically in Fig. 7.10. By using only a thin gold layer on the expansion matching plate, problems of embrittlement of the

Principles of Soldering and Brazing

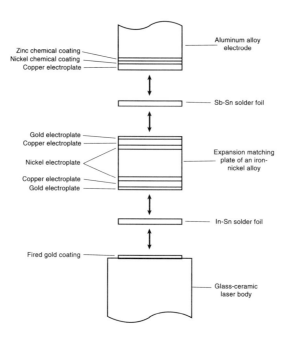

Fig. 7.10 Schematic of the structure of the electrode/laser body assembly of an RLG

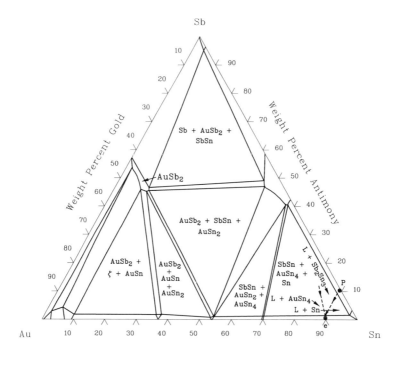

Fig. 7.11 Isothermal section of the Au-Sb-Sn system at 25 °C. A liquidus projection is given for the tin-rich corner (i.e., the dashed lines ep, etc. represent liquidus valleys). Source: After Prince, Raynor, and Evans [1990]

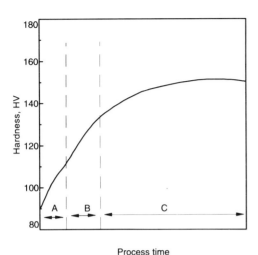

Fig. 7.12 Aging response in terms of hardness of the aluminum alloy during fabrication of an RLG. A, pre-soldering heat treatment at 155 °C (310 °F); B, Sn-Sb soldering, 250 °C (480 °F); C, In-Sn soldering, 180 °C (355 °F) (A,B, and C are on different scales)

Fig. 7.13 Section through a simulated laser assembly of an RLG

Fig. 7.14 CCD camera chip. Package width, ≅15 mm (0.6 in.) Left: empty CCD package. Right: bonded CCD assembly.

joints made with the Sb-95Sn solder, due to the formation of gold-tin intermetallic compounds, are avoided. The maximum thickness of the gold layer permitted was determined by applying the construction procedure described in section 3.2.2 to the gold-antimony-tin ternary phase diagram (Fig. 7.11). From this it was established that the concentration of gold dissolved in the solder must not exceed 10%; this cannot occur with the gold/solder ratio used (see Table 7.2). Other schemes for joining together components that have a significant expansivity mismatch are described in section 4.3.

The process temperatures specified for use with the two solder alloys were sufficiently high to induce precipitation strengthening of the aluminum alloy body. Thus, the total solution encompassed specifying the starting condition of the aluminum alloy and then precisely controlling the joining process temperatures, dwell times, and heating and cooling rates in order to achieve the required aging state in the fabricated RLG. The resulting aging of the aluminum alloy as it is processed through the heating cycles has been confirmed by hardness measurements and is shown in Fig. 7.12.

A sectioned, prototype RLG assembly is shown in Fig. 7.13. Although there is a large thermal expansivity mismatch between the metal expansion matching plate and the aluminum alloy, the operating temperature range of the RLG is relatively narrow and, as the laser generates virtually no heat, the thermal cycles are essentially those of the slowly changing ambient environment. The Sb-95Sn solder is therefore not subjected to severe straining, and thermal cycling tests verified the integrity of the joining solution.

7.3.2 Designing Soldered Joints for Hermetic Seals

7.3.2.1 *Specifying the Functional Requirements*

Soldered joints are frequently required to provide hermetic seals with high mechanical integrity between components. For some applications, the seals must also offer high thermal and electrical conductance. These requirements are most often encountered in the electronics industry.

A representative example of an assembly requiring a hermetic seal is a charge coupled device (CCD), which is the active solid-state component in a television camera. This device receives an optical image through a transparent window and converts it into an electrical signal for subsequent processing and transmission. An example of a packaged CCD chip is shown in Fig. 7.14. The CCD component is susceptible to degradation through oxidation and corrosion if exposed to the ambient atmosphere and needs to be protected within a hermetic package. Therefore, the glass window must be hermetically joined to the package containing the device, which, for similar reasons, is usually a dense, sintered ceramic.

For many large-volume consumer applications, adhesive attachment of the window is normally quite adequate. However, for more demanding applications, such as security cameras that may be required to be installed permanently outdoors, adhesives cannot guarantee the long-term reliability demanded, Consequently, soldered seals are generally preferred.

7.3.2.2 Design Guidelines

In order to achieve a reliable hermetic sealing process, it is necessary to satisfy the conditions relating to joint geometry and the bonding cycle that were identified in sections 1.3, 1.4, and 4.4.1. These may be phrased in the following set of instructions:

1. Make one dimension of the joint area small and preferably below 2 mm (0.08 in.), thereby minimizing the risk of voids developing in the joint due to trapped gas (section 4.4.1.1).
2. Use a solder with a low liquid-solid volume change, again to minimize the level of voids formed in the joint due to the shrinkage contraction of the filler metal (section 4.4.1.2).
3. Carry out the joining operation at low excess temperatures above the melting point of the solder, typically less than 50 °C (90 °F). This condition helps to restrict the extent of flow by the molten filler, thereby maximizing the quantity of solder in the joint gap to effect the seal (section 1.3.4).
4. Endeavor to confine the molten solder to the joint region. If this condition can be met, then the above constraint of having to use low excess temperatures may be relaxed.
5. Ensure that the component surfaces can be readily wet by the molten solder; in particular, they must be clean and dry (section 1.4.2.7). Areas of localized dewetting and voids formed by desorbed species are obviously not desirable when endeavoring to maximize joint filling. Fluxes can be used to provide a degree of surface cleaning, provided that their use is permitted in the application and that the residues can be satisfactorily removed (section 5.3.1).
6. Ensure that the surfaces of the solder preform are clean and dry and that the alloy is internally as free from oxides and trapped volatile constituents as possible, for the reasons given above.
7. Additionally, it is prudent to design the joint so as to encourage the formation of a fillet. Then, even if the joint contains voids, for whatever reason, the fillet will serve to seal the joint, because there is a higher probability that where fillet formation is promoted, these tend to be continuous and void free.

7.3.2.3 Solution

An arrangement that has proved satisfactory for sealing windows on CCD devices comprises a window that is marginally smaller in area than the rim of the package onto which it is mounted (Fig. 7.15). The resulting ledge around the glass window provides a step on which a solder fillet can form (condition 7). The areas to be soldered are coated with a metallization that the solder will wet, and the solder itself must be fully compatible with this metallization. Additional constraints in this application are that the temperature of the package cannot exceed 175 °C (345 °F) without degrading the sensitive electronic circuit and that fluxes cannot be tolerated for the same reason.

Accordingly, the choice fell on the Bi-33In eutectic solder (melting point = 109 °C, or 228 °F) and gold metallizations. This solder is compatible with gold metallizations for the reasons discussed in the previous example. The liquid-solid contraction of the bismuth-indium solder is essentially zero, compared with 3% for common lead-tin solder (condition 2); further details on this point can be found in section 4.4.1.2. The annular width of the seal was maintained at 1.5 mm (0.06 in.) to comply with condition 1 above and, by selectively metallizing only the area to be joined, spreading by the molten solder was constrained (condition 4). A low excess temperature above the melting point of the solder, with the process temperature fixed at 130 °C (265 °F), enabled condition 3 to be met, while simultaneously respecting the upper exposure temperature of the semiconductor device.

The metallizations were achieved using processes described in greater detail in section 4.2.2. A solderable gold-loaded glass frit (thick-film metallization) was applied selectively to the rim of the ceramic package by screen printing and then fired at high temperature (800 °C, or 1470 °F) in order

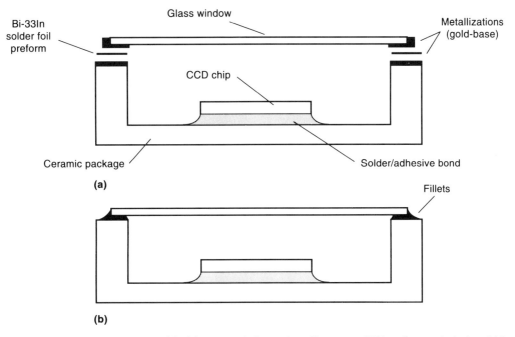

Fig. 7.15 Schematic representation of the fabrication of a hermetic seal between a CCD package and window. (a) Before soldering. (b) After soldering

to fuse the glassy constituent with the ceramic. A sputter metallization of chromium overlaid with gold was selectively deposited onto the edges of the window. A two-layer sputter metallization is necessary, because the chromium provides the reactive bond to the nonmetal and also a metallic surface for the solder to wet, while the gold serves to prevent the chromium from oxidizing, thus ensuring a readily wettable surface.

For the gold coating to be effectively pore free and therefore confer a reasonable solderability shelf life to the underlying chromium, it should be applied by a vapor deposition technique, such as sputtering. The shelf life provided by sputtered gold coatings is superior to that of electroplated gold, as can be seen from the data presented in Fig. 7.16 [Humpston and Jacobson 1990]. Figure 7.16 shows the wettability, in terms of the wetting force measured on a wetting balance, of copper and chromium coupons coated with different thicknesses of gold by the two methods, as a function of the storage time in an open laboratory environment. It is apparent that electroplated gold coatings must be significantly thicker than sputtered gold coatings to provide the same shelf life and that relatively thick coatings are necessary if a shelf life of more than 1 year is required. Where economic and size factors dictate that electroplating is the most suitable coating method, the level of gold in the joint can be greatly reduced by wicking off the coating with solder immediately prior to the soldering operation. The excess gold can be reclaimed from the discarded solder.

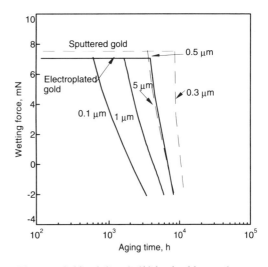

Fig. 7.16 Solderability shelf life of gold-coated components

The bonding of windows on the CCD packages was carried out in a dry nitrogen atmosphere with an oxygen level below 1 ppm in order to en-

Fig. 7.17 Metallographic section through the soldered joint between a CCD package and window. 7.8×

Fig. 7.18 Printed circuit board containing electronic components

sure that the molten solder did not oxidize, which would impede wetting. A metallographic section through the resulting glass-ceramic joint is shown in Fig. 7.17. Note that it is free of voids and possesses smooth, well-formed fillets. No component failures due to leaks in the solder seal have been experienced during 10 years of manufacture and service.

7.3.2.4 *Postscript: Sputtered Chromium Metallizations*

Chromium, or a mixture of nickel with chromium (Nichrome), is frequently recommended as the reactive metal when applying wettable/conductive coatings to nonmetals by sputtering [Holloway 1980] and was used for the windows on the CCD devices. Although more reactive metals, such as titanium, are sometimes used on the premise that they form a more adherent bond to the nonmetal, chromium is preferred because of the nature of the reaction.

Chromium reacts with many nonmetallic compounds to form complex chromates. These are not only strongly bonded to the nonmetal but also act as a barrier to further interaction taking place by diffusion between chromium and the nonmetal [Mattox 1973]. Therefore, it is possible to use chromium as a high-integrity metallization on glass and intrinsically stable ceramics, such as alumina. The benefit of this type of chemical bonding is evident when metallizing relatively unstable ceramics, such as zinc oxide. This material is widely used in electrical voltage surge suppressors, known as varistors [Leite, Varela, and Longo 1992; Wersing 1992]. Heating a zinc oxide ceramic metallized with titanium or zirconium causes the metallization to blister and spall off due to the oxidizing reaction:

$$2ZnO + Ti = TiO_2 + 2Zn$$

At 300 °C (573K), the Gibbs free energy of this reaction is

$$\Delta G_{2ZnO + Ti}(573K) = -272 \text{ kJ}$$

accompanied by the free zinc volatilizing beneath the coating. On the other hand, a chromium metallization is entirely benign toward zinc oxide, even on heating up to 400 °C (750 °F) for prolonged periods, due to the presence of the intervening chromate barrier layer.

7.3.3 Development of an Improved Solder for Die Attachment in Electronics Fabrication

7.3.3.1 *Background*

Electronic circuitry is normally assembled from components connected together on a printed circuit board, an example of which is shown in Fig. 7.18. Today, most of the components are themselves elaborate circuits processed in small silicon and gallium arsenide chips, typically between 1 and 500 mm^2 (0.0016 and 0.8 in.^{2}) in area. These semiconductor chips or dies are housed in hermetic packages to protect them from moisture, contamination, and other sources of degradation and damage. The package is usually made of an electrically insulating ceramic or plastic, and the chips are attached to the metallized base of the package using either a polymeric adhesive, a glass, or a solder. A package containing a chip,

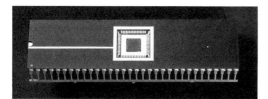

Fig. 7.19 Silicon chip mounted in a ceramic package. The component is completed by soldering on a lid to hermetically seal the electronic circuit.

prior to the attachment of its lid, is shown in Fig. 7.19. Each package can contain several chips. It is normally necessary to achieve good thermal and electrical contact between the chips and the base of the package. For this reason, the polymer or glass die bonding materials are loaded with metal powder, often silver, but these are inevitably inferior to solders with regard to electrical and thermal conductivity, and the achievable bond strength tends to be lower.

7.3.3.2 *Formulating the Problem*

The two solders that are commonly used for die attachment are gold alloys—namely, the eutectic Au-3Si and Au-20Sn compositions. The phase diagrams of these alloy systems are given in Fig. 2.18 and 2.19, respectively. They are chosen because their melting points are relatively high, ensuring that the chip or die does not dislodge from the package when it is soldered to the circuit board using the lead-tin industry standard solder. This operation typically requires a soldering temperature of 220 °C (430 °F). By comparison, the Au-3Si solder melts at 363 °C (684 °F), and the Au-20Sn solder has a melting point of 278 °C (532.4 °F). The two solders also have the advantage of being eutectic alloys, which means that they melt at a single temperature and confer other benefits enumerated in section 3.2.1.

Fluxes are frequently used to promote soldering operations. These chemicals are not permitted in semiconductor packaging, because they are corrosive and therefore incompatible with the sensitive nature of circuitry on semiconductor chips. In the absence of a flux, the wetting and spreading characteristics of the Au-3Si solder are especially poor, on account of the siliceous dross that accumulates on the surface of the molten filler. Even without the inhibiting effect of the dross, this solder is inherently sluggish. This can be seen by comparing the spreading behavior of the Au-3Si alloy, as a function of temperature above its melt-

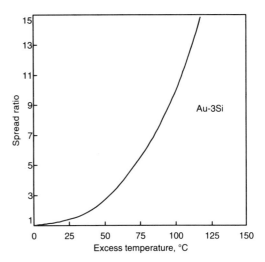

Fig. 7.20 Spreading behavior of Au-3Si solder as a function of the excess temperature above its melting point

ing point (Fig. 7.20), with similar data for other common solder alloys (Fig. 1.9). Poor fluidity of the solder leads to inadequately filled joints, which mar the performance and reliability of the packaged devices.

A further problem, which applies to both the Au-20Sn and Au-3Si solders, is that neither will directly wet bare silicon. Silicon is a relatively refractory element, which, when exposed to air, becomes covered with a stable, continuous, and extremely adherent layer of silica. For this reason, conventional solder alloys are unable to wet silicon in the absence of fluxes. In order to achieve wetting, it is necessary either to coat the back of the precleaned die with a solderable metallization, such as gold, or to manually scrub the molten filler between the die and the package and thereby physically displace the silica layer (see section 5.3.5). The first option is usually not compatible with other stages of semiconductor fabrication, and so scrubbing is resorted to, using a protective shroud of inert gas. This processing step usually has to be performed manually and is therefore expensive. It also introduces the element of human variability, which affects product yield.

There is a clear need for improved die-attach solders that overcome the deficiencies of the existing alloys but that do not require significant alterations to the joining process. Solutions involving changes to the solder composition that have been devised by the authors are described below.

7.3.3.3 Improving the Spreading Characteristics of the Au-2Si Solder

A number of alloying additions are known to be capable of promoting solder spreading [Humpston and Jacobson 1990]. Several of these were assessed by adding identical amounts of each element in turn to the Au-3Si solder and evaluating the effect on the spreading characteristics of the alloy. The most effective promoter of spreading of the Au-3Si solder was found to be tin. Results of spreading tests conducted under identical conditions but with different concentrations of tin in the solder showed that increasing the level of tin gave a progressive improvement in solder spread, as illustrated in Fig. 7.21. Hardness measurements revealed that this was accompanied by a softening of the alloy, with its hardness decreasing by more than 150 HV, a welcome feature that makes the alloy more amenable to mechanical working into foil and wire for solder preforms.

Fig. 7.21 Spreading behavior of Au-2Si solder containing tin. Left to right: 0%, 4%, 8% Sn

It was desirable that any new solder have a melting point similar to the binary Au-2Si composition (363 °C, or 685 °F), so as not to upset the associated manufacturing steps. Unlike a pure metal, which has a single melting point, most alloys melt progressively over a finite temperature range. The narrower this melting range, the better the spreading behavior of the alloy. This was another major consideration in the selection of a new solder composition. The simplified phase diagram of the gold-silicon-tin alloy system, reproduced in Fig. 7.22, indicates that additions of more than 8% Sn to the gold-silicon eutectic alloy result in the formation of two gold-tin intermetallic compounds, $\zeta(Au_5Sn)$ and AuSn. These phases react together with silicon, causing the alloy to melt at 275 °C (527 °F). By restricting the concentration of tin to 8% and below, the melting temperature of the alloy is instead determined by the reaction among gold, silicon, and the Au_5Sn intermetallic compound, which occurs at 356.5 °C (673.7 °F). An alloy of the composition Au-2Si-8Sn was judged to satisfy the combination of requirements sought from a new solder: namely, a melting range of 356.5 to 358.0 °C (673.7 to 676.4 °F), combined with improved spreading characteristics and mechanical properties.

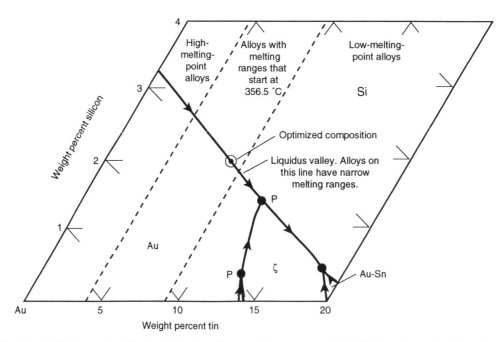

Fig. 7.22 Gold-rich portion of the Au-Si-Sn phase diagram. The system can be divided into three regions by melting point. The circle indicates the alloy composition selected as the solder. Source: After Prince, Raynor and Evans [1990]

7.3.3.4 Improving the Wetting of Silicon by Gold Solders

A successful method of promoting wetting of oxides and oxidized metal surfaces by brazes is to incorporate elements into the filler that are capable of reacting with the oxide to form an adherent bond [Crispin and Nicholas 1984; Xian and Si 1991]. There is a minimum temperature at which a reactive filler is effective in this manner, which, for those that are available commercially, is approximately 800 °C (1470 °F) (see section 5.4.1). Clearly, these are inappropriate for die-attachment because the electronic circuitry cannot withstand such high temperatures. Gold solders containing reactive elements that are effective on bare silicon below 400 °C (750 °F), without the aid of chemical fluxes, have not been reported previously.

Of the elements that are capable of reactively bonding to silica, titanium is preferred because it does not form brittle phases with the constituents of many solders and brazes. In order to evaluate its suitability in gold solders, additions of titanium were made to both the Au-20Sn and the Au-2Si-8Sn alloys, and wetting tests were performed on silicon wafers.

It was found that the addition of titanium to the gold solders was highly effective in promoting wetting and spreading on bare silicon. At the same time, concentrations of titanium of up to 2% were found to have little effect on the melting ranges of the alloys and to be beneficial in reducing their hardness. Figure 1.16 shows a silicon die soldered into a gold-metallized package using a gold solder with the composition Au-2Si-8Sn-1Ti, the joint made by heating in a nitrogen atmosphere without scrubbing. Wetting of the ceramic package and the silicon die by the solder was found to be highly uniform.

7.3.4 Problem Solving: Improving Reliability by Modifying Joint Geometry

7.3.4.1 Introduction

When it comes to problem solving, a major hurdle is to properly formulate the problem. Problems, even mundane ones, that are brought to the laboratory are often not clearly identified by the originator and thus are often incorrectly diagnosed. As an example, the authors were contacted by a company that was experiencing considerable difficulty in the solder wetting of a relatively large component coated with a pure gold electroplate, despite using flux. They supposed that the problem lay with the electroplate or contaminated solder, whereas it transpired that the temperature reached by the gold surface during the joining operation was always below the melting point of the solder!

Formulating the problem correctly is an absolutely essential first step toward achieving a solution. It may require consideration of linked fabrication steps, perhaps involving other parts of the manufacture that are responsible for some of the initial design and processing decisions. Once the cause of the problem has been correctly identified, the cure is often straightforward, the choice of solution being made on the basis of technical merits closely coupled with economic and other practical considerations. A typical example drawn from the authors' experience is described below.

7.3.4.2 Defining the Problem

The geometry of soldered and brazed joints has a major influence on the mechanical integrity of assemblies because it affects the distribution of stresses. With certain designs, stress concentrations can result in the assembly failing at relatively low applied stresses (see section 4.4.3.1). Conversely, by improving the joint configuration, stresses can be better distributed and load-bearing capability thereby enhanced. In a case involving electronic circuit board assembly, it was found possible to raise joint reliability to within the acceptance level by introducing a simple change to the shape of the joints. This problem-solving activity will now be described.

Electronic circuits, principally comprising ceramic body, flat-pack modules soldered to copper tracks on PCBs, were failing in service due to weak joints between the Kovar leads extending out of the packages and the mating pads on the tracks. The failures were manifested by the package leads detaching from the PCB due to cracks propagating along the length of the soldered joints. A schematic illustration of a typical assembly is shown in Fig. 7.23.

Kovar is one trade name, Invar being another, of a family of iron-nickel alloys that contain between 28 and 50% nickel and sometimes also cobalt, chromium, and titanium. These alloys are unusual in that they can be fabricated to have a controlled thermal expansivity and/or elastic modulus over the temperature range of –40 to 350 °C (–40 to 660 °F). By adjusting the composition and metallurgical condition of the material (i.e., grain size, precipitate size, degree of cold work, etc.), the thermal expansivity can be varied from essentially zero to $10^{-6}/°C$ [International Nickel 1965; Angus 1978; Russel and Smith 1989]. Controlled-expansion alloys are widely used in ceramic-metal seals, as the alloy can be tailored to have a thermal expansivity that is close to that of a particular ceramic, over a certain range of temperature, and hence ensure high-reliability joints that can sustain repeated thermal cycling.

Possible causes of the joint weakness were poor wetting by the solder due to the use of insufficiently active fluxes and/or the formation of brittle intermetallic phases. Because the Kovar leads were adequately wetted by the solder, the problem was attributed to, and subsequently confirmed as being due to, a weak interface between the Kovar leads and the solidified lead-tin solder used. The formation of brittle tin-base phases in joints made to iron-nickel alloys using tin-base solders is well known [Kang and Ramachandran 1980; Vianco, Hosking, and Rejent 1990]. The mechanism of formation of these phases is, in principle, identical to the reaction between lead-tin solder and copper that is described in section 3.2.2.

The circuit assembly had already been brought on-line, and so a solution to the problem was required rapidly. Furthermore, due to the large scale of production and considerable setup costs involved, there was a strong incentive not to alter the fabrication process in any significant manner. These considerations precluded the use of metallurgical solutions of the type described in section 4.2.1—namely, of applying metallizations to the Kovar leads (e.g., electroplating with copper) that would provide a readily wettable surface and prevent interaction between tin and Kovar. Another potential solution to the problem would be to use a solder that contains a lower concentration of tin and so reduce the formation of brittle interfacial phases. The data given in Table 7.5 show that the joint strength appears to increase as the proportion of tin in the solder decreases. However, such a radical change to the process was not permitted, as the circuit boards would then have to be requalified. The failure rate was only a subpercentage of joints on all the circuit boards being assembled. A statistical analysis indicated that by doubling the joint strength the problem would be effectively solved.

7.3.4.2 Identification of a Viable Solution

The first step in the search for a solution was to quantify the strength of the substandard joints. The clue to a possible cure was provided by the form of the peel strength profile of joints that were subjected to mechanical testing. The force needed to initiate peel was consistently higher than that required to propagate the delamination, as shown in Fig. 7.24. The initial peak in the peel strength correlated with the fillets at that end of the joints. The load-bearing capability of a simple lap joint benefits greatly from fillets with large, smooth radii, for the reasons given in section 4.4.3.1. Therefore, it was considered that, by increasing the radius of curvature of the fillets, the stress required

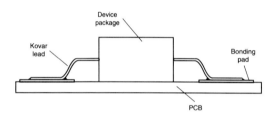

Fig. 7.23 Original joint geometry of a flat-pack module on a circuit board

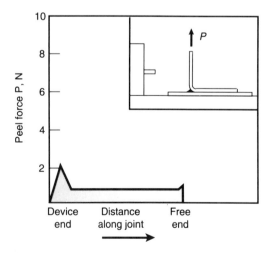

Fig. 7.24 Typical peel force (P) profile of the joints in Fig. 7.23. The peaks in peel strength are associated with the fillets at each end of the joint.

Table 7.5 Effect of tin content on the peel strength of joints to Kovar

Solder alloy	Average joint peel strength, gf/mm (width)
Ag-96Sn	94 ± 45
Pb-60Sn	152 ± 56
Ag-97.5Pb-1Sn	377 ± 164

to initiate peel would be correspondingly increased. This was verified in tests; the roughly linear relationship between fillet height and load required to initiate peel is demonstrated in Fig. 7.25. A similar finding is reported by Hoyt [1987]. Consequently, the Kovar leads were kinked to provide a pair of high-angle joints, which promoted the formation of large fillets when the volume of solder used for each joint was increased (Fig. 7.26). The higher resistance to peel failure of the modified leads can be contrasted with those of the original lead geometry by comparing Fig. 7.27 with Fig. 7.24.

The reshaping of the leads was an operation that could be carried out prior to circuit assembly and thus did not impinge in any way on the production process. This modification was implemented rapidly, within one production shift, causing minimal disruption to manufacture. As predicted, the mean time between circuit failures then met the customer's specification. An additional benefit of this design change was that the joints could be nondestructively tested by applying a set proof stress and taking advantage of the high stress required to initiate peel, as shown in Fig. 7.27. Admittedly, this change constituted a design "palliative" rather than a proper metallurgical solution, although it proved itself fit-for-purpose and reduced the incidence of joint failures in the field to negligible levels.

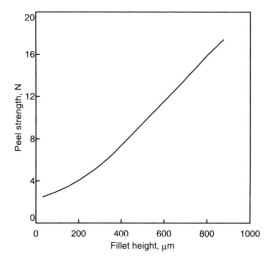

Fig. 7.25 Experimentally derived relationship between the stress required to initiate peel fracture and the height of the solder fillet

Fig. 7.26 Modified joint geometry of the flat-pack module on a circuit board

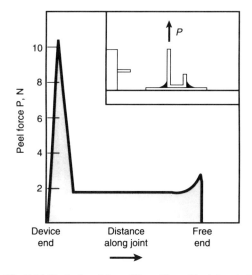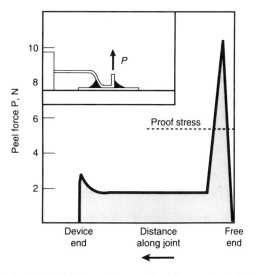

Fig. 7.27 Typical peel force (*P*) profiles of the joints in Fig. 7.26. The joint strength is considerably superior to that in Fig. 7.24.

7.3.5 Product Development: Improving the Fabrication of Silicon Power Device Assemblies

7.3.5.1 Introduction

This is an example of a major research and development exercise, where the solution of an assembly problem impinges on associated fabrication operations and also on the functional characteristics of the product. At the outset, there was no clearly advantageous route for addressing the requirement, and various options had to be considered. The wide-ranging nature of the problem required a collaborative project between industrial partners; this was undertaken within the framework of an EEC-sponsored BRITE Programme. More extensive details of this activity are described by Crees *et al*. [1988] and by Humpston *et al*. [1992].

7.3.5.2 Defining the Problem

A silicon power device comprises a disk of silicon that contains the requisite diffused-in junctions and measures about 0.5 mm (0.02 in.) thick by up to 125 mm (5 in.) in diameter. A detailed description of this type of semiconductor component and its manner of operation is analyzed in detail by Taylor [1987]. During operation, heat at a flux density of typically 1 W/mm^2 is generated by the device and must be removed (i.e., about 2 kW from a 50 mm, or 2 in., diam device and nearly 8 kW from a 100 mm, or 4 in., diam device) to prevent failure of the semiconductor through overheating. The requisite cooling is usually achieved by attaching the silicon disk to a metal plate that acts as a heat sink. A schematic section through a silicon power device is shown in Fig. 7.28. The significant mismatch between the coefficients of thermal expansion of silicon (CTE = 2.6×10^{-6}/°C at 25 °C) and high-conductivity metals, notably copper (CTE = 16.7×10^{-6}/°C at 25 °C), prevents large-area components of these materials from being bonded together. Instead, the standard practice is to braze the silicon disk to an intermediate plate of molybdenum (CTE = 5.0×10^{-6}/°C at 25 °C) using a 50 µm (2000 µin.) thick foil of the Al-13Si eutectic composition braze (melting point = 577 °C, or 1071 °F) to form what is known as a "basic unit." Conventional soldering cannot be used to make this joint because, under certain operating conditions, the temperature of the silicon component can exceed 350 °C (660 °F). Mechanical pressure contacts are then made to provide an electrical and thermal path between the molybdenum and copper plates and also to effect the electrical connection to the top of the device.

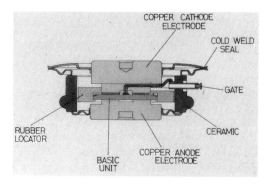

Fig. 7.28 Schematic illustration showing the construction of a typical high-power semiconductor device

Aluminum and silicon form a low-melting-point eutectic at Al-12.6Si, as shown in Fig. 2.16, so that the Al-12Si composition is suitable for use as a braze and is metallurgically compatible with the silicon wafer. Wetting of the molybdenum plate by this alloy results in the formation of a continuous, but strong and relatively compliant, interfacial layer, usually assumed to be molybdenum disilicide (MoSi$_2$), provided that process temperatures above about 650 °C (1200 °F) are used. At lower brazing temperatures, the layer is discontinuous and the joint to the molybdenum is correspondingly inferior [Crees *et al*. 1988]. The attribution of the interfacial compound as MoSi$_2$ does not accord with the constitution of the aluminum-molybdenum-silicon system [Brukl, Nowotny, and Benesovsky 1961]. It can be seen from the isothermal section of the aluminum-molybdenum-silicon phase diagram given in Fig. 7.29 that within the aluminum- and silicon-rich portion of the system, aluminum and silicon solid solutions exist together with the intermetallic compound Mo(Al,Si)$_2$, which is represented by the interfacial layer.

The method of fabrication described is satisfactory for devices with a diameter of less than about 75 mm (3 in.). However, the drive toward ever higher power ratings requires not simply improved heat sinking but, more importantly, an increase in the size of devices, because the power-handling capability is approximately proportional to device surface area. As the diameter of the device increases, the thermal mismatch stresses introduced by the existing joining method become excessive to the point that the silicon component fractures, and other solutions must be sought.

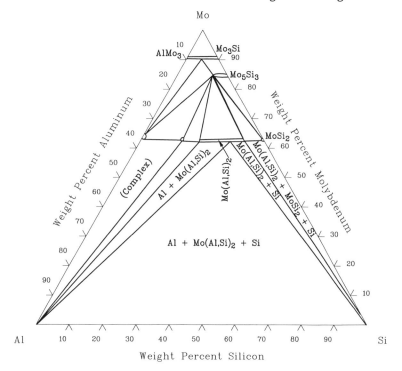

Fig. 7.29 Triangulation of the Al-Mo-Si ternary system. Aluminum and silicon enter into ternary equilibrium with the compound Mo(Al,Si)$_2$. Source: After Brukl, Nowotny, and Benesovsky [1961]

Additional problems that have arisen out of the evolution of power devices also needed to be addressed in the project. These were the heavy and nonuniform erosion of the silicon wafer by the existing braze and precipitation, on cooling, of a layer of silicon at the interface of the silicon device. Uneven erosion of the surface of the silicon wafer is clearly evident in Fig. 7.30, and a regrowth layer can be seen in Fig. 7.31. The precipitated silicon layer can be detrimental to the performance of these types of semiconductors, because the silicon has a different doping level to that carefully engineered into the surface of the silicon wafer and thereby alters its electronic characteristics. Nonuniform erosion of the silicon/braze interface means that this junction then also can have nonuniform electrical characteristics, which limits the maximum rating of the device.

The mechanisms of erosion and reprecipitation are explained in section 3.2.1. They occur because the aluminum-silicon braze is unsaturated with respect to silicon when heated above the eutectic temperature, so that on heating during the brazing cycle, silicon is dissolved from the wafer. On cooling, the solubility of silicon in the braze decreases, and thus the excess dissolved silicon reprecipitates on the wafer.

Modifications to the fabrication of power device/heat sink assemblies that did not require significant changes to the fabrication technology and production facilities will be referred to as incremental changes to distinguish these from the more radical changes that were devised.

Fig. 7.30 Cross section through a joint made between a silicon wafer and a molybdenum disk using Al-12Si braze and a peak process temperature of 680 °C (1255 °F). The braze has wet and dissolved the silicon wafer in a nonuniform manner that is detrimental to the electrical performance of devices. 1000×

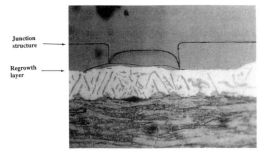

Fig. 7.31 Etched section through a brazed joint to a silicon power device, made using Al-12Si braze, revealing junctions in the silicon wafer and the silicon regrowth layer. 585×

Fig. 7.32 Microsection of a joint between a silicon wafer and a molybdenum backing plate made by replacing the Al-12Si braze foil with a 15 µm (600 µin.) thick vapor-deposited coating of the same alloy applied to a clean the silicon surface. 420×

7.3.5.3 *Solutions by Incremental Changes to the Fabrication Process*

Two incremental changes were based on optimizing the control of the process to give more reproducible metallurgical reactions. They are discussed below.

The first involved applying the Al-12Si braze as a thin (10 to 20 µm, or 400 to 800 µin.), vapor-deposited coating to the silicon wafer after the wafer had been cleaned in the deposition chamber maintained under high vacuum. This resulted in more uniform wetting of the silicon semiconductor by the braze, and the reduced volume of the aluminum alloy led to reduced erosion and subsequent precipitation.

The improvement in the uniformity of wetting was a direct consequence of the cleaning procedure, due to the removal of the native silica layer from the surface of the wafer. This layer acts as a barrier to wetting when the brazing is carried out using foil preforms. The highly aggressive fluxes that would be required in order to remove silica are not permitted in this application. By applying the braze alloy directly to the silicon, there is no intermediate oxide layer between it and the wafer, which would otherwise impede wetting.

The second change involved applying a coating to the molybdenum that is readily wetted by the aluminum alloy braze when it is molten. This change enabled substantially lower brazing temperatures to be used, while still forming a continuous interfacial layer of Mo(Al,Si)$_2$, owing to the absence of an interfering layer of refractory molybdenum oxide. The use of reduced process temperatures also led to reduced erosion and precipitation, because the solubility of silicon in the braze is a function of the excess temperature above its melting point.

Coatings that proved successful were MoSi$_2$ and the metals palladium and rhodium. To avoid changing the joint metallurgy through introducing new elements, the coating chosen was MoSi$_2$. A choice of alternative methods for applying the MoSi$_2$ coatings was identified, with chemical vapor deposition (CVD) proving to be the most rapid and cost effective that was capable of achieving adequate quality. It was found that submicron layers of palladium and rhodium, applied by electroplating, were equally effective. They were readily wetted by the braze and dissolved to form fine dispersions of the intermetallic compounds of either Al$_2$Rh or Al$_3$Pd.

By implementing these changes together, the electrical performance of power devices was significantly improved [Humpston *et al.* 1992]. A cross section through a typical joint made using the modified brazing process is shown in Fig. 7.32 and indicates a more regular interface between the silicon and the braze.

7.3.5.4 *Solutions by Radical Changes to the Fabrication Process*

There remained to be solved the fundamental problem of stress in the device assembly that arises from the thermal expansion mismatch between silicon and molybdenum and that is introduced when the assembly cools from the melting point of the braze at 577 °C (1071 °F). Of the range of options explored, diffusion soldering yielded early advances and was developed through to production.

The first approach involved evaluating alternative low-melting-point aluminum alloy brazes for joining silicon devices to molybdenum plates, with both sets of components suitably metallized. Germanium and silicon occupy neighboring positions in the same group of the periodic table and

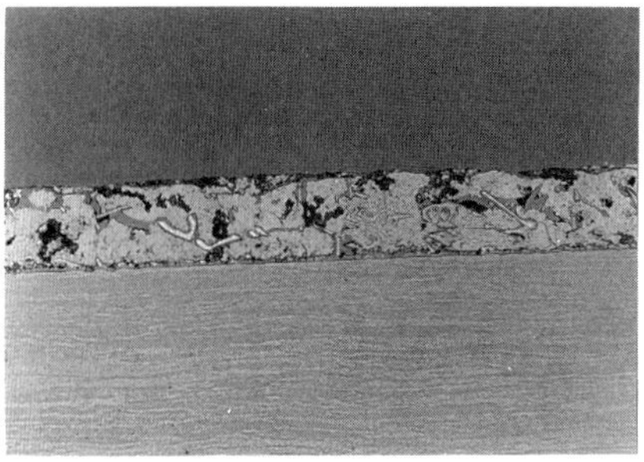

Fig. 7.33 Microsection through a joint made between a silicon wafer and a molybdenum backing plate using an aluminum-germanium-base braze. 380×

Fig. 7.34 Diffusion-soldered joint made using layers of silver and indium between the silicon and molybdenum components of a high-power semiconductor device. 380×

have closely similar chemical properties. Both form low-melting-point eutectics with aluminum, but the aluminum-germanium eutectic point is at 420 °C (788 °F), as compared with 577 °C (1071 °F) for aluminum-silicon. Figure 7.33 shows a cross section through a joint made using a braze based on aluminum-germanium. The lower melting point of the braze reduces the distortion arising from the residual thermal expansivity mismatch by 20%.

Although brazing with aluminum-germanium alloys was demonstrated to be practicable, the braze foil is exceedingly brittle, having a germanium content of 43%, and cannot be produced by conventional metalworking methods [Illgen et al. 1991]. Consequently, it was necessary to rely on rapid solidification technology to produce the preforms (see section 1.4.2.2). Because the available casting facility could not produce foil widths greater than 50 mm (2 in.), attention then shifted to diffusion soldering.

The principles of the diffusion soldering process are outlined in section 4.4.2. The silver-tin and silver-indium systems were found to be eminently suitable for this application because of the relatively low bonding temperatures that could be used and the suitably high remelt temperatures that apply [Marconi Electronic Devices Ltd. 1990]. Moreover, the mechanical properties of the joint that could be achieved were perfectly adequate, and the excellent filling that was obtained ensured a high thermal conductivity path to the heat sink. A typical joint is shown in Fig. 7.34.

Finding a suitable metallization for the silicon wafer represented a significant part of the exercise. This had to be metallurgically compatible with the diffusion soldering process and also had to provide an adherent ohmic contact layer to the silicon device. The metallization that was eventually selected comprised three layers. First, aluminum was applied and sintered to the silicon by a heat treatment, thereby forming an ohmic contact. Over this was sputter-deposited a submicron layer of titanium to provide a reactive bond to the aluminum, followed by a layer of silver of adequate thickness for the diffusion soldering process. The silver layer needed to be sufficiently thick to ensure that all of the molten solder was transformed to silver-rich intermetallic phases [Jacobson and Humpston 1992]. A layer of silver 10 μm (400 μin.) thick was found to be perfectly adequate.

A third possible option for reducing the mismatch was explored. This involved substitution of the molybdenum backing plate by one of a metal having a thermal expansion coefficient closer to that of silicon. The only metal that offers a reduction in the mismatch stress and a simultaneous increase in thermal conductivity is tungsten (CTE = $4.5 \times 10^{-6}/°C$ at 25 °C). A further merit of tungsten as a substitute for molybdenum is that the same brazing and diffusion soldering processes can be used. However, disks of tungsten are at least three times the cost of equivalent molybdenum components, and they represent a significant proportion of the overall cost of manufacture. Therefore, this option was not pursued.

Other heat sink configurations were identified that could achieve a closer thermal expansion coefficient match to silicon than molybdenum and tungsten. One of these involved a composite heat sink of graduated layers of metal designed to distribute mismatch stresses away from the joint with the silicon wafer. A component of this type is shown in Fig. 4.7. Another scheme examined was a filamentary structure of high compliance that would readily absorb differences in thermal expansivity and not generate significant stress levels. A plate of this type, made using copper filaments, is shown in Fig. 4.10. This approach is being exploited by other companies [Glascock and Webster 1983]. Both of these engineering solutions are discussed further in section 4.3. Although composite heat sinks were found to be technically feasible, these would be significantly more expensive than molybdenum plates and would require the joining processes to be suitably retailored.

7.3.5.5 Outcome

All of the incremental changes to the fabrication process described above were implemented in production, with considerable benefits in terms of process yield and device performance for the existing product range. Commercial considerations dictated that, of the more radical changes to the fabrication process, the diffusion soldering route in conjunction with molybdenum plates be adopted. Because of the greater complexity of this joining process, it was reserved for the largest and most critical devices in the product range, where the technical benefits allowed the additional process costs to be amply covered.

References

Ackroyd, M.L., Mackay, C.A., and Thwaites, C.J., 1975. Effect of Certain Impurity Elements on the Wetting Properties of 60% Tin-40% Lead Solders, *Met. Technol.*, Vol 2, p 73-85

Angus, H.C., 1978. Advances in Magnetic Low-Expansion Alloys, *Metall. Mater. Technol.*, Vol 2, p 80-84

BBI 1991. New Bismuth-Bearing Free-Machining Copper Alloy for Pb-Free Plumbing Fittings, *Bull. Bismuth Inst.*, Vol 62, p 9

Becker, G., 1991. Impurities in Solder—Their Impact on Solderability and Corrosion, *Solder. Surf. Mount Technol.*, Vol 7 (No. 2), p 24-33

Bornemann, A., 1956. Tin-Disease in Solder Type Alloys, *Proc. Symp. Solder*, STP 189, ASTM, p 129-149

Boughton, J.D. and Sloboda, M.H., 1970. Embrittling Effects of Trace Quantities of Aluminum and Phosphorus on Joints Brazed in Steel, *Weld. Met. Fabr.*, Vol 8, p 335-339

Brukl, C., Nowotny, H., and Benesovsky, F., 1961. Untersuchungen in der Dreistoffsystemen: V-Al-Si, Nb-Al-Si, Cr-Al-Si, Mo-Al-Si bzw. Cr(Mo)-Al-Si, *Monatsh. Chem.*, Vol 92, p 967-980

Crees, D.E. *et al.*, 1988. Silicon/Heat-Sink Assemblies for High Power Device Applications, *GEC J. Res.*, Vol 6 (No. 2), p 71-79

Crispin, R.M. and Nicholas, M.G., 1984. Reactive Brazing of Alumina to Metals, *Brazing Soldering*, Vol 6 (No. 1), p 37-39

EP&P 1992. Reduction in Antimony Puts Solder Joints at Risk, *Electron. Packaging Prod.*, Vol 32 (No. 2), p 20

Evans, D.S. and Prince, A., 1982. The Effect of Au on the Pb-1.5%Ag-1%Sn Solder, *Mater. Res. Bull.*, Vol 17, p 681-687

Feller, H.G. *et al.*, 1984. Lead Corrosion in Aqueous Solution and Tap Water, *Z. Metallkd.*, Vol 75 (No. 8), p 619-624

Glascock, H.H. and Webster, H.F., 1983. Structured Copper: A Pliable High Conductance Material for Bonding to Silicon Power Devices, *IEEE Trans. Components Hybrids Manuf. Technol.*, Vol 6 (No. 4), p 460-466

Holloway, P.H., 1980. Gold Chromium Metallizations for Electronic Devices, *Solid State Technol.*, Vol 2, p 109-115

Hoyt, J., 1987. Influence of Leg Shape and Solder Joint Metallurgy on Surface Mount Solder Joint Strength, *Brazing Soldering*, Vol 13 (No. 3), p 10-19

Humpston, G. and Jacobson, D.M., 1990. Solder Spread: A Criterion for Evaluation of Soldering, *Gold Bull.*, Vol 23 (No. 3), p 83-95

Humpston, G. *et al.*, 1992. Recent Developments in Silicon/Heat-Sink Assemblies for High-Power Applications, *GEC Rev.*, Vol 7 (No. 2), p 67-78

Illgen, L. *et al.*, 1991. Preparation of Ductile Al-Ge Soldering Foils by PFC Technique, *Mater. Sci. Eng.*, Vol A133, p 738-741

International Nickel, 1965. "Controlled Expansion and Constant Modulus Nickel-Iron Alloys," Publ. No. 2934

Jacobson, D.M. and Humpston, G., 1990. Assessment of Fluxless Solderability With Reference to Three Tin-Based Solders, *GEC J. Res.*, Vol 8 (No. 1), p 21-31

Jacobson, D.M. and Humpston, G., 1992. Diffusion Soldering, *Solder. Surf. Mount Technol.*, Vol 10 (No. 2), p 27-32

Jarman, R.A., Myles, J.W., and Booker, C.J.L., 1973. Interfacial Corrosion of Brazed Stainless Steel Joints in Domestic Tap Water, *Br. Corros. J.*, Vol 8 (No. 1), p 33-37

Jarman, R.A., Linekar, G.A.B., and Booker, C.J.L., 1973. Interfacial Corrosion of Brazed Stainless Steel Joints in Domestic Tap Water. II. Metallographic Aspects, *Br. Corros. J.*, Vol 10 (No. 3), p 150-154

Kang, K.S. and Ramachandran, V., 1980. Growth Kinetics of Intermetallic Phases at the Liquid Sn and Solid Ni Interface, *Scr. Metall.*, Vol 14 (No. 4), p 421-424

Kuhn, A.T., Rawlings, R.D., and May, R., 1984. A Potentiometric and Microstructural Study of the Corrosion of Silver-Brazed Stainless Steel Joints, *Brazing Soldering*, Vol 6 (No. 1), p 14-20

Leite, E.R., Varela, J.A., and Longo, E., 1992. Barrier Deformation of ZnO Varistors by Current Pulse, *J. Appl. Phys.*, Vol 72 (No. 1), p 147-150

Macek, W.H. and Davis, D.T.M., Jr., 1963. Rotation Sensing With Travelling-Wave Ring Lasers, *Appl. Phys. Lett.*, Vol 2, p 67-68

Marconi Electronic Devices Ltd., 1990. "Method of Joining Components," U.K. Patent Appl. GB 2 235 642 A

Mattox, D.M., 1973. Thin Film Metallization of Oxides in Microelectronics, *Thin Solid Films*, Vol 18, p 173-186

Mei, Z. and Morris, J.W., Jr., 1992. Characterization of Eutectic Sn-Bi Solder Joints, *J. Electron. Mater.*, Vol 21 (No. 6), p 599-607

MacIntosh, R.M., 1968. Tin in Cold Service, *Tin Uses*, Vol 72, p 7-10

Nielsen, K., 1984. Corrosion of Soldered and Brazed Joints in Tap Water, *Br. Corros. J.*, Vol 19 (No. 2), p 57-63

Prince, A., Raynor, G.V., and Evans, D.S., 1990. *Phase Diagrams of Ternary Gold Alloys*, Institute of Metals, London

Rogers, R.R. and Fydell, J.F., 1953. Factors Affecting the Transformation to Gray Tin at Low Temperatures, *J. Electrochem. Soc.*, Vol 100, p 383-387

Russel, K.C. and Smith, D., Ed., 1989. Physical Metallurgy of Controlled 'Invar' Type Alloys, TMS

Schmitt-Thomas, Kh.G. and Becker, R., 1988. Impurities in Solder Baths, *Brazing Soldering*, Vol 15 (No. 3), p 43-47

Schultze, W. and Schoer, H., 1973. Fluxless Brazing of Aluminum Using Protective Gas, *Weld. J.*, Vol 52 (No. 9), p 644-651

Stone, K.R. *et al.*, 1983. Mechanical Properties of Solders and Soldered Joints, *Brazing Soldering*, Vol 4 (No. 1), p 20-27

Taylor, P.D., 1987. *Thyristor Design and Realisation*, John Wiley & Sons

Thwaites, C.J. and McDowall, D., 1984. A Study To Determine the Effect on the Mechanical Strength of Soldered Joints Made to Brass of the Presence of Antimony in 40% Tin-60% Lead Solder, *Brazing Soldering*, Vol 6 (No. 1), p 32-36

Tomlinson, W.J. and Bryan, N.J., 1986. The Strength of Brass/Sn-Pb-Sb Solder Joints Containing 0 to 10%Sb, *J. Mater. Sci.*, Vol 21, p 103-109

Tomlinson, W.J. and Collier, I., 1987. The Mechanical Properties and Microstructures of Copper and Brass Joints Soldered With Eutectic Tin-Bismuth Solder, *J. Mater. Sci.*, Vol 22, p 1835-1839

Tomlinson, W.J. and Cooper, G.A., 1986. Fracture Mechanism of Brass/Sn-Pb-Sb Joints and the Effect of Production Variables on the Joint Strength, *J. Mater. Sci.*, Vol 21, p 1730-1734

Vianco, P.T., Hosking, F.M., and Rejent, J.A., 1990. Solderability Testing of Kovar With 60Sn-40Pb Solder and Organic Fluxes, *Weld. J.*, Vol 69 (No. 6), p 230s-240s

Wersing, W., 1992. Improving Ceramics Through Multilayer Technology, *Met. Mater.*, Vol 8 (No. 6), p 326-331

Williams, W.L., 1956. Gray Tin Formation in Soldered Joints Stored at Low Temperatures, *Proc. Symp. Solder*, STP 189, ASTM, p 149-157

Xian, A. and Si, Z., 1991. Wetting of Tin-Based Active Solder on Sialon Ceramic, *J. Mater. Sci. Lett.*, Vol 10 (No. 22), p 1315-1317

Yost, F.G., Ganyard, F.P., and Karnowsky, M.M., 1976. Layer Growth in Au-Pb/In Solder Joints, *Metall. Trans.*, Vol 7A (No. 8), p 1141-1148

Index

A

Abietic acid, used in soldering of fluxes, 157

Accelerated salt-spray test, 202

Activation energy for diffusion, 71

Active hydride process, 116

Adhesion tests, 184

Adhesive bonding, 1
 characteristic features, 3

Aging, accelerated, of components, 193-194

Alkoxypropanol, 172-174

Alloy constituents, conversion between weight and atomic fraction, 103

Alloy constitution, 71-72

Alloy J, as solder, applications, 59

Alloy Phase Diagram International Commission (APDIC), 73

Alumina, 139, 161, 167
 chromium metallizations, 244
 copper-alumina joints, 170, 171

Aluminum
 as addition to gold-bearing filler metals, 44
 boiling/sublimation temperature at 10^{-10} atm pressure, 151
 copper-aluminum alloy ($CuAl_2$), 195
 critical concentration threshold for embrittlement, 230
 electrode potential at 25 degrees C, 51
 fluxless brazing, 155-156
 furnace brazing, 146
 impurity element addition affecting braze, 229
 in aluminum-bearing filler metals, 52
 in brazes, 5, 16
 in brazes, bismuth and beryllium effect, 231
 in composition of commercially available rapidly solidified filler metals, 49
 in composition of zinc-bearing solders, 52, 53
 in metallization for silicon semiconductors, 253
 lowest impurity concentration producing detrimental effects in Pb-60Sn solder, 231
 reduction of surface tension when melted, as produced by additions, 114
 solidification shrinkage (% of solid), 127
 X-ray maps for concentration shown in SEM image, 201

Aluminum alloys
 aging behavior of that used for a ring laser gyroscope, 238, 240
 joining using brazes, 34
 linear expansivity at room temperature, 19
 some "unsolderable" in air, 157
 substrate for metal combinations used for diffusion brazing/soldering, 129

Aluminum-bearing brazes
 applications, 52
 melting range for various compositions, 52
 problems associated with use on aluminum alloy components, 51-52

Aluminum-bearing filler metals, as brazes, 50-52

Aluminum-copper-silicon alloys, brazes, 201-202

Aluminum-copper-silicon alloys, specific types
 Al-4Cu-10Si, melting point of brazes, 5
 68Al-27Cu-5Si, melting point and problems encountered, 7

Aluminum fluoride, 162

Aluminum-germanium alloys, eutectic point, 95, 253

Aluminum-germanium alloys, specific types
 48Al-52Ge, melting point and problems encountered, 7

Aluminum-magnesium alloys, specific types
 36Al-37Mg, melting point and problems encountered, 7

Aluminum-molybdenum-silicon phase diagram, isothermal section showing equilibrium, 250, 251

Aluminum nitride, 166

Aluminum oxide
 Gibbs free energy of formation at room temperature, 147
 oxidation, 150

Aluminum-silicon alloys
 eutectic point, 253
 phase diagram, 50, 51

Aluminum-silicon alloys, specific types
 Al-12Si, 250-252
 Al-12Si, filler alloy and surface roughness, 14
 Al-12Si, metallographic examination, 195
 Al-12Si, silicon wafer eroded by, 251-252
 Al-13Si, additions which promote wetting, 155

Al-13Si, cleaning agents, 155
Al-13Si, eutectic composition alloy, 50
Al-13Si, joined with aluminum alloys, 32
Al-13Si, melting range, 49
Al-13Si, structure, 49
Al-13Si, typical applications, 49
Al-13Si, wetting tungsten in example, 80
Aluminum-silver alloys, for determining phase diagrams, 91-92
Aluminum-zinc alloys, phase diagram, 52, 53
Aluminum-zinc alloys, specific types
  Al-94Zn, zinc-bearing solder, 52
  6Al-94Zn, melting point and problems encountered, 7
Amine hydrohalides, 158
Annealing, stainless steels, 28
Annular joint, 208
Antimony
  as addition to lead-tin solders, 67
  additions causing embrittlement in brazes, 128
  boiling/sublimation temperature at 10^{-10} atm pressure, 151
  energy of formation with solid copper, 11
  gold-antimony alloys as solders, 56, 58
  gold-antimony-tin system, isothermal section at 25 degrees C, 240, 241
  in composition of commercially available rapidly solidified filler metals, 49
  in composition of tin alloy solders, 58
  in composition of zinc-bearing solders, 53
  in solders, 66
  in solders, causing voids in brass components, 230
  lead-antimony-tin solders, 63, 230
  lowest impurity concentration producing detrimental effects in Pb-60Sn solder, 231
  solidification shrinkage (% of solid), 127
  volume expansion on freezing, 127
  ZnSb pest acceleration, 232
Antimony-tin alloys, phase diagram, 62
Antimony-tin alloys, specific types
  Sb-95Sn, gold addition effect on solder, 234
  Sn-5Sb solder, 240-241
Area-of-spread measurement (tests), 186, 187, 231
Arrhenius relationship, 212
Arrhenius-type rate relationships, 17
Arsenic
  as addition to brazing alloy to promote wetting, 155
  gallium arsenide, soldering of, 117
  lowest impurity concentration producing detrimental effects in Pb-60Sn solder, 231
Atmospheres. See *Joining atmospheres*
Atomic fraction, conversion from weight of constituents of alloys, 103
Axially loaded butt joints, 133, 134, 135, 138

B

Backscatter of beta particles, 184
Barium
  addition to brazing alloy to promote wetting, 155
Basic unit, 250
Bending tests, 209, 210, 229-230
Berthoud equation, 17
Beryllium
  activating agent for fluxless brazing of aluminum, 155
  as addition to lead-tin solders, 68
  contact angle reaction with brazing alloys, 11
  effect on wettability of aluminum- and silicon-containing brazes, 231
  joined in a vacuum furnace, 147
  silver as braze for bonding, 33
  "unsolderable" in air, 157
Beryllium oxide, Gibbs free energy of formation at room temperature, 147
BHN scale, 217
Binary solder alloys, spread characteristics, 15
Bismuth
  additions causing embrittlement in brazes, 128
  additions to brazing alloy to promote wetting, 155
  as addition to lead-tin solders, 67
  boiling/sublimation temperature at 10^{-10} atm pressure, 151
  effect on wettability of aluminum- and silicon-containing brazes, 231
  in composition of commercially available rapidly solidified filler metals, 49
  in composition of tin alloy solders, 58
  in low-melting-point eutectic composition alloys used as solders, 58
  in solders, 64-65, 66
  in solders, applications, 65
  lowest impurity concentration producing detrimental effects in Pb-60Sn solder, 231
  solidification shrinkage (% of solid), 127
  tin pest retardation, 232
  volume expansion on freezing, 127
Bismuth-containing solders, and wetting of nonmetals, 115
Bismuth-containing solders, specific types
  Bi-In, principal intermetallic phases formed by alloying between the common binary solders and engineering parent metals and metallizations, 64
  Bi-33In eutectic solder, 242
  Bi-Pb, principal intermetallic phases formed by alloying between the common binary solders and engineering parent metals and metallizations, 64
  Bi-44Pb, joint fill ratio as a function of joint width, 124
  Bi-Pb-Sn system, liquidus surface, 63

Index

silver-bismuth (Ag-Bi) alloys, as solder, 64

Bismuth-tin solders
   mechanical properties of soldered joints to copper, 236
   phase diagram, 60
   principal intermetallic phases formed by alloying between the common binary solders and engineering parent metals and metallizations, 64
   shear strength of soldered joints to brass, 236

Bismuth-tin solders, specific types
   Bi-42Sn, 212, 213
   Bi-43Sn, 127-128
   Bi-43Sn, form of the filler metal, 22
   Bi-43Sn, gold addition effect on solder, 234
   Bi-43Sn, joint fill ratio as a function of joint width, 124
   Bi-43Sn, melting range, 49
   Bi-43Sn, structure, 49
   Bi-43Sn, typical applications, 49

Blister test, 184

BNi-1, nickel-base braze, 47

Boiling/sublimation temperature of selected elements at 10^{-10} atm pressure, 151

Bolometers, 24

Borates, 159

Borides, formation in joint gap of nickel-bearing filler metal brazes, 48

Boron
   as addition to gold-nickel brazes, 46
   as addition to nickel-bearing filler metals, 47
   as addition to pure copper brazes, 33
   depressing melting point of brazes to increase joint strength, 128
   in composition of commercially available rapidly solidified filler metals, 49
   incorporated into metallic coatings in wet plating, 140
   nickel-boron braze, 127, 129, 130

Boron nitride, 166

Borosilicates, 159

Bow distortion, 118-119, 121

Brass components, antimony-containing solders causing voids, 230

Brasses
   as brazes, 36, 37
   degradation in a vacuum, 150

Brazes
   aluminum-bearing filler metals, 50-52
   compared to solders, 5-8
   copper, pure, 33-34
   copper-manganese-tin alloys, 50
   copper-nickel-tin-phosphorus alloys, 50
   derivation of term, 5
   distinguishing features, 7
   gold-bearing filler metals, 43-46
   gold-copper, 44-45
   gold-nickel, 44, 45-46
   gold-palladium, 44, 46
   important consequences, 7
   low-melting point, 7
   nickel-bearing filler metals, 47-50
   palladium-bearing filler metals, 46-47
   principal braze alloy families and their melting ranges, 6
   rapid solidification casting, 48
   replacement with solder at a joint, 119
   silver, pure, 33
   silver-copper eutectic compositions, 34-36
   silver-copper-zinc-cadmium alloys, 39-42
   silver-copper-zinc-tin alloys, 42-43
   temperature range for applications, 8
   zinc-bearing solders, 52-54

Braze-"tinned" surfaces, 116

Brazing, aluminum, 155-156

Brinell hardness tests, 217, 218

Buoyancy, 191

Butt and lap joints, mechanical testing, 202-203, 211

Butt joints, 202-203, 211
   axially loaded, 133, 134, 135, 138
   corrosion in copper pipes conveying tap water, 233
   fatigue curves, 215

C

Cadmium
   banned for filler alloys and plumbing fittings for drinking water appliances, 232
   boiling/sublimation temperature at 10^{-10} atm pressure, 151
   electrode potential at 25 degrees C, 51
   energy of formation with solid copper, 11
   in composition of silver-copper-zinc-cadmium brazing alloys, 41
   in composition of zinc-bearing solders, 52, 53
   in fluxes and fillers, joining atmosphere, 25
   in solders, health hazards, 58
   in solders and brazes, 5
   lowest impurity concentration producing detrimental effects in Pb-60Sn solder, 231
   silver-copper-zinc-cadmium alloys as brazes, 39-42, 77, 135, 215, 229-230
   solidification shrinkage (% of solid), 127
   volatile materials in metallic components and filler metals, 126

Cadmium containing alloys, specific types
   62Cd-38Cu, melting point and problems encountered, 7

Cadmium-free carat gold "solders", melting range of various compositions, 43

Calcium, activating agent for fluxless brazing of aluminum, 155

Principles of Soldering and Brazing

Calorimetric thermal analysis (CTA), 96, 104, 106-107, 108
    rate of heating or cooling of a specimen while it is undergoing a phase change, 108
    time/temperature heating curves, 107

Capillary forces, 10-11, 14-15, 21

Carat gold brazing alloys, melting range of various compositions, 43

Carat gold "solders," 43

Carbide cutting tip, brazed in steel tool, 204, 205

Carbides
    formation in joint gap of nickel-bearing filler metal brazes, 48
    gold-nickel brazes, 45

Carbon
    as addition to nickel-bearing filler metals, 47
    in composition of commercially available rapidly solidified filler metals, 49

Carbon fibers, chopped, incorporated into fillers, 139

Carbon monoxide gas atmospheres, 151, 152

Carburizing surface treatment, 115

Cast iron
    graphite inclusions and wetting of metals, 114
    linear expansivity at room temperature, 19
    silver-copper-zinc alloys for brazes with manganese, 38
    wetting constraints, 117

Cathodic sputtering, 140

Cemented carbides, silver-copper-zinc alloys for brazes, 38

Ceramics
    chromium metallizations, 244
    engineering, joined using reactive filler alloys, 164
    interlayers used to reduce stress from thermal expansion mismatch, 120-121
    mechanical constraints, 119
    metallographic examination, 196
    sintered, for packaging of charge coupled devices, 242-243
    solders for, 7
    spreading, 11
    substrate for finding surface tension of molten filler, 192
    wetting by electroplating, 116
    wetting of, 11, 164-166

CFC-113, 174

Chadwick test, 210, 211

Chamfering, 2

Charge coupled device (CCD), 241-242

Charpy impact test, 210

Chemical displacement, 140

Chemical etching, 172

Chemical fluxes, 155-164
    aluminum brazing fluxes, 161-162
    aluminum gaseous fluxes, 162
    aluminum soldering fluxes, 161
    application methods, 158
    brazing fluxes, 159-160
    chloride-based aluminum brazing fluxes, 161-162
    chloride-base fluxes, 161
    classification of soldering fluxes using the method adopted by the International Organization for Standardization, 159
    commercial designation of fluxes used for soldering, 158
    fluoride-based aluminum brazing fluxes, 162
    fluxes for aluminum and its alloys, 160-162
    organic fluxes, 161
    self-fluxing filler alloys, 162-163
    shelf life, 156
    soldering fluxes, 157-159
    ultrasonic fluxing, 163-164

Chemical vapor deposition (CVD), 140
    advantages and disadvantages, 142
    characteristic features, 141
    coating quality parameters, 142
    coatings for nonmetallic components, 115
    for applying $MoSi_2$ coating to silicon semiconductors, 252
    process parameters, 142
    to apply wettable metallizations, 114-115

Chill-block melt spinning, nickel-bearing filler metal brazes, 48

Chlorofluorocarbons (CFCs), 154-155
    to remove flux residues, 157

Chromium
    as addition to gold-bearing filler metals, 44
    as addition to gold-nickel brazes, 45, 46
    as addition to pure copper brazes, 33
    boiling/sublimation temperature at 10^{-10} atm pressure, 151
    in composition of commercially available rapidly solidified filler metals, 49
    in composition of palladium-bearing filler metals, 47
    in solders to form a eutectic alloy for nonmetals, 115
    insoluble in most solders, 13
    metallization for CCD devices, on glass and alumina, 244
    metallizing of a nonmetallic material, 115
    to help wetting of engineering ceramics, 164
    "unsolderable" in air, 157

Chromium oxide
    Gibbs free energy of formation at room temperature, 147
    oxidation, 150
    reduction, 152

Clamping, 1, 2-3

Classification schemes, chemical fluxes, 158

Clausius' theorem, 176

Index

Climatic test, typical conditions used in accelerated aging assessments, 194

Cobalt
- addition to brazing alloy to promote wetting, 155
- addition to pure copper brazes, 33
- boiling/sublimation temperature at 10^{-10} atm pressure, 151
- in composition of commercially available rapidly solidified filler metals, 49
- in composition of palladium-bearing filler metals, 47
- substrate material for copper-lead-tin solder, 86

Cobalt alloys, substrate for metal combinations used for diffusion brazing/soldering, 129

Cobalt alloys, specific types
- Co-19Cr-19Ni-8Si-1B, melting range, 49
- Co-19Cr-19Ni-8Si-1B, structure, 49
- Co-19Cr-19Ni-8Si-1B, typical applications, 49

Coefficients of thermal expansion (CTE), 237, 238

Coffin-Manson formula, 213

Columbium
- addition to Ag-Cu-Ti alloys, 169, 171
- joined in a vacuum furnace, 147

Compliant structures, to reduce stress from thermal expansion mismatch, 121-123

Congruently melting compounds, 95

Conservation of Energy, Principle of, 175

Constitution, of alloys, 71-72

Contact angle, 9-12, 145, 164-166, 168
- between molten filler and testpiece from surface tension and wetting force, 192
- calculation of, 187
- of droplets, 226
- effect on fillet formation and joint filling, 12
- effect on spread ratio and spread factor, 189-190
- equilibrium, 12
- measurement of, 189
- nonequilibrium, 12
- time dependence of, 11

Cooling rate, relationship to grain size in cast alloys, 97

Copper
- addition to aluminum-bearing brazes, 51
- addition to gold-nickel brazes, 46
- boiling/sublimation temperature at 10^{-10} atm pressure, 151
- brazes, pure, 33-34
- electrode potential at 25 degrees C, 51
- filler metal for aluminum alloys, diffusion brazing/soldering, 129
- high-purity, for interlayers in brazed joints, 120
- in aluminum-bearing filler metals, 52
- in composition of cadmium-free carat gold "solders," 43
- in composition of commercially available rapidly solidified filler metals, 49
- in composition of industrial gold brazing alloys, 43
- in composition of palladium-bearing filler metals, 47
- in composition of silver-copper-zinc-cadmium brazing alloys, 41
- in composition of silver-copper-zinc-tin brazing alloys, 42
- in composition of tin alloy solders, 58
- in composition of zinc-bearing solders, 52, 53
- joined by lead-tin solders, 67
- joined with soldering fluxes, 157
- lowest impurity concentration producing detrimental effects in Pb-60Sn solder, 231
- melting point, 33
- metallization, substrate for metal combinations used for diffusion brazing/soldering, 129
- metallization of components examined metallographically, 195
- metallization systems based on, 141
- silver-copper eutectic brazes, 34-36
- silver-copper-phosphorus brazing alloys, 5
- silver-copper-phosphorus brazing alloys, corrosion of joints in tap water, 232-233
- silver-copper-zinc brazing alloys, 36-40
- silver-copper-zinc brazing alloys, corrosion of joints in tap water, 232, 233
- silver-copper-zinc brazing alloys, for joints in stainless steel pipes, 233
- silver-copper-zinc-cadmium alloys as brazes, 39-42
- silver-copper-zinc-tin alloys as brazes, 42-43
- solderability rate in presence of impurity elements, 231
- solidification shrinkage (% of solid), 127
- solid solutions formed with silver-lead solders, 5
- soluble in lead-tin solders, 83
- substrate material for copper-lead-tin solder, 86
- surface coating to help diffusion soldering, 130
- tin pest acceleration, 232
- used as wettable metallization, 114
- water installations and corrosion of joints, 232-233
- X-ray maps for concentration shown in SEM image, 201

Copper abiet, 158

Copper alloys
- brazes of self-fluxing filler alloys, 162
- linear expansivity at room temperature, 19
- substrate for metal combinations used for diffusion brazing/soldering, 129

Copper-alumina joints, 170, 171

Copper-aluminum alloys, specific types
- Al-4Cu-10Si, 5
- $CuAl_2$, 195

Copper-antimony alloys, specific types
- 24Cu-76Sb, melting point and problems encountered, 7

Copper-gold alloys, as brazes, 44-45

Copper-lead-tin alloys, 83-84
- phase diagram, 84

Copper-manganese-tin alloys, as brazes, 50

261

Copper-managanese-tin alloys, specific types
　　Cu-10Mn-30Sn, melting range, 49
　　Cu-10Mn-30Sn, structure, 49
　　Cu-10Mn-30Sn, typical applications, 49
Copper-nickel alloys, 5
Copper-nickel alloys, specific types
　　Cu-50Ni, filler metal for aluminum alloys, diffusion brazing/soldering, 129
Copper-nickel-tin-phosphorus alloys, as brazes, 50
Copper-nickel-tin-phosphorus alloys, specific types
　　Cu-10Ni-4Sn-8P, melting range, 49
　　Cu-10Ni-4Sn-8P, structure, 49
　　Cu-10Ni-4Sn-8P, typical applications, 49
Copper oxide
　　chemical reactions with soldering fluxes, 157-158
　　reduction in vacuum, 150
Copper-phosphorus brazes, 162
　　phase diagram, 33, 34
Copper-phosphorus brazes, specific types
　　Cu-15P, melting range, 49
　　Cu-15P, structure, 49
　　Cu-15P, typical applications, 49
Copper-silver alloys
　　brazes, 165
　　eutectic alloys with titanium, 168
Copper-silver alloys, specific types
　　Ag-28Cu eutectic braze, 75, 76, 81, 139
　　Ag-Cu-Au system, 81-82
　　Ag-Cu-Cd-Zn braze, 135
　　Ag-Cu-Hf braze, 169
　　Ag-Cu-Ti, phase diagram, 167
　　Ag-Cu-5Ti, 166
　　Ag-27Cu-2Ti, 165, 166, 167
　　Ag-66Cu-22Ti, 167
　　Ag-Cu-Zn-Cd brazes, 77, 215
　　Ag-Cu-Zn-Cd brazes, impurity element effect, 229, 230
Copper-tin alloys, 132
　　corrosion of solders at joints in tap water, 233
　　diffusion soldering, 130-131
　　intermetallic compounds, 85-86, 204
　　phase diagram, 83
Copper-tin alloys, specific types
　　Cu-20Sn, melting range, 49
　　Cu-20Sn, structure, 49
　　Cu-20Sn, typical applications, 49
Copper-tin-phosphorus braze, 162
Copper-titanium brazes, specific types
　　Cu-5Ti, reaction layer thickness as function of brazing temperature for Si_3N_4, 169
　　Cu-5Ti, reaction layer thickness as function of brazing time for Si_3N_4, 168
Copper-tungsten components, interlayers to reduce mismatch stress concentration, 121
Copper-zinc alloys
　　brazes, 36, 37
　　melting point, 36
　　phase diagram, 36, 37
Coring, 74, 75
Corrosion, 194, 202, 216
　　aluminum-bearing brazes resistant, 50, 51
　　ammonia-containing solder fluxes for brass components, 156
　　avoided by use of soldering fluxes, 157
　　carbide cutting tip brazes in a steel tool, 204, 205
　　CCD component, 242
　　fluxless brazing of aluminum, 155
　　gold-bearing filler metals resistant, 44
　　gold-copper-brazes, 45
　　in tap water, of soldered and brazed joints, 232-233
　　palladium-bearing filler metals resistant, 47
　　scanning electron microscopy, 201
　　zinc-bearing solders, 54
　　zinc resistant to aluminum alloys, 52
Corrosion fatigue, 216
Crack, maximum stress at end of, 135
"Cracked" or dissociated ammonia gas atmospheres, 151
Creep, 211-212, 213
Creep tests, 182, 183, 212
Crimping, 1, 2
Cuprous oxide, Gibbs free energy of formation at room temperature, 147
Cupric oxide, Gibbs free energy of formation at room temperature, 147
Cycles to failure, 215
Cyclohexanol, boiling, thermal cycling tests, 206

D

Damp heat aging test, typical conditions used in accelerated aging assessments, 194
Data sheet, for a defluxing agent, listing health, safety, and environmental information, 172-174
Design life, 79
Developer, 219
Dewetting, 115, 116
Dew point, 152-153, 154
Diamond/silica/carbide fiber, linear expansivity at room temperature, 19
Differential scanning calorimetry (DSC), 104, 105-106
　　time/temperature heating curves, 105
Differential thermal analysis (DTA), 104-105, 106
　　time-temperature heating curves, 104-105
Differential thermocouple

Index

to detect departure by specimen from thermal equilibrium for DTA and DSC, 104, 105
to maintain a constant temperature gradient across the barrier in CTA, 106

Diffusion, 75
    activation energy for, 71
    rate of, 71

Diffusion bonding, 1, 127, 128
    characteristic features, 4
    transient-liquid phase, 5

Diffusion brazing, 47, 128-132

Diffusion coefficients, 71

Diffusion soldering, 127, 128-132
    silicon semiconductors, 252, 253, 254

Dip coatings, of gold, 115

Dip method of brazing, 162

Dipping methods, to apply tin coatings, 114

Dissociation pressure of oxide, 148

Dissolution, 74

Dissolution rate
    engineering parent metals and metallizations in lead-tin solder with temperature, 84
    reduced by preloading the filler with parent metal, 117
    silver in silver-gold-tin alloys as solders, 89-90
    of solid metal in a molten metal, 17

Doping level, Al-12Si braze for a silicon wafer, 251

Dowanol PX-16S, data sheet, 172-174

Drying agents, 154

Dry heat test, typical conditions used in acclerated aging assessments, 194

Dwell stages, 28

E

Ecology data, 173

Edge dip method, 186, 188

Elastic limits, 118, 208

Electrical current, for welding, 4

Electrode potential, selected elements at 25 degrees C, 51

Electroless plating, 140
    advantages and disadvantages, 142
    characteristic features, 141
    coating quality parameters, 142
    process parameters, 142

Electron beam bombardment, 140

Electron beams, for welding, 4

Electroplating, 22, 26, 195
    advantages and disadvantages, 142

application method for lead-tin solders, 67
ceramics, 116
characteristic features, 141
coating quality parameters, 142
for applying submicron layers of palladium and rhodium to silicon semiconductors, 252
of gold, 115
gold and nickel layers for ring laser gyroscope, 239
in flip-chip bonding process, 123
process parameters, 142
pure copper brazes, 33-34

Ellingham diagrams, 148-154, 162

Elongation to failure, 208

Embrittlement
    antimony additions as cause in brazes, 128
    bismuth additions as cause in brazes, 128
    brazed joints by impurity elements, 229-230
    exact stoichiometric compounds, 89
    gold-lead-tin alloys for solders, 86, 88
    gold-silicon alloys as solders, 57
    of gold wettable metallizations, 115
    in presence of standard gas atmospheres, 147
    intermetallic, caused by diffusion of constituents, 181
    nickel-bearing filler metals as brazes, 47-48, 49
    silver-copper-zinc-tin brazes, 43
    tin alloy solders, 66
    tin-base solders, 117

Emission reduction/recycling/disposal, 174

Endurance limit, 215-216

Energy-dispersive analyzer, 199 200

Energy-dispersive X-ray analysis (EDAX), 199, 200, 201
    to identify phases present, 93

Energy of formation, 11

Entropy, 175-176

Entropy change, 176

Entropy increment, 148

Equal liquidus slope, method of, 92, 95, 96, 99-100, 101

Equilibrium, gold-nickel phase diagram, 74

Equilibrium constant, 178

Equilibrium contact angle, 12

Erosion, 74
    germanium, of a molten silver-aluminum-germanium alloy, 102
    of gold, by molten tin as a function of reaction time and temperature, 79
    indium-base solders, 66
    lead-tin solders, 67
    nickel-bearing filler metal brazes, 48
    as process constraint, 117-118
    silicon wafer with an Al-12Si braze, 251-252
    of silver, by molten tin as a function of reaction time and temperature, 78

of silver by tin in silver-gold-tin alloys, 89
silver-tin eutectic solder, 77
to physically remove a surface coating, 157
Eutectic alloys, 5
high-melting-point solders and low-melting-point brazes, 7
Eutectic composition, filler metals, characteristics for preference, 76-77
Eutectic point, prediction of composition in high-order systems, 95-101
Eutectiferous solidification, 91
Evaporation, 22
Exact stoichiometric compounds, 89

F

Fast atom bombardment, 140
Fatigue, 212-216
cracking, 200-201
Fatigue failure, 212
Fatigue life, 202, 213
Fatigue strength, 212-216
Fatigue tests, 182, 183, 202
Ferric oxide, Gibbs free energy of formation at room temperature, 147
Ferrous oxide, Gibbs free energy of formation at room temperature, 147
Field-assisted brazing, 167-168
Filler alloys, 31-69
features influencing manner and extent of flow into the joint, 8
solder vs. braze systems, 32-33
spreading characteristics, 13-14
storage shelf life, 32
Filler metals, for welding, 4
Fillet formation, hermetic seals, 242
Fillets, 3
contact angle effect on formation, 12
influence of, 205-206
Fill ratio, 124
Finite element analysis (FEA), soldering and brazing applications, 111-112, 113
First Law of Thermodynamics, 175
Flash point, 172
Flip-chip bonding process, 122-123
Fluid flow, 12-13
mean velocity of, 13
Fluoborate fluxes, 159-160
Flux carrier, 158

Fluxes, "no clean," 159
Fluxes, true, 155
Flux-free joining, 154
Fluxless brazing processes
aluminum-bearing brazes, 52
clad materials, 51
Fluxless joining process, 91-92
Focal spot size, 220, 222
Focusing cup, 220
Formation, energy of, 11
Free energy change, 148, 149
Free energy curves, 148
Free-machining steels, wetting constraints, 117
Frost point, 26
Furnace joining
advantages, 146
disadvantages, 146-147

G

Gallium
addition to Au-20Sn solder, 117
in composition of cadmium-free carat gold "solders," 43
Gallium arsenide
soldering using Au-20Sn with 3.4% Ga by preloading filler, 117
tensile strength of soldered joints, 206
Gas Law, 177
GEC Meniscograph solderability tester, 192, 193
Geometric unsharpness, 220, 222
Germanium
aluminum-germanium alloys, eutectic point, 95, 253
aluminum-silver-germanium alloys, determination of phase diagrams, 91-101
components metallographically examined, 195
contact angle on silicon carbide, 11
gold-germanium alloys, as solders, 56, 57, 58
Gibbs free energy, 11, 91, 94-95, 152
change as measure of strength of metal-oxygen chemical bond, 147-148, 149
driving force of a chemical reaction, thermodynamic function, 175
gaseous constituents, 177-178
metallization of zinc oxide and oxidizing reaction, 244
reactions possible between silicon nitride and titanium, 165
Gibbs free energy change, 153, 165, 177-178
Gibbs free energy function, dependence on pressure, 176-178

Gibbs free energy of formation, wetting of nonmetals, 115

Glass, linear expansivity at room temperature, 19

Glass carrier, 159

Glass frits, fired-on, 115

Gold. See also *Gold-bearing filler metals*
  addition to the silver-tin eutectic solder, 118
  boiling/sublimation temperature at 10^{-10} atm pressure, 151
  coatings that dissolve in solder bath and wetting balance testing, 193
  crystal structures and primitive rhombohedral cells of selected phases in Au-Pb-Sn system, 94
  effect on common tin-base eutectic solders, 234-235
  electrode potential at 25 degrees C, 51
  erosion by molten tin as a function of reaction time and temperature, 79
  filler metal for aluminum alloys, diffusion brazing/soldering, 129
  in composition of cadmium-free carat gold "solders," 43
  in composition of commercially available rapidly solidified filler metals, 49
  in composition of industrial gold brazing alloys, 43
  in solders, 5
  metallization systems based on, 141
  silver-gold, 5
  silver-gold-copper system, 81-82
  silver-gold-palladium alloys, 5
  solidification shrinkage (% of solid), 127
  sputter deposition, 115
  as surface coating for fluxless joining processes, 147
  surface coating to help diffusion soldering, 130
  used as wettable metallization, 114
  wettable metallization overlay for nonmetals, 116

Gold alloys, microhardness profile across a brazed joint, 218

Gold-antimony alloys
  phase diagram, 56
  as solders, 56, 58

Gold-antimony alloys, specific types
  Au-25Sb, eutectic temperature, 54
  75Au-25Sb, melting point and problems encountered, 7
  Au-Sb-Sn system, isothermal section at 25 degrees C, 240, 241

Gold-bearing filler metals
  applications, 43-44
  as brazes, 43-46
  color of various compositions, 43
  families, 44
  properties showing superiority to base-alloy brazes, 44
  as solders, 54-58
  as solders, melting range, 54

Gold-bearing solders, compositions and eutectic temperatures, 54

Gold-copper alloys
  as brazes, 44-45
  melting point of brazes, 44-45
  phase diagram, 44

Gold-copper alloys, specific types
  Au-20Cu braze, 44

Gold-copper-manganese alloys, specific types
  Au-34Cu-16Mn-10Ni-10Pd, gold-nickel braze, 45

Gold-copper-nickel alloys, specific types
  Au-16.5Cu-2Ni braze, 45

Gold-germanium alloys
  phase diagram, 56
  as solders, 56, 57, 58

Gold-germanium alloys, specific types
  Au-12Ge, eutectic temperature, 54
  88Au-12Ge, melting point and problems encountered, 7

Gold-indium alloys
  intermetallic phase continued growth, 79
  phase diagram, 65

Gold-indium alloys, specific types
  58Au-42In, melting point and problems encountered, 7

Gold-indium-tin ternary phase diagram, isothermal section, 238-239

Gold-lead-tin alloys
  applications, 86
  crystal structure and primitive rhombohedral cells, 93-94
  in solders, 86-90
  liquidus projection, 87
  phase diagram determination, 94, 95
  vertical section between eutectic tin-lead composition and gold, 87

Gold metallizations, substrate for metal combinations used for diffusion brazing/soldering, 129

Gold-nickel alloys
  applications, 45
  brazes, 44, 45-46, 91
  melting range of brazes, 45
  phase diagram, 45, 46, 73, 74
  target properties of brazes defined, 46
  with nickel components, 73

Gold-nickel alloys, specific types
  Au-18Ni, 76
  Au-18Ni, gold-nickel braze, 46
  Au-38Ni, 74
  Au-48Ni, level rule applied to solidification of Au-70Ni alloy, 75
  Au-50Ni, as braze, 73, 74
  Au-54Ni, lever rule applied to solidification of Au-70Ni alloy, 75
  Au-61Ni, lever rule applied to solidification of Au-70Ni alloy, 75
  Au-66Ni, 74

Au-70Ni, 74
Au-70Ni, lever rule applied to solidification of Au-70Ni alloy, 75
Au-74Ni, lever rule applied to solidification of Au-70Ni alloy, 75
Au-78Ni, lever rule applied to solidification of Au-70Ni alloy, 75
Au-84Ni, lever rule applied to solidification of Au-70Ni alloy, 75

Gold oxide, Gibbs free energy of formation at room temperature, 147

Gold-palladium alloys
  as brazes, 44, 46
  melting point, 46

Gold-silicon alloys, 163
  decomposition, 57
  phase diagram, 55
  as solders, 54-57

Gold silicon alloys, specific types
  Au-2Si solder-based, for silicon chips, 28
  Au-3Si solder, eutectic temperature, 54
  Au-3Si solder, for die attachment, 245
  Au-3Si solder, gold-bearing filler metal solder, 43
  Au-3Si solder, melting point, 5, 245
  Au-3Si solder, melting range, 49
  Au-3Si solder, spreading, 245, 246
  Au-3Si solder, spreading characteristics improvement, 246
  Au-3Si solder, structure, 49
  Au-3Si solder, typical applications, 49
  Au-3Si solder, wetting, 245
  97Au-3Si, melting point and problems encountered, 7

Gold-silicon-tin alloys, phase diagram, 246
Gold-silicon-tin alloys, specific types
  Au-2Si-8Sn, titanium addition promoting wetting and spreading on bare silver, 247

Gold-tin alloys
  crystal structures and primitive rhombohedral cells of selected phases in Au-Pb-Sn systems, 94
  phase diagram, 55
  as solders, 55, 57-58
  as solders, applications, 57-58

Gold-tin alloys, specific types
  Au-20Sn, 117
  Au-20Sn, eutectic solder, 57
  Au-20Sn, eutectic temperature, 54
  Au-20Sn, for die attachment, 245, 247
  Au-20Sn, gold-bearing filler metal solder, 43
  Au-20Sn, melting point, 245
  Au-20Sn, melting range, 49
  Au-20Sn, structure, 49
  Au-20Sn, titanium addition promoting wetting and spreading on bare silicon, 247
  Au-20Sn, typical applications, 49

Grain size, relationship to cooling rate in cast alloys, 97

Graphite
  gold-nickel brazes, 45
  linear expansivity at room temperature, 19

Gray cast iron, chemical pretreatment before wetting, 117

Gray tin, 49, 231-232

H

Hafnium
  addition to fillers for ceramics, 11
  addition to solders for materials containing glass or ceramic phases, 91
  filler metal for brazes, 36
  to help wetting of engineering ceramics, 164

Hafnium-containing brazes, Ag-Cu-Hf alloy, 169

Hafnium nitride, 169

Halide reduction, 140

Hardness measurements, 216, 217-219

HB scale, 217

Health hazard data, 173

Heat-affected zone (HAZ), in welding, 4

Heating cycles, 26-28

Heat-resistant alloys, brazes of self-fluxing filler alloys, 162

Heat-resistant steels, gold-nickel brazes, 46

Heat sinks, 253-254

Heavy metal deposition, 161

Hermetic seals
  applications, 241
  design of soldered joints, 241-244

High homologous temperatures, 212, 213

Homologous temperature (T/Tm), 212, 216

HV scale, 217-218

Hybrid joining processes, 128

Hydrazine hydrochloride, 158

Hydrogen, electrode potential at 25 degrees C, 51

Hydrogen fluoride, 162

Hydrogen gas atmospheres, 151-152

Hydrostatic force, 10-11

Hydrostatic tension, 133

I

IA (inorganic acid), 158

Image analysis, 189, 196, 197

Image analysis system, 100

Impact test, 205, 209, 210
Impurity element effects on properties of filler metals and joints, 229-232
    melting points, 234-235
    metallurgical reactions that vary with solder composition, 235-236
    solder composition influence on mechanical properties of soldered joints, 236
Inboard wire peel test, 184, 185
Indium
    addition to silver-copper-titanium alloys, 167
    boiling/sublimation temperature at 10^{-10} atm pressure, 151
    filler metal for aluminum alloys, diffusion brazing/soldering, 129
    gold-indium intermetallic phase continued growth, 79
    gold-indium phase diagram, 65
    gold-indium-tin ternary phase diagram, isothermal section, 238-239
    in composition of commercially available rapidly solidified filler metals, 49
    in composition of tin alloy solders, 59
    in low-melting-point eutectic composition alloys used as solders, 58
    in solders, 64, 65-66
    in solders, applications, 65-66
    silver-indium phase diagram, 60
    silver-indium solder, for semiconductors, 253
    solidification shrinkage (% of solid), 127
Indium-containing alloys, specific types
    Ag-In, 64
    Ag-97In, 124
    Bi-In, 64
    Bi-33In, eutectic solder, 242
Indium-lead alloys
    phase diagram, 80
    principal intermetallic phases formed by alloying between the common binary solders and engineering parent metals and metallizations, 64
    as solders, 59, 80
Indium-lead alloys, specific types
    In-50Pb, as a solder, 235
    In-50Pb, joint fill ratio as a function of joint width, 124
Indium-lead-tin alloys, specific types
    In-18Pb-70Sn, gold addition effect on solder, 235
Indium oxide, Gibbs free energy of formation at room temperature, 147
Indium-tin alloys
    phase diagram, 59
    principal intermetallic phases formed by alloying between the common binary solders and engineering parent metals and metallizations, 64
Indium-tin alloys, specific types
    In-48Sn, gold addition effect on solder, 234
    In-48Sn, joint fill ratio as a function of joint width, 124
    In-48Sn, as a solder, 238
Induction heating, 23
Industrial gold brazing alloys, melting range of various compositions, 43
Interlayers
    applications for reducing stress from thermal expansion mismatch, 121
    to reduce stress from thermal expansion mismatch, 120-121
Intermetallic compounds, silver-tin solder's reaction with silver, 90
Intermetallic phases, 79, 163, 167, 195, 204-205
    Cu_6Sn_5, 204
    and erosion effect on process constraints, 117
    metallization of silicon semiconductors, 253
    nickel-boron, 129-130
    slowing down rate of corrosion of copper-tin and silver-tin solders, 233
Internal energy, 175, 176
Invar, lead material for PCBs, 247-248, 249
Ion-aided process, 140
Ion beam deposition, 140
Ion implantation, 140
Ionized-cluster beam deposition, 140
Ion plating, 140
Iron
    addition to brazing alloy to promote wetting, 155
    boiling/sublimation temperature at 10^{-10} atm pressure, 151
    copper brazing, phase diagram, 73
    electrode potential at 25 degrees C, 51
    in composition of commercially available rapidly solidified filler metals, 49
    substrate material for copper-lead-tin solder, 86
Iron alloys, linear expansivity at room temperature, 19
Iron alloys, specific types
    Fe-12Cr-4B, filler metal for aluminum alloys, diffusion brazing/soldering, 129
Iron-nickel alloys, lead material for PCBs, 247-248, 249
Iron oxide, Gibbs free energy of formation at room temperature, 147
Iron-tin alloys, specific types
    $FeSn_2$, 5
Isothermal conditions, 9, 13
Isothermal solidification, 128

J

Jethete, gold-nickel brazes, 46
Joining atmospheres

bonding of windows on CCD packages, 243-244
chemically active gas atmosphere, 146
chemically inert gas atmosphere, 146
Ellingham diagram, joining in inert atmospheres and vacuum, 150-151
fluxless brazing of aluminum, 155-156
halide, 154-155
reducing atmospheres, 151-155
reducing atmospheres of common industrial metals, 154
and reduction of oxide films, 147
silicon chip die attachment under shroud of inert gas, 245
thermodynamic aspects of oxide reduction, 147-150

Joining cycle, duration of, 79

Joining environment, 145-178
chemical fluxes, 155-164
health, safety, and environmental aspects of soldering and brazing, 172-174
joining atmospheres, 145-156
reactive filler alloys, 164-172

Joining methods, 1-4

Joint area, strength as a function of, 123-128

Joint filling, 13

Joint fill ratio, 124

Joint gap, 13

Joint quality assessment, 181-226
mechanical testing of joints, 202-216
metallization quality evaluation, 183-186
metallographic examination, 195-196
microscopic examination of joints, 194-202
nondestructive evaluation of joints and subassemblies, 216-225
scanning acoustic microscopy (SAM), 195, 196-198, 199
scanning electron microscopy (SEM), 195, 198-202
schematic flow chart for a joining process, 182
sequence recommended within framework of qualifying a new bonding process, 181-183
wetting assessment, 186-194

K

Kelvin-Planck statement, Second Law of Thermodynamics, 176

Kirkendall effect, 127

Kirkendall porosity, 4, 204-205

Kirkendall voids, 181

Knoop hardness test, 218

Kovar, 139
leads material for PCBs, 247-248, 249
tin content effect on peel strength of joints, 248

L

Lap joints, 202-203, 209, 211
longitudinally loaded, 135-138
peel strength, 248-249

Lapping, 172

Laser heating, 23

Lasers, for welding, 4

Lead
banned for filler alloys and plumbing fittings for drinking water appliances, 232
boiling/sublimation temperature at 10^{-10} atm pressure, 151
crystal structures and primitive rhombohedral cells of selected phases in Au-Pb-Sn system, 94
electrode potential at 25 degrees C, 51
in composition of commercially available rapidly solidified filler metals, 49
in composition of zinc-bearing solders, 53
in lead-tin solders, 66-69
in low-melting-point eutectic composition alloys used as solders, 58
in solders, 5, 58-59, 64, 65-66
in solders, applications, 65
in solders, health hazards, 58
solidification shrinkage (% of solid), 127
tin pest retardation, 232

Lead-antimony-tin solders, 230
liquidus surface, 63

Lead-antimony-tin solders, specific types
Pb-40Sn-2Sb solder, 230

Lead-containing alloys
bismuth-lead alloys, specific types
Bi-Pb, 64
Bi-44Pb, 124
bismuth-lead-tin alloys
liquidus surface, 63
gold-lead-tin alloys
in solders, 86-90
phase diagram determination, 94, 95
indium-lead solders, 80
phase diagram, 80
indium-lead solders, specific types
In-Pb, 64
In-50Pb, 124, 235
indium-lead solders, specific types
In-18Pb-70Sn, gold addition effect, 235
silver-lead solders, 64
phase diagram, 62
silver-lead solders, specific types
Ag-97.5Pb-1Sn, tin content effect on peel strength in joints to Kovar, 248
tin-lead solders, 64, 117, 163, 182

Lead-copper-tin system, 83-84

Lead-indium-silver alloys, specific types
Pb-5In-2.5Ag, melting range, 49
Pb-5In-2.5Ag, structure, 49

Pb-5In-2.5Ag, typical applications, 49

Lead oxide, Gibbs free energy of formation at room temperature, 147

Lead-palladium alloys, specific types
  75Pb-25Pd, melting point and problems encountered, 7

Lead-tin solders, 32, 77
  additions, 67
  advantages, 66-67
  applications, 67-68
  cleaning method effect on wetting behavior, 192, 193
  composite solders, 68-69
  copper-tin intermetallic compound growth on a copper substrate varying with time, thickness, and/or temperature, 85-86
  degradation modes, 68
  dispersion strengthening for fatigue resistance, 68-69
  dissolution rate of engineering parent metals and metallizations with temperature, 84
  grain refinement for fatigue resistance, 68, 69
  health hazards, 58
  mechanical properties of solder joints to copper, 236
  metallographic examination, 195
  phase diagram, 61
  reaction with copper substrate following heat treatment, 85
  shear strength of soldered joints to brass, 236
  tensile strength of cast bars, 77
  used with copper components, 83-86
  viscosity at 50 degrees C, 76

Lead-tin solders, specific types
  Pb-4Sn, solder and jigging of the components, 21
  Pb-5Sn solder, 123
  Pb-60Sn, 83-84, 86-87, 89, 132, 139, 211
  Pb-60Sn, eutectic solder, very dense, 53
  Pb-60Sn, joint fill ratio as a function of joint width, 124
  Pb-60Sn, lowest impurity concentrations producing detrimental effects, 230-231
  Pb-60Sn, spreading tests, 190
  Pb-60Sn, tensile strength of cast bars, 76-77
  Pb-60Sn, tin content effect on peel strength in joints to Kovar, 248
  Pb-62Sn, melting range, 49
  Pb-62Sn solder, gold addition effect on solder, 234
  Pb-62Sn, structure, 49
  Pb-62Sn, typical applications, 49

Lever rule, 74, 75, 81-83, 90, 100

Liquid metal embrittlement (LME), 21

Liquid penetrant inspection, 216, 219, 225

Liquidus temperature
  filler alloy, 112-113
  of solder or braze, 31

Lithium
  activating agent for fluxless brazing of aluminum, 155
  addition to lead-tin solders, 68
  filler metal for brazes, 36
  fluxing agent for self-fluxing filler alloys, 162

Load/extension curve, 208, 211

Low-cycle fatigue, 213

Low-expansion alloys (Fe-Ni-base), linear expansivity at room temperature, 19

M

Magnesia, 161

Magnesium
  activating agent for fluxless brazing of aluminum, 155
  addition to aluminum-bearing brazes, 51
  boiling/sublimation temperature at 10^{-10} atm pressure, 151
  disadvantages of use as an activating agent, 155
  electrode potential at 25 degrees C, 51
  in solders and brazes, 5
  "unsolderable" in air, 157
  volatile materials in metallic components and filler metals, 126

Magnesium oxide, Gibbs free energy of formation at room temperature, 147

Magnesium vapor, from furnace brazing of aluminum, 146

Magnetic particle inspection, 219

Magnetron sputtering, 140

Manganese
  addition to gold-bearing filler metals, 44
  addition to gold-nickel brazes, 45
  addition to silver-copper-zinc-nickel alloys, 38-39
  addition to silver-copper-zinc-tin brazes, 43
  boiling/sublimation temperature at 10^{-10} atm pressure, 151
  in aluminum-bearing filler metals, 52
  in brazes, joining atmosphere, 25
  in composition of commercially available rapidly solidified filler metals, 49
  in composition of palladium-bearing filler metals, 47
  in composition of silver-copper-zinc brazing alloys, 39
  in composition of silver-copper-zinc-cadmium brazing alloys, 41
  in composition of silver-copper-zinc-tin brazing alloys, 42
  metallization systems based on, 141
  volatile materials in metallic components and filler metals, 126

Materials, role in defining process constraints, 111-143

Maximum exposure limit, 172

Maximum inhalable quantity, 172

Maximum load to fracture, 208

Mechanical agitation, gold-tin alloys, 58

Mechanical fastening, characteristic features, 1-3

Mechanical fluxing process, gold-tin alloys, 58

Mechanical properties, scope of, 181

Mechanical shock test, 182

Mechanical testing of joints
creep rupture strength, 211-212
devising test to furnish usable data, 202-203
establishing the magnitude of the stresses to which the joint will be subjected in service, 203-204
fatigue strength, 212-216
impact resistance, bending, and peeling strength, 209-211
influence of fillets, 205-206
reproducing the conditions under which the manufactured products are joined, 205
shear strength, 208-209
tensile strength, 203, 206-208
thermal history replication, 204-205

Melting point (T_m), 71
changes introduced by impurity element additions, 234-235

Meniscograph, 13

Meniscus rise as a function of time, 186, 187, 188

Meniscus rise end value determination, 186, 188

Mercury, filler metal for aluminum alloys, diffusion brazing/soldering, 129

Metallizations, 164
copper, alloys examined metallographically, 195
for hermetic seals of charge coupled devices, 242
for ring laser gyroscope, 239
for silicon semiconductors, 253
quality evaluation, 183-186
techniques, 140-143
titanium application, 166

Metal-loaded glass frits, fired on nonmetal component surfaces for metallization, 141-143

Metallographic examination, 195-196, 230
corrosion of copper pipework soldered and brazed joints in tap water, 232-233
joint between a CCD package and window, 244

Metalloids, in nickel-bearing filler metals, 50

Metal-matrix composites (MMCs), 133
nonmetallic components and wetting of metals, 114
reactive filler alloys, 164
wetting constraints, 117

Metal oxides, Gibbs free energy of formation at room temperature, 147

Metals, wetting of, 114-115

Method of equal liquidus slope, 92, 95, 96, 99-100, 101

Mica, water wetting on, surface energies, 10

Mild environment test, typical conditions used in accelerated aging assessments, 194

Mild steel
bending test, 209
copper brazing, phase diagram, 73
fatigue curves, 215
impurity element effect on braze filler alloy, impact strength, 230
lead-tin eutectic solder wetting behavior, 193
pure copper for joining brazes, 33
substrate for impact test on brazed T-joints, 205
zirconia/mild steel joints with Ag-Cu-3Ti, 170

Military specifications, MIL-STD-883, 183

Minimum practicable joining temperature, 112-113

Mischmetal, activating agent for fluxless brazing of aluminum, 155

Molar ratio, 152

Molybdenum
aluminum-molybdenum-silicon phase diagram, equilibrium shown, 250, 251
boiling/sublimation temperature at 10^{-10} atm pressure, 151
diffusion soldering of backing plates for silicon semiconductors, 250-254
joined in a vacuum furnace, 147
joined with aluminum alloy brazes, 51
metallization systems based on, 141

Molybdenum components, gold-plated, 81

Molybdenum disilicide, 250, 252

N

National Standards, 183, 192

Nernst-Shchukarev equation, 17

Nichrome, as metallization, 244

Nickel
addition to brazing alloy to promote wetting, 155
addition to gold-bearing filler metals, 44
addition to gold-copper brazes, 45
addition to pure copper brazes, 33
addition to silver-copper eutectic brazes, 36
addition to silver-copper-zinc brazes, 38, 39
addition to silver-copper-zinc-tin brazes, 43
application to prevent reaction with lead-tin solder, 117
boiling/sublimation temperature at 10^{-10} atm pressure, 151
effect on Ag-Cu-Zn brazed stainless steel pipe water installations, 233
electrode potential at 25 degrees C, 51
in composition of cadmium-free carat gold "solders," 43
in composition of commercially available rapidly solidified filler metals, 49

Index

in composition of industrial gold brazing alloys, 43
in composition of palladium-bearing filler metals, 47
in composition of silver-copper-zinc brazing alloys, 39
in composition of silver-copper-zinc-cadmium brazing alloys, 41
in composition of silver-copper-zinc-tin brazing alloys, 42
in composition of tin alloy solders, 58
in composition of zinc-bearing solders, 53
iron-nickel alloys, lead material for PCBs, 247-248, 249
metallizing of a nonmetallic material, 115
provides adhesion for alloys in ring laser gyroscope, 239
pure, in composition of nickel-bearing filler metals, 47
substrate material for copper-lead-tin solder, 86

Nickel alloys
   copper-nickel, 5
   gold-nickel braze, 91
   linear expansivity at room temperature, 19
   substrate for metal combinations used for diffusion brazing/soldering, 129

Nickel alloys, specific types
   Ni-4B braze, 129, 130
   Ni-4B braze, filler metal for aluminum alloys, diffusion brazing/soldering, 129
   Ni-15Cr-3B, melting range, 49
   Ni-15Cr-3B, structure, 49
   Ni-15Cr-3B, typical applications, 49
   Ni-14Cr-5Si-5Fc-3B, melting range, 49
   Ni-14Cr-5Si-5Fe-3B, structure, 49
   Ni-14Cr-5Si-5Fe-3B, typical applications, 49
   Ni-32Pd-8Cr-3B-1Fe, melting range, 49
   Ni-32Pd-8Cr-3B-1Fe, structure, 49
   Ni-32Pd-8Cr-3B-1Fe, typical applications, 49
   Ni-41Pd-9Si, melting range, 49
   Ni-41Pd-9Si, structure, 49
   Ni-41Pd-9Si, typical applications, 49
   Ni-10P, melting range, 49
   Ni-10P, structure, 49
   Ni-10P, typical applications, 49
   Ni-12P, filler metal for aluminum alloys, diffusion brazing/soldering, 129

Nickel-base superalloys, diffusion brazing, 129

Nickel-bearing filler metals
   applications of brazes, 47-48
   as brazes, 47-50
   melting range, 50
   melting ranges of various compositions, 47
   metalloids, 50

Nickel-boron braze, 127
   phase diagram, 47, 48

Nickel-containing alloys, specific types
   Au-18Ni, 76

Nickel oxide, Gibbs free energy of formation at room temperature, 147

Niobium
   addition to Ag-Cu-Ti alloys, 169, 171
   joined in a vacuum furnace, 147

Nitride/carbide, linear expansivity at room temperature, 19

Nitrogen gas, silicon nitride with nickel-containing brazes, 91

Nocolok flux, 162

Nondestructive evaluation of joints and subassemblies; see also *Nondestructive testing (NDT) methods*
   evaluation of fabricated products, 225
   hardness measurements, 216, 217-219
   liquid penetrant inspection, 216, 219, 225
   technique selection, 225
   ultrasonic inspection, 216, 222-225
   visual inspection and metrology, 216-217, 225
   X-radiography, 216, 219-222, 225

Nondestructive testing (NDT) methods, 183, 216-225; see also *Nondestructive evaluation of joints and subassemblies*

Nonequilibrium contact angle, 12

Noneutectic alloys, 5

Nonmetals, wetting of, 115-117

O

OA (organic acid), 158

Optical-fiber light pipes, 217

Oxidation, 140, 194
   CCD component, 242
   gold-bearing filler metals resistant, 44
   gold-nickel brazes resistant, 45, 46
   of metal, equation, 147-148
   palladium-bearing filler metals resistant, 47
   phosphorus in silver-copper eutectic brazes, 36
   rate of, and temperature, 150
   titanium and its alloys, 154

Oxide, linear expansivity at room temperature, 19

Oxygen potential, 152

P

Palladium
   addition to gold-copper and gold-nickel alloys, 45, 46
   boiling/sublimation temperature at 10^{-10} atm pressure, 151
   coating for molybdenum for silicon semiconductors, 252

in composition of commercially available rapidly solidified filler metals, 49
in composition of industrial gold brazing alloys, 43
in composition of palladium-bearing filler metals, 47
metallization systems based on, 141
silver-gold-palladium alloys, 5
silver-palladium, 5

Palladium alloys, specific types
Pd-38Ni-8Si, melting range, 49
Pd-38Ni-8Si, structure, 49
Pd-38Ni-8Si, typical applications, 49

Palladium-bearing filler metals
as brazes, 46-47
melting range of various compositions, 47
properties conferred on joints, 47

Palladium oxide
joining in inert atmospheres and vacuum, 150
reduction in vacuum, 150

Partial pressures, 150, 151, 152, 178
and dew point, 153
of gases, 25

Partial wetting, 191

Peak joining temperature, 205

Peel strength, joints profiled on printed circuit boards, 248-249

Peel tests, 209, 210, 211

Peel-type debonding, 136-137

Peritectic reaction, 80

Peritectic solders, 80-81

Peritectic transformation, 80

Peritectic transformation temperature, 80-81

Peritectiferous solidification, 91

Phase diagrams
application to soldering and brazing, 71-108
information not revealed by, 72-73
information provided by, 72
methodology for determining, 91-102
methodology for determining literature search, 92
methodology for determining metallographic examination, 92-93, 94
methodology for determining, predicting the composition of eutectic points in high-order systems, 95-96, 99-100, 101
methodology for determining, quantitative metallography, 99-102
methodology for determining, subdivision of high-order systems, 93-95
methodology for determining, thermal analysis, 96-99

Phase formation, as process constraint, 118

Phase stability, 81

Phosphides, formation in joint gap of nickel-bearing filler metal brazes, 48

Phosphorus
addition to nickel-bearing filler metals, 47
addition to pure copper brazes, 33
addition to silver-copper eutectic brazes, 34-36
critical concentration threshold for embrittlement, 230
depressing melting point of brazes to increase joint strength, 128
fluxing agent for self-fluxing filler alloys, 162
impurity element addition affecting braze, 229
in composition of commercially available rapidly solidified filler metals, 49
incorporated into metallic coatings in wet plating, 140
lowest impurity concentration producing detrimental effects in Pb-60Sn solder, 231
silver-copper-phosphorus brazes, 5
silver-copper-phosphorus brazing alloys, corrosion of joints in tap water, 232-233

Phosphorus-containing brazes, 162-163

Phosphorus pentoxide, 153-154

Physical vapor deposition (PVD), 140
coatings for nonmetallic components, 115

Plasma arcs, for welding, 4

Platinum
addition (limited by high cost) to brazing alloys, 47
metallization systems based on, 141

Plumbing, self-fluxing filler metal applications, 36

Polytetrafluoroethylene (PTFE), 192

Porcelain/clay, linear expansivity at room temperature, 19

Potential energy, 9

Power cycling, 215

"Pressure variation" method, 126-127

Printed circuit boards (PCBs), 200-201, 214, 236
improving reliability by modifying joint geometry, 247-248

Process constraints
erosion, 117-118
imposed by components and solutions, 123-139
joints to strong materials, 133-139
materials systems approach to joining process development, 111, 112
mechanical constraints and solutions, 118-123
metallurgical constraints and solutions, 113-118
phase formation, 118
wetting of metals, 114-115

Process design examples, 236-254
die attachment in electronics fabrication, development of an improved solder, 244-247
improving reliability by modifying joint geometry, 247-250
improving the fabrication of silicon power device assemblies by product development, 250-254

mitigating thermal expansion mismatch in bonded assemblies, 237-241
Process temperature, 164
"Proof" testing, 206
Protection (personal) information, 172, 173
Pseudoeutectic composition, silver-copper-zinc-cadmium alloys, 41
Pull-off test, 184
Pulse-echo ultrasonic inspection, 222-223, 224-225
Pyrometers, 23-24

Q

Quality assessment of joints. See *Joint quality assessment*
Quantitative image analysis, 100-101

R

R(rosin), 158, 159
RA (rosin activated), 158
Rapid solidification casting
 brazes, 48
 nickel-bearing filler metal brazes, 48-49
 solders, 48
Rapid solidification techniques, 166
 nickel-bearing filler metals, 48-50
 to produce preforms for aluminum-germanium brazes, 253
Rapidly solidified filler metals, commercially available
 applications, 49
 compositions, 49
 melting range, 49
 structure, 49
Reaction, rate of, 71
Reaction product zone, 168-169
Reaction zone, 201
Reactive filler alloys, 164-172
 formation and nature of the reaction products, 168-169
 mechanical properties of joints, 169-172
 reactive constituent concentration influence, 166-168
Reactivity, 173
Reduction, 148, 150-155
Reduction reaction, 162
Reflow operation, 28

Refractory metals, joined using reactive filler alloys, 164
Reprecipitation, Al-12Si braze for a silicon wafer, 251-252
Resistance heating, 140
Reverse bias sputtering, 140
Rheological agent, 158
Rhodium, coating for molybdenum for silicon semiconductors, 252
Ring laser gyroscope (RGL), 237-241
RMA (rosin mildly activated), 158, 159
Rockwell hardness test, 218-219
Rosin fluxes, 157
Rotary dip method, 186
Rotary dip tests, 231
Rotary fatigue testing machine, 213, 214
Rubbers, linear expansivity at room temperature, 19
Rub or scrub soldering, 163

S

SA (synthetically activated), 158-159
Salt bath brazing, 162
Salt-spray test, accelerated, 202
Saturation limit, 17
Scandium, activating agent for fluxless brazing of aluminum, 155
Scanning acoustic microscopy (SAM), 195, 196-198, 199, 216, 225
Scanning electron microscopy (SEM), 195, 198-202
Scarfing, 137
Scarf joints, 203, 204
Screen printing, 158, 242-243
Second Law of Thermodynamics, 175-176
Self-fluxing filler metals, 34, 35-36
 copper-nickel-tin-phosphorus alloys, 50
Semiconductor, construction of typical high-power device, 250, 251-252
Semicrystalline, linear expansivity at room temperature, 19
Sessile drop test, 165, 186, 187-188, 192
Severe environment test, typical conditions used in accelerated aging assessments, 194
Shear stress, 135
Shear tests, 182, 183, 202, 208-209
Shelf life

evaluation of, 186
gold-coated components of hermetic seal for charge coupled
devices, 243

"Shortlist" of filler alloys, 113

Sheradizing surface treatment, 115

Silica gel, 153-154

Silicates, 159

Silicon
  addition to nickel-bearing filler metals, 47
  aluminum-molybdenum-silicon phase diagram, equilibrium shown, 250, 251
  aluminum-silicon alloys, eutectic point, 253
  chip mounted in a ceramic package, 245
  gold-silicon alloys, 54-57, 163
  gold-silicon-tin alloy system, phase diagram, 246
  in aluminum-bearing filler metals, 52
  in brazes, 5
  in brazes, bismuth and beryllium effect, 231
  in chloride-base fluxes, 161
  in composition of commercially available rapidly solidified filler metals, 49
  molybdenum disilicide, 250, 252
  tin pest acceleration, 232
  wetting impossible in the absence of fluxes, 245
  wetting improvement by gold solders, 247
  X-ray maps for concentration shown in SEM image, 201-202

Silicon carbide, 169

Silicon-containing alloys, specific types
  Al-4Cu-10Si, 5
  Al-12Si braze, 250-252
  Al-12Si braze, metallographic examination, 195
  Al-12Si braze, silicon wafer eroded by, 251-252
  Al-13Si braze, 155
  Au-3Si solder, for die attachment, 245-247

Silicon nitride, 165, 166, 168-169
  with nickel-containing brazes, 91

Silicone oil, in thermal cycling tests, 206

Silver
  addition to brazing alloy to promote wetting, 155
  addition to gold-copper brazes, 45
  addition to lead-tin solders, 67
  addition to tin/lead eutectic solders, 65
  aluminum-silver-germanium alloys, for determination of phase diagrams, 91-101
  boiling/sublimation temperature at 10^{-10} atm pressure, 151
  brazes, pure, 33
  dissolution by tin-lead solders, 117
  electrode potential at 25 degrees C, 51
  erosion by molten tin as a function of reaction time and temperature, 78
  erosion by tin in silver-gold-tin alloys, 89
  in composition of cadmium-free carat gold "solders," 43
  in composition of commercially available rapidly solidified filler metals, 49
  in composition of industrial gold brazing alloys, 43
  in composition of palladium-bearing filler metals, 47
  in composition of silver-copper-zinc brazing alloys, 39
  in composition of silver-copper-zinc-cadmium brazing alloys, 41
  in composition of silver-copper-zinc-tin brazing alloys, 42
  in composition of zinc-bearing solders, 53
  in composition of tin alloy solders, 58
  in low-melting-point eutectic composition alloys used as solders, 58
  in metallization for silicon semiconductors, 253
  in solders, 64, 66
  in solders, applications, 64
  melting point, 33
  metallization systems based on, 141
  solidification shrinkage (% of solid), 127
  stress level when a filler in a joint to high-strength steel, 133, 134
  substrate material for copper-lead-tin solder, 86
  surface coating to help diffusion soldering, 130
  used as wettable metallization, 114

Silver alloys, substrate for metal combinations used for diffusion brazing/soldering, 129

Silver-aluminum-germanium alloys, phase diagram determination, 91-101

Silver-antimony alloys, specific types
  56Ag-44Sb, melting point and problems encountered, 7

Silver-bismuth alloys, principal intermetallic phases formed by alloying between the common binary solders and engineering parent metals and metallizations, 64

Silver-cadmium alloys, specific types
  5Ag-95Cd, melting point and problems encountered, 7
  23Ag-53Cd-24Cu, melting point and problems encountered, 7

Silver-carbon alloys, specific types
  Ag-40C, melting range, 49
  Ag-40C, structure, 49
  Ag-40C, typical applications, 49

Silver-copper alloys
  brazes to wet titanium silicide, 165
  eutectic brazes, 34-36
  eutectic melting point, 34
  hardness when containing titanium, 166
  phase diagram, 35, 75, 76
  to join copper components, 75

Silver-copper alloys, specific types
  Ag-28Cu, eutectic braze, 75, 76, 139
  Ag-28Cu, eutectic braze for fluxless brazing of gold-plated molybdenum components, 81
  Ag-28Cu, gold coatings, 81, 82
  Ag-30Cu, filler metal for aluminum alloys, diffusion brazing/soldering, 129
  Ag-Cu-Cu$_3$P partial ternary system, liquidus surface, 35, 36

Index

Silver-copper-hafnium alloys, reaction layer thickness of braze as function of concentration of active metal for SiC, 169

Silver-copper-lithium braze, 162

Silver-copper-phosphorus brazing alloys, 162
  corrosion of joints in tap water, 232-233
  reaction with steels, 5

Silver-copper-titanium alloys
  braze with nickel interlayer for joint between alumina and steel, 120
  phase diagram, 167

Silver-copper-titanium alloys, specific types
  Ag-Cu-5Ti, melting range, 166
  Ag-27Cu-2Ti, 167
  Ag-27Cu-2Ti, brazing temperature effect on wetting of Si_3N_4, 165
  Ag-27Cu-2Ti, brazing time effect on wetting of Si_3N_4, 166
  Ag-38Cu-5Ti, melting range, 49
  Ag-38Cu-5Ti, structure, 49
  Ag-38Cu-5Ti, typical applications, 49
  Ag-66Cu-22Ti, 167

Silver-copper-zinc alloys
  applications, 38, 39
  as brazes, 32, 36-39
  corrosion of joints in tap water, 232, 233
  hallmarking limits on silver content for brazes, 38
  isoelongation (%) contours of as-cast alloys, 40
  isohardness (Brinell) contours of as-cast alloys, 40
  isostrength (MPa) contours in the cast condition, 39
  joints in stainless steel pipes serving as water installations, 233
  liquidus temperature and liquidus surface, 36-38
  melting point, 36
  melting range of brazes, 39
  melting range vs. silver content, 41

Silver-copper-zinc alloys, specific types
  Ag-15Cu-15Zn, filler metal for aluminum alloys, diffusion brazing/soldering, 129

Silver-copper-zinc-cadmium alloys, 39-42, 77, 135, 215
  applications, 41
  cadmium addition benefits, 39-41
  health hazards, 42
  impurity element effect, 229, 230
  liquidus temperature of alloy compositions, 41, 42
  melting ranges of different braze compositions, 41
  melting range vs. silver content, 41
  pseudoeutectic composition, 41

Silver-copper-zinc-cadmium alloys, specific types
  42Ag-17Cu-16Zn-25Cd, 43

Silver-copper-zinc-tin alloys
  as brazes, 42-43
  melting point, 42-43
  melting ranges of braze compositions, 42
  melting range vs. silver content, 41
  tin content effects, 43

Silver-copper-zinc-tin alloys, specific types
  55Ag-21Cu-22Zn-2Sn braze, 43

Silver-gold alloys, 5

Silver-gold-copper alloys, 81-82
  isothermal projection, 82
  liquidus surface, 45
  phase diagram, 81, 82

Silver-gold-copper alloys, specific types
  45Ag-29Au-26Cu, 45, 81

Silver-gold-palladium alloys, 5

Silver-gold-tin alloys
  liquidus projection, 88
  phase diagram, 88-89
  vertical section between eutectic tin-silver alloy and gold, 88

Silver-indium alloys
  phase diagram, 60
  principal intermetallic phases formed by alloying between the common binary solders and engineering parent metals and metallizations, 64
  solders for semiconductors, 253

Silver-indium alloys, specific types
  Ag-97In, joint fill ratio as a function of joint width, 124

Silver-lead alloys
  phase diagram, 62
  principal intermetallic phases formed by alloying between the common binary solders and engineering parent metals and metallizations, 64
  solid solutions formed with copper, 5

Silver-lead-tin alloys, specific types
  Ag-97.5Pb-1Sn, tin content effect on peel strength in joints to Kovar, 248

Silver metallizations, substrate for metal combinations used for diffusion brazing/soldering, 129

Silver oxide
  Gibbs free energy of formation at room temperature, 147
  joining in inert atmospheres and vacuum, 150
  reduction in vacuum, 150

Silver-palladium alloys, 5

Silver-tin alloys
  corrosion of soldered joints in tap water, 233
  diffusion soldering, 131-132
  eutectic solder, 90
  for semiconductors, 253
  phase diagram, 61, 77
  phase diagram, tin-rich portion, 78
  principal intermetallic phases formed by alloying between the common binary solders and engineering parent metals and metallizations, 64

Silver-tin alloys, specific types
  Ag_3Sn, 77, 78
  Ag-96Sn, 16, 64, 77-78, 88-89, 90, 132

Ag-96Sn, applications, 77
Ag-96Sn, gold addition effect on solder, 234
Ag-96Sn, joint fill ratio as a function of joint width, 124
Ag-96Sn, mechanical properties as a function of gold addition, 88-89
Ag-96Sn, solder radiograph, 16
Ag-96Sn, tin content effect on peel strength in joints to Kovar, 248

Silver-tin-antimony alloys, specific types
  25Ag-65Sn-10Sb, solder applications, 59

Silver-zinc alloys
  as brazes, 36, 37
  melting range, 36
  phase diagram, 36, 37

Simplex lattice design method, 97-101

S-N fatigue curves, 215

Solderability, 186-187

Soldering and brazing
  characteristic features, 1, 3-4
  cleaning treatment, 26
  coatings applied to surfaces of components, 26
  design and application of processes, 19-28
  dissolution of parent materials by molten fillers, 16-18
  fluid flow, 12-13
  form of the filler metal, 22-23
  functional requirements and design criteria, 19-21
  heating methods, 23
  inert joining atmosphere, 24-26
  jigging of the components, 21-22
  key parameters, 8-19
  oxidizing joining atmosphere, 24-26
  postjoining treatments, 28
  processing aspects, 21-28
  reducing joining atmosphere, 26
  significance of the joint gap, 18-19
  surface energy and surface tension, 8-9
  surface roughness, 14-16
  temperature measurement, 23-24

Soldering fluxes
  basic ingredients, 157

Solders, 54-69
  activated, 7
  classification scheme for those melting below 300 degrees °C (570 degrees F), 59-64
  compared to brazes, 5-8
  complexities presented by higher-order and nonmetallic systems, 90-91
  derivation of term, 5
  distinguishing features, 7
  gold-bearing filler metals, 54-58
  gold-silicon alloys, 54-57
  gold-tin alloys, 55, 57-58
  high-melting-point, 7
  indium-lead, 80
  lead-tin alloys, 66-69, 83-86, 117
  lead-tin alloys, advantages, 66-67
  lead-tin alloys, applications, 66
  principal solder alloy families and their melting ranges, 6
  rapid solidification casting, 48
  silver-tin vs. lead-tin, 67
  temperature range for applications, 7-8
  titanium additions, 36

Solidification, rate of, 75

Solidification shrinkage, 135
  as process constraint, 127-128

Solid-state diffusion, 205

Solidus temperature, of parent materials, 31

Solubility limit, 167

Space Shuttle, gold-nickel brazes for main engine, 45

Split globule test, 186-187

Spread factor, 189-190, 226

Spreading, 11, 13, 15, 217
  filler characteristics, 13-14
  impurity element effect on properties, 230-231
  molten copper on mild steel, 154
  nonmetallic materials, 145

Spreading test, 184, 188-191

Spread ratio, 189-190, 226

Sputtering (sputter-deposition), 140, 239
  advantages and disadvantages, 142
  characteristic features, 141
  chromium overlaid with gold metallization, 243, 244
  coating quality parameters, 142
  metallization of silicon semiconductors, 253
  process parameters, 142
  to apply wettable metallizations, 114-115
  wetting of nonmetals, 115-116

Stainless steels
  annealing, 28
  brazes for self-fluxing filler alloys, 162
  joined with aluminum alloy brazes, 51
  linear expansivity at room temperature, 19
  pure copper brazes for joining, 33
  silver-copper-zinc alloys as brazes with manganese, 38-39
  soldering fluxes, 157
  substrate for metal combinations used for diffusion brazing/soldering, 129
  water installations in place of copper pipework, 233
  Zircaloy 2 joined by diffusion brazing, 130

Steady-state strain rate, 212

Steam aging test, typical conditions used in accelerated aging assessments, 194

Steel
  interaction with silver-copper-phosphorus brazes, 5
  medium-carbon, butt joint fracture stress vs. joint thickness, 134
  solderability rate in presence of impurity elements, 231

substrate for metal combinations used for diffusion brazing/soldering, 129
Stenciling, 158
Step height, 137
Step joints, 137, 203, 204
Strain cycles to failure, 213
Straining, 212
Strap joints, 137-138, 203, 204
Stress
　due to thermal expansion mismatch, 118-123
　scanning electron microscopy, 201
Stress corrosion, 202, 216
Stress-corrosion cracking, silver-zinc alloys, 36
Stress relaxation, 211-212
Stress relief, 28, 119
Stress-relief heat treatments, to reduce stress on metallization layer of nonmetals, 116
Stress rupture, 211-212
Strontium
　activating agent for fluxless brazing of aluminum, 155
　addition to brazing alloy to promote wetting, 155
Sulfur, lowest impurity concentration producing detrimental effects in Pb-60Sn solder, 231
Superheat, 7, 14
Superplastic forming, 128
Surface energy, 8-9, 10, 13
　of liquid, 9
　MKS unit, 9
　origin of, 8
　of solid, 9
Surface-mounted electronic devices (SMDs), adapted wetting balance for solderability testing, 192
Surface roughness, soldering and brazing, 14-16
Surface tension, liquid, 9
Surface tension force, 191, 192
Surfactants, 158
Surroundings, defined, 145
Syringe dispensing, 158
System, 175

T

Tantalum, joined in a vacuum furnace, 147
Taper, angle of, 138
Taper sections, 196

Tap water, corrosion of soldered and brazed joints, 232-233
Temperature, brazing, effect on wetting of Si_3N_4 by Ag-27Cu-2Ti alloy, 165
Temperature cycles, typical profiles, 27
Tensile force, liquid, 9
Tensile stress, 135
　applied, 133, 134
Tensile tests, 182, 183, 202, 206-208
　soldered joints of GaAs samples, 206
Ternary alloy systems, 81-90
Thermal analysis
　methods and their characteristics, 104-108
　to determine temperature of eutectic transformation, 92, 94, 96-99, 101
Thermal cycling, 215
Thermal cycling screening test, 182
Thermal cycling tests, 206, 207, 241
Thermal expansivity, reduction brought about by fiber additions, 139
Thermal expansivity (expansion) mismatch, 2, 170-172, 184, 202, 213-214, 238
　aluminum alloy of ring laser gyroscope and metal expansion matching plate, 241
　between aluminum-germanium braze and silicon wafers, 253
　between silicon and molybdenum, 252
Thermal shock test, 206, 207
Thermal soak test, 195
Thermocouples, 23-24
Thermodynamics, 175-176
Thermogravimetric analysis (TGA), 104, 107, 108
Thermomechanical analysis (TMA), 104, 107-108
Thermomechanical fatigue testing, 203
Thick-film metallization techniques, 115, 140-141
　advantages and disadvantages, 142
　characteristic features, 141
　coating quality parameters, 142
　process parameters, 142
Threshold limit value (TLV), 174
Tie line, 82, 83, 95
Time, brazing, effect on wetting of Si_3N_4 by Ag-27Cu-2Ti alloy, 166
Tin
　addition effects on silver-copper-zinc-tin alloys, 43
　addition to promote spreading of Au-2Si solder, 246
　addition to silver-copper-titanium alloys, 167
　boiling/sublimation temperature at 10^{-10} atm pressure, 151

content effect on peel strength of joints to Kovar, 248
crystal structures and primitive rhombohedral cells of selected phases in Au-Pb-Sn system, 94
electrode potential at 25 degrees C, 51
energy of formation with solid copper, 11
filler metal for aluminum alloys, diffusion brazing/soldering, 129
in composition of commercially available rapidly solidified filler metals, 49
in composition of silver-copper-zinc-tin brazing alloys, 42
in composition of solders, 7, 58-66
in composition of zinc-bearing solders, 52, 53
in low-melting-point eutectic composition alloys used as solders, 58
solidification shrinkage (% of solid), 127
used as wettable metallization, 114

Tin-antimony alloys, phase diagram, 62
Tin-antimony alloys, specific types
Sn-5Sb solder, 240-241
Sb-95Sn, 234

Tin-antimony-lead alloys, 230
liquidus surface, 63

Tin-antimony-lead alloys, specific types
Pb-40Sn-2Sb solder, 230

Tin-bismuth alloys, 64
mechanical properties of soldered joints to copper, 236
phase diagram, 60
shear strength of soldered joints to brass, 236

Tin-bismuth alloys, specific types
Bi-42Sn solder, 212, 213
Bi-43Sn solder, 124, 127-128, 234

Tin-bismuth-lead alloys, liquidus surface, 63

Tin-copper alloys, 132
corrosion of joints in tap water, 233
Cu_6Sn_5 intermetallic phases, 204
for diffusion soldering, 130-131
phase diagram, 83

Tin-copper-lead system, 83-84

Tin-gold alloys
as solders, 55, 57-58

Tin-gold alloys, specific types
Au-20Sn solder, 117
Au-20Sn solder, for die attachment, 245, 247

Tin-gold-antimony system, isothermal section at 25 degrees °C, 240, 241

Tin-gold-indium ternary phase diagram, isothermal section, 238-239

Tin-gold-lead alloys
in solders, 86-90
phase diagram determination, 94, 95

Tin-gold-silicon alloy system, phase diagram, 246

Tin-indium alloys, 64

Tin-indium alloys, specific types
In-48Sn, 124, 234
In-49Sn solder, 238

Tin-indium-lead alloys, specific types
In-18Pb-70Sn solder, gold addition effect, 235

Tin-lead solders, 66-69, 77, 83-86, 117, 182
cleaning method effect on wetting behavior, 192, 193
mechanical properties of soldered joints to copper, 236
metallographic examination, 195
phase diagram, 61
principal intermetallic phases formed by alloying between the common binary solders and engineering parent metals and metallizations, 64
reaction with a copper substrate following heat treatment, 85
shear strength of soldered joints to brass, 236
ultrasonic fluxing, 163

Tin-lead solders, specific types
Pb-5Sn, 123
Pb-60Sn, 83-84, 86-87, 89, 132, 139, 211
Pb-60Sn, joint fill ratio as a function of joint width, 124
Pb-60Sn, lowest impurity concentrations producing detrimental effects, 230-231
Pb-60Sn, spreading tests, 190
Pb-60Sn, tensile strength of cast bars, 76-77
Pb-60Sn, tin content effect on peel strength in joints to Kovar, 248
Pb-62Sn, 234

Tin oxide, Gibbs free energy of formation at room temperature, 147

Tin pest, 67
beneficial effect of prevention in solders, 231-232

Tin-silver alloys, 64
corrosion of joints in tap water, 233
eutectic solder, 90
for diffusion soldering, 131-132
for semiconductors, 253
phase diagram, 61
phase diagram, tin-rich portion, 78

Tin-silver alloys, specific types
Sn-3.5Ag, melting range, 49
Sn-3.5Ag, structure, 49
Sn-3.5Ag, typical applications, 49
Ag_3Sn, 77, 78
Ag-96Sn, 77-78, 88, 90, 124, 234
Ag-96Sn, eutectic solder, 132
Ag-96Sn, tin content effect on peel strength in joints to Kovar, 248

Tin-silver-antimony alloys, specific types
Sn-25Ag-10Sb, melting range, 49
Sn-25Ag-10Sb, structure, 49
Sb-25Ag-10Sb, typical applications, 49

Tin-silver-copper-zinc alloys, as brazes, 42-43

Tin-silver-lead alloys, specific types

Index

Ag-97.5Pb-1Sn, tin content effect on peel strength in joints to Kovar, 248

Titanium
- active constituent of brazes for nonmetals, 115
- addition to fillers for ceramics, 11
- addition to improve wetting of silicon by gold solders, 247
- addition to nickel-bearing filler metals, 50
- addition to silver-copper-base brazes to help wetting of engineering ceramics, 164
- addition to silver-copper eutectic brazes, 36
- addition to solders, 36
- addition to solders for materials containing glass or ceramic phases, 91
- boiling/sublimation temperature at 10^{-10} atm pressure, 151
- concentration effect on wetting of nitride ceramics, 165
- Gibbs free energy of possible reactions with silicon nitride, 165
- in composition of commercially available rapidly solidified filler metals, 49
- in metallization for silicon semiconductors, 253
- in tin-containing solders, 7
- joined in a vacuum furnace, 147
- metallization for zinc oxide, 244
- metallization systems based on, 141
- metallizing of a nonmetallic material, 115, 116
- oxidation, 154
- silver as braze for bonding, 33
- "unsolderable" in air, 157
- used in active hydride process, 116

Titanium-activated brazes, 168

Titanium alloys
- linear expansivity at room temperature, 19
- substrate for metal combinations used for diffusion brazing/soldering, 129

Titanium-copper alloys, specific types
- Cu-5Ti braze, 168-169

Titanium-copper-nickel alloys, specific types
- Ti-15Cu-15Ni, melting range, 49, 50
- Ti-15Cu-15Ni, structure, 49
- Ti-15Cu-15Ni, typical applications, 49

Titanium-copper-silver alloys
- phase diagram, 167

Titanium-copper-silver alloys, specific types
- Ag-27Cu-2Ti, 165, 166, 167
- Ag-Cu-5Ti, 166
- Ag-66Cu-22Ti, 167

Titanium dioxide, dissociation pressure, 154

Titanium-nickel-zirconium alloys, specific types
- Ti-20Zr-20Ni, melting range, 49
- Ti-20Zr-20Ni, structure, 49
- Ti-20Zr-20Ni, typical applications, 49

Titanium nitride, 165

Titanium oxide, Gibbs free energy of formation at room temperature, 147

Titanium silicide, 165

T-joints, 205

T-peel test, 210, 211

Transient liquid phase bonding, 129

Transition reactions, 81

Transmission electron microscopy, for precipitates of reaction of lead-tin solder on copper substrate, 85

Trapped gas, 135, 242
- as source of voids in soldered and brazed joints, 124-127

Triaxial (hydrostatic) tension, 133, 134

Trifoils, titanium sheet with silver-copper braze, 36

Triode sputtering, 140

True fluxes, 155

Tungsten
- boiling/sublimation temperature at 10^{-10} atm pressure, 151
- brazed in air under cover of mild fluxes, 147
- joined with aluminum alloy brazes, 51
- metallization systems based on, 141
- not cost-effective for molybdenum in semiconductors, 253
- powder compact for infiltrating molten copper, 121
- wetted by Al-12Si eutectic braze, 79

Tungsten/molybdenum alloys, linear expansivity at room temperature, 19

U

Ultimate tensile strength, 208

Ultrasonic agitation, for diffusion bonding, 4

Ultrasonic fluxing, 163-164

Ultrasonic inspection, 195, 196-198, 216, 222-225

Ultrasonic vibration, for diffusion bonding, 4

Ultrasound, 222, 223, 224, 225

Unsharpness, 220, 222

V

Vacuum bakeout, 125

Vacuum evaporation
- advantages and disadvantages, 142
- characteristic features, 141
- coating quality parameters, 142
- process parameters, 142

Vanadium
- in composition of commercially available rapidly solidified filler metals, 49
- joined in a vacuum furnace, 147

Van der Waal forces, 10

Vapor deposition techniques, 22, 143, 239
 chromium overlaid with gold metallization, 243
 in flip-chip bonding process, 123
 to apply wettable metallizations, 114

Vapor phase deposition process, to metallize germanium window with silver, 102

Vapor-phase technique, gold-silicon alloys as solders for silicon components, 57

Vickers hardness test, 218

Viscosity
 of lead-tin alloys at 50 degrees C, 76
 liquid, 13
 molten eutectic alloys, 76

Visual inspection and metrology, 216-217, 225

V-notch Charpy impact test, 229

Void clusters, 195

Voids, 213, 220
 generation of, compromising strength of soldered joints, 230
 hermetic seals, 242

Volatilization, 150

Volume energy, 9

W

Wave soldering, 186

Weight, conversion to atomic fraction of constituents of alloys, 103

Weight fraction of braze not melted under equilibrium conditions, 74

Welding, 1 characteristic features, 4

Wet coating, 143

Wet plating, 140, 143
 aluminum coating with silver, 102
 coatings for nonmetallic components, 115
 to apply wettable metallizations, 114-115

Wetting, 9-12, 13, 182, 217
 assessment of, 186-194
 ceramics, 164-166
 impurity element effect on properties, 230-231
 metals, 114-115
 nonmetals, 115-117, 145
 self-fluxing filler metals, 163

Wetting angle, 182

Wetting balance solderability testing, 191-193

Wetting equation, 9, 10, 11, 13

Wetting force measurements, as function of time, 186

White tin, 231-232

WS (water soluble), 158-159

X

X-radiography, 198, 216, 219-222, 225

X-ray fluorescence, 184

Y

Yield point, 208

Yttrium, activating agent for fluxless brazing of aluminum, 155

Z

Zinc
 addition to aluminum-bearing brazes, 51
 boiling/sublimation temperature at 10^{-10} atm pressure, 151
 copper-zinc alloys as brazes, 36, 37
 electrode potential at 25 degrees C, 51
 in composition of cadmium-free carat gold "solders," 43
 in composition of silver-copper-zinc brazing alloys, 39
 in composition of silver-copper-zinc-cadmium brazing alloys, 41
 in composition of silver-copper-zinc-tin brazing alloys, 42
 in composition of tin alloy solders, 58
 in solders and brazes, 5
 lowest impurity concentration producing detrimental effects in Pb-60Sn solder, 231
 solidification shrinkage (% of solid), 127
 volatile materials in metallic components and filler metals, 126

Zinc alloys, linear expansivity at room temperature, 19

Zinc-aluminum alloys, specific types
 Zn-6Al, melting range, 53
 Zn-7Al-4Cu, melting range, 53
 Zn-5Al-2Ag-1Ni, melting range, 53

Zinc-bearing solders
 as brazes, 52-54
 factors limiting their adoption, 54
 potential benefits, 53-54

Zinc-cadmium alloys, specific types
 Zn-10Cd, melting range, 53
 Zn-25Cd, melting range, 53

Zinc chloride, a salt added to a flux to adjust corrosivity, 158

Zinc-copper alloys, specific types
 Zn-1Cu, filler metal for aluminum alloys, diffusion brazing/soldering, 129
 Zn-1Cu, melting range, 53

Zinc-copper-silver alloys

as brazes, 36-40
corrosion of joints in tap water, 232, 233
for joints in stainless steel pipes, 233

Zinc-copper-silver-antimony alloys, specific types
Zn-30Cu-3Sb-1Ag, melting range, 53

Zinc-copper-silver-cadmium alloys
as brazes, 39-42, 77, 135, 215
impurity element effect, 229, 230

Zinc-copper-silver-tin alloys, as brazes, 42-43

Zinc-nickel alloys, specific types
Zn-2Ni, melting range, 53

Zinc oxide, titanium or zirconium metallization prone to blistering, 244

Zinc-tin alloys, specific types
Zn-70Sn, melting range, 53

Zinc-tin-lead alloys, specific types
Zn-10Sn-1Pb, melting range, 53

Zircaloy 2, diffusion brazing for joining to stainless steel, 130

Zirconia, 168

Zirconia/mild steel joints, Ag-Cu-3Ti filler alloys, 170

Zirconium
addition to fillers for ceramics, 11
addition to solders for materials containing glass or ceramic phases, 91
filler metal for brazes, 36
in composition of commercially available rapidly solidified filler metals, 49
joined in a vacuum furnace, 147
metallization for zinc oxide, 244
metallization systems based on, 141
used in active hydride process, 116

Zirconium-nickel alloys, specific types
Zr-17Ni, melting range, 49
Zr-17Ni, structure, 49
Zr-17Ni, typical applications, 49

Zirconium-titanium-vanadium alloys, specific types
Zr-16Ti-28V, melting range, 49
Zr-16Ti-28V, structure, 49
Zr-16Ti-28V, typical applications, 49